CAMBRIDGE LIBRARY COLLECTION

Books of enduring scholarly value

Monographs of the Palaeontographical Society

The Palaeontographical Society was established in 1847, and is the oldest Society devoted to study of palaeontology worldwide. Its primary role is to promote the description and illustration of the British fossil flora and fauna, via publication of an authoritative monograph series. These monographs cover a wide range of taxonomic groups, from microfossils, trilobites and ammonites through to Coal Measure plants, mammals and reptiles, and from all ages from Cambrian to Pleistocene. They form a benchmark for understanding the past life of the British Isles and many include the original descriptions of numerous key species. The first monograph (on the Crag Mollusca) was published in March 1848 and the Society still continues this work today. Notable authors in the series include Charles Darwin (fossil barnacles) and Richard Owen (dinosaurs and other extinct reptiles). Beginning in 2014, the Cambridge Library Collection and the Society are collaborating to reissue the earlier publications, focusing on monographs completed between 1848 and 1918.

A Monograph of the Crag Mollusca

The Pliocene–Pleistocene Crags of East Anglia are an incredibly rich source of fossil shells, many belonging to extant Boreal and Mediterranean genera. Dominated by marine gastropods and bivalves, the deposits also contain evidence of terrestrial and non-marine gastropods and bivalves, brachiopods, and extensive epifauna including bryozoans. Published between 1848 and 1879 in four volumes, the latter two being supplements with further descriptions and geological notes, this monograph by Searles Valentine Wood (1798–1880) covers more than 650 species and varieties of fossil mollusc. For each species Wood gives a synonymy, diagnosis (in Latin), full description, dimensions, occurrence and remarks. The supplements also provide a breakdown of the species and their current distribution. The detailed plates were prepared by the conchologist George Brettingham Sowerby and his namesake son. Volume 3 (1872–4) comprises the first supplement, covering univalves and bivalves, and includes an important map of the Crag district.

Cambridge University Press has long been a pioneer in the reissuing of out-of-print titles from its own backlist, producing digital reprints of books that are still sought after by scholars and students but could not be reprinted economically using traditional technology. The Cambridge Library Collection extends this activity to a wider range of books which are still of importance to researchers and professionals, either for the source material they contain, or as landmarks in the history of their academic discipline.

Drawing from the world-renowned collections in the Cambridge University Library and other partner libraries, and guided by the advice of experts in each subject area, Cambridge University Press is using state-of-the-art scanning machines in its own Printing House to capture the content of each book selected for inclusion. The files are processed to give a consistently clear, crisp image, and the books finished to the high quality standard for which the Press is recognised around the world. The latest print-on-demand technology ensures that the books will remain available indefinitely, and that orders for single or multiple copies can quickly be supplied.

The Cambridge Library Collection brings back to life books of enduring scholarly value (including out-of-copyright works originally issued by other publishers) across a wide range of disciplines in the humanities and social sciences and in science and technology.

A Monograph of the Crag Mollusca

With Descriptions of Shells from the Upper Tertiaries of the East of England

Volume 3: Supplement,
being Descriptions of Additional Species
(Univalves and Bivalves)

Searles V. Wood

CAMBRIDGE
UNIVERSITY PRESS

University Printing House, Cambridge, CB2 8BS, United Kingdom

Cambridge University Press is part of the University of Cambridge.
It furthers the University's mission by disseminating knowledge in the pursuit of
education, learning and research at the highest international levels of excellence.

www.cambridge.org
Information on this title: www.cambridge.org/9781108076906

This edition first published 1872–4
This digitally printed version 2014

ISBN 978-1-108-07690-6 Paperback

THE

PALÆONTOGRAPHICAL SOCIETY.

INSTITUTED MDCCCXLVII.

LONDON:

MDCCCLXXIV.

SUPPLEMENT

TO THE

MONOGRAPH OF THE CRAG MOLLUSCA,

WITH

DESCRIPTIONS OF SHELLS

FROM THE

UPPER TERTIARIES OF THE EAST OF ENGLAND.

BY

SEARLES V. WOOD, F.G.S.

VOL. III.

UNIVALVES AND BIVALVES.

WITH

AN INTRODUCTORY OUTLINE OF THE GEOLOGY OF THE
SAME DISTRICT, AND MAP.

BY

S. V. WOOD, Jun., F.G.S., and F. W. HARMER, F.G.S.

LONDON:

PRINTED FOR THE PALÆONTOGRAPHICAL SOCIETY.

1872—1874.

PRINTED BY
J. E. ADLARD, BARTHOLOMEW CLOSE.

INTRODUCTION.

WHEN the 'Crag Mollusca' was passing through the press, during the years from 1846 to 1856, our knowledge of the Upper Tertiaries was not only very restricted, but some erroneous ideas existed as to the identity of beds now found to be very different in age. Geologists had begun to consider the Clyde beds as newer than the Crag; but the Bridlington deposit was regarded as coeval with the Norwich Crag, while the relation of this Crag to the Red was but vaguely ascertained. Moreover, the thick beds of sand which overlie the Red Crag, but which are now found to have no connexion with it, were then regarded as the "unproductive sands of the Crag," while the Glacial beds were known but dimly under the term "Northern Drift," or as the "Boulder Clay," under which term two very different deposits, the Boulder Till of the Cromer Cliff and the wide-spread Boulder Clay of the East Anglian uplands, were confounded together. Further, the Post-glacial freshwater beds of Clacton, Ilford, Grays, Stutton, and other places, were thought at this time by many, myself included, to belong to the age of the Crag, which, indeed, so far as any light their freshwater testaceous remains would throw on the question, there was every reason to suppose was the case.

The advance of geology during the years that have since elapsed has thrown new light on the relations borne by these various newer tertiary beds to each other, and I have been able in consequence to refer the species noticed in the present work to the formations in which they occur, according to what I believe to be their true geological sequence, and accordingly several new formations and localities are referred to throughout the Supplement that are not to be found in the original work.

Under these circumstances I have thought it desirable to confine this Supplement to fossils from that portion of England in which the succession of the various beds has been systematically worked out, and to avail myself of the labours of my son and his co-adjutor, Mr. F. W. Harmer, of Norwich, who have for several years past occupied themselves in studying these formations in the field, and in regularly mapping them on the one-inch scale Ordnance maps. The area over which they have carried, and now very nearly completed this mapping, is that which extends from the Norfolk coast, as far west as the Wash, to the River Thames, comprising nearly all of the three counties, Essex, Suffolk, and Norfolk; so that they have had the opportunity of tracing the beds over a wide area, and of thus guarding themselves against the not unlikely error of

b

regarding as separate formations, beds that differ only from local peculiarities, but are in fact identical in age and continuity.

From the map so constructed by them, the lithographic map which accompanies this Supplement, and which, although it embraces the entire Crag area, comprises only a portion of that embraced by the original, has been taken by reduction to the scale of one fourth (linear) of the original; and the sections which accompany it (and without which it would be difficult to understand either the structure of the country or the sequence of the deposits) have been carefully prepared by them for its illustration. All the formations quoted for Mollusca in my work will be found in this map, with the following exceptions; viz. the Bridlington bed, which is on the Yorkshire coast; the Kelsea Hill gravel, which is near the same coast; the gravel of Hunstanton, which is at the north-western extremity of the Norfolk Coast; the Brick-earth of the Nar, which is in the same neighbourhood; and the gravel of March, which is in the midst of the great fen district of Cambridgeshire.

The geological position of the beds thus beyond the limit of the map will, however, be explained in the following outline of the geology of the Upper Tertiaries of East Anglia, which has been written by my son and Mr. Harmer as embodying their views of the subject. The actual area from which the Mollusca described or referred to in my Supplement were obtained, is that which lies east of a line drawn from London to Bridlington, and including the latter place.

I defer some remarks which I propose to make on the fauna of the several Upper Tertiaries embraced in this Supplement, until the remaining part of it, which will contain the Bivalvia; and I propose then to give a tabular list of the species.

S. V. WOOD.

AN OUTLINE OF THE GEOLOGY OF THE UPPER TERTIARIES OF EAST ANGLIA.

By S. V. WOOD, junior, F.G.S., and F. W. HARMER, F.G.S.

THE CORALLINE CRAG.

The oldest of the Upper Tertiaries of East Anglia—the Coralline Crag—is but a fragment of the once continuous deposit that must have spread at least from Tattingstone on the south-west, to Aldborough on the north, nearly the whole of the formation having undergone destruction prior to the accumulation of the Red Crag. It is not improbable that under the highlands formed of Glacial beds other isolated masses of the Coralline Crag may be concealed, but the only fragments which offer themselves to our investigation are a very small one at Tattingstone (see the south-western end of Section A), another at Pettistree Hall, Sutton (see also Section A), another at Ramsholt Cliff (on the Deben River, near Sutton), and the main mass which occupies the parishes of Orford, Sudbourn, Iken, and Aldborough. To these may be added a trace (not shown in the map) at Trimley, where it was observed in the digging of a ditch by the late Mr. Acton.

The Cor. Crag has long been known to consist of two main portions, and a third subordinate bed. The first and lowest of these, (3′ of Section XXII,) consists of a series of calcareous sands, in some places more or less marly, which are rich in Molluscan remains. The second, 3″, consists of a solid bed formed of Molluscan remains, agglutinated with the fronds and fragments of various species of Polyzoa into a rock, so hard as to have been formerly quarried for building. The third and uppermost, 3‴, is a thin subsidiary bed, consisting of a few feet of the abraded material of the rock, reconstructed evidently in very shoal water, probably, indeed, between tide marks, as it is very obliquely bedded.

From the outliers at Tattingstone and Ramsholt this rock bed is absent, but over the Sutton outlier a small cap of it remains. Over the main mass, however, it spreads continuously, and either from a slight northerly dip of the whole formation, or else from a displacement of the underlying shelly sands, this rock bed descends to the sea level at the northern extremity of the area. The thickness of the formation has been estimated at between eighty and ninety feet, but this seems to us to be much in-excess of the fact. The place to test the true thickness of the formation is clearly that where it is in the greatest state of preservation. This is the neighbourhood of Sudbourn, at the southern extremity of the main mass. The London clay, upon which it rests, comes out along the Butley Creek Marshes, and the shelly beds in their full force appear along the slope which fringes those marshes on the eastern side of the creek, where they are exposed in several pits, known as the Gedgrave, the Gomer, the Broom, and the Hall pits, from which the Mollusca of that neighbourhood are obtained. Higher up this slope comes the rock bed, which forms the upper part of the low hills, and this is exposed in numerous pits, that stretch away from those eminences to the northern extremity of the formation. The outcrop of the shelly beds (3′) along this slope is shown in Section XXII. Now, it is clear, as it seems to us, that, the whole formation being thus present in the complete state, we have only to take the elevation of the highest points attained by the Cor. rock bed, 3″, and its overlying reconstructed bed, 3‴, in these parts, above the elevation at which the London clay thus crops out at its base, and the difference, after making allowance for any slight dip there may be, will be the thickness of the entire mass. Estimated in this way, it will be difficult to make out the thickness of the Cor. Crag as exceeding sixty feet.

Mr. Prestwich has attempted to divide the shelly sands, 3′, into constant and determinable horizons, which he thinks might by investigation be identified by special groups of fossils. We doubt the constancy or determinability of such horizons; and, so far from their being characterised by special groups of fossils, the author of the ' Crag Mollusca ' in his long researches has mainly confined himself to one pit at Sutton, affording a vertical range of but a very few feet, and yet from this spot he has obtained specimens of nearly the whole known species of the Cor. Crag, many of these being known to collectors as occurring only at the pits near Orford. Not only this, but so inconstant is the Molluscan facies at any one place, that many species which once occurred at this spot (and some of them abundantly) have not been noticed there for many years. An attempt under such circumstances to group these shelly beds in any order of Molluscan succession would thus evidently be illusory. At this spot, moreover, Foraminifera were once abundant, and from it Mr. Wood collected all the species obtained by him from the Cor. Crag which are described in the Monograph of Messrs. Jones and Parker. No Foraminifera, however, have been found by him there for many years, although very many tons of the Crag from the same spot have been sifted by him for Mollusca during that period.

The depth of water under which this Crag was accumulated has been estimated by Mr. Prestwich at " possibly from 500 to 1000 feet." Against this view there are, it seems to us, objections. A depth of 1000 feet would have carried the Crag sea over the whole of East Anglia, and, indeed, across England,[1] and it is unlikely that all traces of such a sea should have been so completely removed as to present the blank which now exists. At elevations of about 350 feet (*i. e.* fifty-eight fathoms), some pebble beds occur through Southern Essex, which might represent the shingle of the southern shore of a sea having a depth of about sixty fathoms over Suffolk, and these might belong to the age of the Coralline Crag, but these seem better to associate themselves with the Bagshot beds on which they rest than to any later formation.[2] The author of the ' Crag Mollusca ' considers that nothing among the forms of Mollusca yet obtained from the Crag, points to the existence of any greater depth of water for their habitat than thirty-five or forty fathoms ; so that, coupling the physical difficulties with the exigences of the Molluscan evidence, we may, we think, regard the depth of the Cor. Crag sea of Suffolk as under 300 or even 250 feet, rather than as approaching 1000.[3]

[1] Some traces occur on the North Downs of Kent of beds supposed by some to belong to the oldest Crag, or Diestien formation. These are by the Geological Survey assigned to sands beneath the London Clay ; but if they be of Crag age, there has, we contend, been a great upheaval in this part of England in Post-glacial times, which has quite changed the relative levels of the Crag period.

[2] See ' Quart. Journ. Geo. Soc.,' vol. xxiv, p. 464, and also the forthcoming new edition of the Memoir of the Geological Survey on sheets 1 and 7, in which the view of their Bagshot age is adopted.

[3] The line of 600 feet is now distant between 350 and 600 miles from the Suffolk coast. The deepest parts of the North Sea between Suffolk and the Dutch coast are but 180 feet, while the depth of the chief part of that sea ranges between tide marks and 150 feet. Messrs. Jones and Parker, in the introduction to their Monograph (p. 3), say, in reference to the Coralline Crag Foraminifera, that they are best represented by dredgings between 300 and 420 feet south of the Scilly Isles, and from the Mediterranean at 126 feet. This small depth in the Mediterranean is significant, as it is with the Mediterranean that the Molluscan Fauna of the Coralline Crag has its chief affinity.

The *Entomostraca* throw no light on the subject, as Mr. Rupert Jones, according to the revision of the Monograph of himself, Mr. Parker, and Mr. Brady, made by him in the ' Geol. Magazine' (vol. vii, p. 155),

The Red Crag.

1. *Its structure.*—The physical structure of the Red Crag is unlike that of any other formation known to us, ancient or modern. The more considerable portion of it is formed of a succession of beds, varying from two or three, to nearly twenty feet in thickness, each of which consists of laminæ of sand and shells, inclined at a high degree to the horizon. The laminæ planes of each bed form an angle with those of the other beds above and beneath them; and at the base of each bed they change into a slight curl.[1] This structure is altogether different from the well known one of false bedding, which also exists in some parts of the Red Crag, especially in that under which the phosphatic nodules are worked, and the two forms of bedding pass more or less into each other. This oblique lamination may be traced (as, for instance, in Bawdsey Cliff) for a consider-

OBLIQUE RED CRAG IN BAWDSEY CLIFF.

The oblique bedded crag appearing above the Talus is nearly twenty
feet in thickness.

able distance in a constant manner, without shading off into horizontal stratification or passing into false bedding. If we examine a section of any beach or foreshore (at right angles to the line of the shore) we shall find it presents exactly this

describes eighteen species, of which only five are recognised as living, viz. *Cythere punctata, C. sublacunosa, C. trachypora, Cythereis ceratoptera,* and *C. tamarindus;* the first and fourth of which are species of the British coasts, the third and fourth of the Norwegian coast, and the fifth an Atlantic form. Of the Polyzoa their describer Mr. Busk says, in a letter to the author of the 'Crag Mollusca,' "Judging from the habits of existing forms, those of the Crag may have lived at any depth from the surface downwards," to which we may add that the rock bed made up of the remains of Polyzoa is very false-bedded, which is indicative of the reverse of deep water.

[1] Something approaching this oblique lamination may be seen in the Great Oolite of the Great Northern Railway cuttings near Grantham, and on the Yorkshire coast south of Scarborough.

feature. The pebble heaps of the London Clay basement bed about Bickley, in Kent, not unfrequently exhibit this, while the Lower Glacial pebble beds (to be spoken of hereafter) in the neighbourhood of Halesworth and Henham exhibit it yet more distinctly; their oblique bedding, which is constant and parallel throughout sections between twenty and thirty feet in depth, contrasting with the horizontal bedding of equally deep sections of the same pebbles at other localities—the one case representing the accumulations of a foreshore subaërially heaped up under the action of tide and surge, and the other the same accumulations spread out under a shallow sea. This beach or foreshore character pertains to a large part of the Red Crag, and well agrees with some of its other features; as, e. g., the intercalation occasionally of land and fresh-water shells in a truly marine deposit, which presents none of those intermediate features that occur in a fluvio-marine one;[1] the presence of beds of *Cardium* double, as they lived among the tossed-about heaps of other shells; and the bedding up of the deposit against and around a low cliff of Coralline Crag (as at Sutton) which is perforated by lithodomous mollusca. The Red Crag thus presents itself as the remains of an extensive series of banks that were more or less dry at every tide, and that were from time to time partially swept away and reaccumulated; every bed (of which three or four are often to be seen resting on one another) representing some of this destruction and reaccumulation, since the top of every preceding bed is planed off evenly to form a floor for the next above, in the base of which small pebbles and small rolled phosphatic nodules often abound, in some cases forming thin bands. In the channels which permeated these banks there seems to have been accumulated those portions of the Red Crag which exhibit the true features, often very extreme, of false bedding. If we examine any long section of this beached-up Crag where its base is exposed, such as in the cliff at Bawdsey, we do not find beneath it that bed, several inches thick, of phosphatic nodules which occurs in the various pits worked for nodule extraction, nor any of those large angular flints and masses of Septaria so abundant at the base of the Crag, where it is thus worked; while in these pits we find an absence (at least in the beds immediately over the nodule bed) of those foreshore features just described, and in their stead true and often very fantastic false bedding. An inference seems to follow from this that where these flint erratics and Septaria masses occur we have the bottoms of the channels permeating the banks, up which there drove floating ice freighted with flints from some chalk shore; and that these channels afterwards silted up and became part of the banks, for it is very remarkable that none of these erratics occur in the body of the Crag itself. If the main mass of the Crag had been deposited under water, we should naturally expect to find the large flints and masses of Septaria, so abundantly present in places at its base, distributed through the entire mass of the formation.

[1] Mr. Bell, in vol. viii of the 'Geol. Mag.,' p. 451, gives a section at Butley, where a bed of land and fresh-water shells is intercalated between two of these beached-up beds of Red Crag.

2. *The age of the different portions of the Red Crag.*—The physical structure of the Red Crag just discussed assists, we think, in the elimination of the age of its different portions, as indicated by their organic remains. The author of the ' Crag Mollusca' very long ago, and again in 1866,[1] pointed out an affinity between the organic contents of the Walton Naze bed and the Coralline Crag, which distinguished the former from the rest of the Red Crag. In this he was guided both by the predominance at Walton of certain species characteristic of the Coralline, and by the absence of others specially characteristic of the rest of the Red Crag. Thus, it is doubtful whether there has ever occurred at Walton any of the three species of *Tellina, T. obliqua, T. prætenuis,* and *T. lata,* the individuals of which make up the principal part of the shells of the Butley[2] and Scrobicularia Crags, and of the Fluvio-marine and Chillesford beds ; nor has there occurred there the several species of the genus *Leda,* so common in the Crag of Butley ; and particularly is there absent from Walton the shell so highly characteristic of the newer part of the Red Crag, of the Fluvio-marine and Chillesford beds, and of the Lower, Middle, and Upper Glacial formations, *Nucula Cobboldiæ ;* while the dextral form of *Trophon antiquus,* which abounds in the rest of the Red Crag,[3] is unknown at Walton, the sinistral or older form[4] alone occurring there.

In testing the age of the different portions of the Red Crag, we must reverse those rules which make negative evidence of so little value, and positive evidence of so much ; for it is the negative, and not the positive upon which we must rely in this case. Great stress has been laid upon the leavening of the Red Crag fauna by shells derived from the Coralline, but the inclination of modern research is to attach less importance to Coralline Crag derivation than formerly. It is, however, Red Crag derivation that seems to us to be the complicating element. It is obvious that where banks of sand full of shells are swept away and reaccumulated, the dead shells of these banks must be undistinguishably mixed up with those that have but just parted with the living animal ; so that if, as probably was the case, a very material change in the denizens of the Crag sea was taking place between the earliest and latest parts of the Red Crag formation, by the disappearance of certain old forms and the incoming of some new ones, we should get no clear record of it in the beds of Crag formed during this time ; because the existing fauna would in these contemporary accumulations be largely leavened by heaps of semi-fossilized shells of the antecedent period, derived from the destruction of the older banks, such as the Walton bed.[5]

[1] ' Quart. Journ. Geol. Soc.,' vol. xxii, p. 538.

[2] *T. lata* is rare in the newer beds of the Red Crag, but profuse in the Fluvio-marine and Chillesford beds. *T. obliqua* is a Cor. Crag. shell.

[3] Except in that of Bentley, the Crag of which presents a nearer approach to that of Walton than any other, although it has a decidedly newer facies.

[4] See the remarks as to this shell, p. 19 of the ' Supplement to the Crag Mollusca.'

[5] Collectors often clear out from their boxes,in order to make room for other specimens which they value more, shells obtained at other localities. These thrown away specimens may be found by the next comer, and spurious localities thus arise for shells. We anticipate that much confusion will arise from this, and

In this way it is almost impossible to separate the Red Crag into those chronological divisions of it that probably exist; but there are two, besides that of Walton, which may be clearly indicated. The first of these is the Red Crag of Butley, in which the northern forms of Mollusca predominate, and in which *Nucula Cobboldiæ* and the species of *Leda* become very common, and of which the whole fauna, both in individuals and in species, makes a great approach to that of the Fluvio-marine Crag, offering in this respect a great contrast to the Crag of Walton. This approach to the one and contrast with the other would, we are convinced, be greatly enhanced could we eliminate from the evidence those false witnesses, the intermixed shells derived from older banks swept away. The second of these beds needs no special palæontological test for its distinction, as it rests on the Red Crag of Butley, in the section under Chillesford Church (see Section XVII). It consists of Crag, gradually losing both the red colour and the oblique bedding as we ascend in the section, becoming horizontal in the upper layers. This Crag is poor in species, being largely made up of *Tellina prætenuis* and *T. obliqua;* but in it appears *Scrobicularia plana* in some abundance, a shell unknown in the other parts of the Red, but occurring in the Fluvio-marine Crag of Bramerton. Valves of *Mya truncata* also, which are unknown to the Walton bed, and almost so to the Crag of the Deben region, but which become common in the Butley Crag, are very abundant in these Scrobicularia beds, where exposed over the Coralline Crag at Sudbourn in Section XVIII.

Although there is doubtless an intermingling of more than one stage of the Red Crag in that region which is cut by the rivers Orwell and Deben, it would be impracticable to distinguish them further here; and, accordingly, in the sections all this Crag has been grouped under the same symbol as the Crag of Butley, viz. as 4″, although it is probably all, or most of it, older than the Butley bed; the still older bed of Walton being distinguished by the symbol 4′, and the newest, or Scrobicularia Crag, by 4‴.

The Chillesford Beds, and the Correlation of the Red-and Fluvio-marine Crags.

Although for the most part they seem to have been swept off the Red Crag region, yet we find this Crag capped in a few places by some beds that remain more complete over the Fluvio-marine Crag area. These consist of a micaceous sand (5′) in which occurs, though not constantly, the shelly bed *x*. This sand passes up without break into a bed of laminated micaceous clay (5″), which, in some localities, yields a few shells, or their casts. This bed varies from a dark blue tenacious laminated clay, as at Aldeby and

geologists must be on their guard against it. A specimen of *Tellina obliqua* obtained by H. Norton, Esq., of Norwich, at Walton, probably got there in this way. In Mr. Wood's collection in the British Museum is a specimen of *T. prætenuis* labelled Walton, which he thinks must have originally come from some other locality.

Easton Cliff, to a loamy micaceous sand, more or less interbedded with seams of laminated clay, as on the immediate west of Beccles and on the south of Norwich, but it is easily recognisable everywhere.

These beds 5′ and 5″ were first recognised[1] in the pit behind Chillesford Church (see Section XVII), where they occur immediately over the Scrobicularia Crag. Their fauna differs but slightly from that of the Fluvio-marine Crag, and as little, except in its greater richness, from that of the Scrobicularia Crag beneath them. In the well-known pit at Bramerton Common the Chillesford shell bed (x), with true marine facies, and its overlying laminated clay, are exposed, as well as the Fluvio-marine Crag itself, which rests on the Chalk (see Section XVI). The bed x is there divided from the Fluvio-marine Crag, $\overline{4}$, by about twelve feet of unfossiliferous sand (not distinguished in the section by any symbol), which exactly take the place of the Scrobicularia Crag of the Chillesford section. *Scrobicularia plana* occurs in the Fluvio-marine Crag of Section XVI, though rarely; but in another pit, about a quarter of a mile east of that represented in Section XVI, and known to the Norwich collectors as the Scrobicularia pit, a deep section of sand, interspersed with shelly beds, is exposed, part of which answers in position to the sands thus intervening in Section XVI, between $\overline{4}$ and x, and in this pit the shell is common. Comparing thus the section at Bramerton with that at Chillesford, the inference arises that the Fluvio-marine Crag in the former is the equivalent of the Marine Red Crag of Butley in the latter; and that the sands without symbol, separating the Fluvio-marine Crag from the bed x, are represented by the Scrobicularia Crag of the Chillesford section. Or we may even confine our correlation with the Fluvio-marine Crag to the base of the Scrobicularia Crag itself.

We might thus, without much hesitation, arrive at the conclusion that the Fluvio-marine Crag of Bramerton was coeval with the newest parts of the Red, were it not for one conflicting feature, which we have endeavoured to bring out by making Section XVII partly hypothetical. This section represents what we conceive would be presented by an excavation made at right angles to the pit under the Church (Stackyard pit), back to the Chillesford beds pit behind the Church. The section afforded by the Stackyard pit is truly represented by the extreme right of Section XVII; and there we have the Scrobicularia Crag overlaid by a few feet of sand marked ?. This sand is divided from the Scrobicularia Crag by a well-marked line of erosion, which descends in potholes into the latter. In this sand no organic remains have been detected, while that under the Chillesford clay, in the pit behind the Church, has yielded a series of fossils in high preservation. Holes sunk in the bottom of this pit disclosed the upper beds of the Scrobicularia Crag, which are exposed in the Stackyard pit below, but without any signs of such a line of erosion as that appearing on the face of the Stackyard pit. It is obvious that if the unfossiliferous sand marked ?, separated by this line of erosion from the Scorbicularia Crag,

---

[1] By a party of geologists headed by Mr. Prestwich; see 'Quart. Journ. Geo. Soc.,' vol. v, p. 345.

c

be the same as the sand 5′ of the pit behind the church, containing the fossiliferous bed x, the identity of the Chillesford with the Bramerton section fails; and the idea forces itself that the Fluvio-marine Crag of the Bramerton section, and its overlying sands, which pass so uninteruptedly up into the Chillesford beds, are newer than even the Scrobicularia Crag itself.

It, however, appears to us the more probable alternative, that the sands marked *?* are not the Chillesford sands of the pit behind the church, but some later deposit; probably the Middle Glacial sand (8), which, in a pit only a furlong north of the church, occurs under the Boulder clay (9). Figure XVII accordingly represents the section on this hypothesis, and on the assumption that if a clear section were carried down from the Chillesford clay of the pit behind the church into the Red Crag, it would disclose that uninterrupted passage of the formations into each other which is represented on the left side of the figure; the sands marked *?* lying somewhat in the way suggested by the broken line.

It is important to observe that while this probable conformity exists at Chillesford, the Red Crag of Walton is clearly unconformable to the Chillesford beds which overlie it; the sands of those beds (5′) lying on a very irregular surface of the Red Crag and filling up the depressions in it (see Section XXI); while both these sands and the laminated clay (5″) above them overlap the Crag on the south side, and rest there on the London clay (see Section Q).[1] Nothing, therefore, can be clearer, we think, than that the Walton Crag, so distinguishable by its fauna from the newer Red Crag beds 4″ and 4‴, had been denuded before the Chillesford beds overspread it, and is quite disconnected from them.

If the hypothesis presented in Section XVII is true, we see that the northern part of the Red Crag area continued to receive accumulations up to and during the time when the Fluvio-marine Crag was deposited; and that from a depression which then set in, the Fluvio-marine Crag gave place, through the sands shown in the Bramerton section without symbol, to the marine deposit, x, of that place; while the sandbank deposit, 4″, of the Chillesford section gave place through the beds 4‴ to the same overlying bed x; the submergence carrying the sands 5′, in which this bed x occurs, over the Walton Crag.

The other instances which exist of the Chillesford beds over the Red Crag area, such as those on the eastern side of the Deben, opposite Woodbridge, do not afford any section which would show their conformability, or the reverse, to the underlying Red Crag; but the appearances, as far as they go, all point there to an unconformability. How far the sands which, in some of the excavations for phosphatic nodule working, seem to pass down into the Red Crag by thin seams of comminuted shell in their lower part may belong to the Crag, it is difficult to say. They are not, we think, the Chillesford

---

[1] In Section Q this overlap is represented as existing at the northern end also; but as the face of the cliff is there obscured, this is uncertain.

The bed x can be detected in Walton Cliff, but the shells are in too decayed a state for extraction or even recognition.

sands, for they neither contain the shell bed x, nor, though sometimes twenty feet thick, do they present any traces of the Chillesford clay over them. It is not improbable that while the upper part of these sands belong to the Glacial formation, the lower with shell seams belong to the Crag, being the result of the silting up of those channels between the banks, at the bottom of which it is that the great chalk flints and other erratics occur. Accordingly, in the map we have, as the safer course, nowhere over the Red Crag area shown the Chillesford beds unless we could detect the Chillesford clay in the neighbourhood.[1]

The Fluvio-marine Condition of the Chillesford Beds.

In the Sections XVI and XVII first discussed, and in that at Aldeby (No. XIV), the Chillesford bed is marine; but in all the other sections of the beds, such as those shown in Sections VII, IX, X, XI, XII, and XIV, the bed 5' is more or less Fluvio-marine in character, and in the cases of Nos. IX, X, and XI, as much so as is the Fluvio-marine Crag of Bramerton itself; and, accordingly, the beds shown in the last-mentioned sections have always been regarded by collectors as the Fluvio-marine Crag. The sands, however (5'), which in these sections underlie the laminated clay (5″), and contain the shelly seam, x, are but very few feet in thickness. The more likely view seems to us to be that the sites of Sections IX, X, and XI, which were land during the formation of the Red and Fluvio-marine Crag, became, by the depression which at Bramerton caused the Fluvio-marine conditions to disappear, and the marine bed x to occur, covered by the estuary, and received a Fluvio-marine fauna; the clay 5″ eventually spreading over all. In some places in the valley of the Tese, near Saxlingham, the bed 5' is in the condition of compact pebbles overlain by the clay 5″, which is somewhat thin (see Section XII). This pebbly condition of the bed 5' seems to have extended across from the Tese valley to Bungay; for at Ditchingham, near that place, a patch of it full of shells occurs at the edge of the Waveney, in the garden of a house a mile north of the line of Section J. It is too small to be shown in the map, and is overlain by the Middle Glacial sand. This line of pebbles doubtless marked part of the south-western shore of the estuary portion of the Upper Crag sea.

[1] While this introduction is passing through the press, the Fluvio-marine Crag has been discovered by Mr. W. M. Crowfoot at the base of Dunwich Cliff, having been cleared of the talus covering by the storms of the winter of 1871-2; and various Crag shells, inclusive of *Cyrena fluminalis*, have been obtained by him from it. The Crag here consists of a deep red bed with shells, overlain by micaceous sands with threads or thin seams of clay. These sands seem to be those intervening between the Fluvio-marine Crag and the Chillesford beds, and they have accordingly been represented in the section of this cliff (R) under the same shading as the Crag $\overline{4}$. This shading should have been continued in the section to near the southern end of the cliff. An irregular line of denudation parts these micaceous sands from the red (or orange) sand marked b? As to the capping beds (10) of this cliff, see note, page xxix.

The limits of the Chillesford clay were extensive, for we find what seems to be this deposit as far north-west as Needham Market,[1] showing that the valley of the Gipping had come into existence prior to the Crag period; while well-preserved remnants of it extend northwards to near Burgh and Oxnead, on the Bure, and to Barton Turf on the Ant.

THE MAMMALIAN REMAINS OF THE CRAG.

Before leaving the subject of the Upper Crag, the bed of rough flints which occurs, not only at the base of the Fluvio-marine Crag, but also at that of the Chillesford sands, where these rest on the Chalk, should be noticed. The mammalian remains (chiefly molar teeth) which have come from this bed have been regarded by some as belonging to the fauna of the Upper Crag period; while by others this bed is regarded as an old land surface, in which the remains of animals that lived upon it are imbedded. From both these views, however, we dissent. Land surfaces are not to be looked for in a bed of rough flints, without any traces of a peat or soil covering, but in such beds as those of the forest series presently to be described; while it is not in pure land surfaces that the remains of the mammalia that lived on it are preserved (for in these the bones perish into dust by atmospheric agencies), but in the sediment of the pools, lakes, swamps, and rivers of the time; and even in these we find the remains in a more connected form than that of bouldered teeth, and portions of the more solid parts of the bones, such as have come from this bed of rough flints. On the other hand, these are just the remains which, when an anterior deposit yields to cliff waste, escape destruction from the waves, and are preserved in the estuarine and coast deposits of the subsequent period. Accordingly, we believe the origin of all the mammalian remains found in this bed to be derivative equally with the rough flints themselves; and we may add that marine Mollusca sometimes occur in the bed, though we never heard of a land shell having been found in it.

With the exception of the connected cetacean vertebræ which occurred a few years since in the Chillesford clay at Chillesford, it therefore seems to us open to much question whether any of the mammalian remains obtained from the Fluvio-marine Crag, or from the Chillesford beds, belonged to individuals which lived during the accumulation of these deposits. In the case of the Red Crag we venture to affirm that all such remains (even those of the Ziphioid Cetaceans, notwithstanding their occurrence at Antwerp) are derivative; since, independently of their consisting only of isolated fragments, more or less bouldered and perforated by lithodomous Mollusca, they do not occur in the body of the Red Crag itself, but only in the nodule bed at its base, which is, *par excellence*, a bed of erratics of all sorts. Of derivative origin also, are, as it seems to us, the mammalian remains occurring at the base of the Coralline Crag.

[1] In a pit on the north of the town by the Windmill. One side of the section shows the laminated clay forced up into a vertical position, apparently by the pressure of land ice during the subsequent Glacial period. There is a pit of micaceous sand three furlongs north-west of Easton Church (six miles north of Woodbridge) that may belong to these Chillesford beds, but we have not ventured to show it as such in the map.

The fact that no trace of the mastodon has yet occurred among the abundant elephantine remains from the Hasboro' Forest bed and the other beds of that series, of itself raises the doubt whether that animal lived during the age of the Upper Crag; since the Hasboro' Forest bed, if it be not actually coeval with the Upper Crag beds containing mastodon remains, is evidently separated from the latest [1] of them by an interval of time too slight (and accompanied apparently by no change of climate) satisfactorily to account for the disappearance of this great proboscidean genus. The Mastodon teeth found in the Red and Fluvio-marine Crags, and in the Chillesford beds, have, we think, been derived from destroyed freshwater deposits intermediate between the Coralline Crag and the Red; [2] while the Cetacean remains in the Red Crag have, as the author of the 'Crag Mollusca' long since suggested, been probably derived from the destroyed Coralline Crag itself.

The Forest Beds.

The several deposits of this series have been distinguished in the sections by the letters A, B, C, D, and E.

The Green clay of Bacton (A) occurs upon the beach about half a mile south of that place. It was described by the Rev. Chas. Green in 1842, [3] and consists of a greenish sandy mud, which has yielded numerous mammalian remains and freshwater Mollusca, as well as leaves and fragments of wood. It is associated with some blue and brown clay, but, like the other deposits of this series, is too much obscured by the shingle of the beach and the cliff talus to exhibit any section.

The Forest bed of Hasboro' (B) occurs at the base of the cliff, and on the beach. It consists of a ferruginous gravel indurated into a pan, which is associated with a bed of blue clay. In and on this occur the remains of a forest growth. This bed has also yielded abundantly mammalian remains, fir cones, and land and freshwater shells.

Beds A and B may not improbably be identical, while the traces of this old forest growth are said to occur at the following other places along the beach, but they are usually so covered by the modern beach shingle that it is only at rare intervals they are exposed. These places are three quarters of a mile south of Overstrand Gap, at Mundesley, [4] at Trimingham, and at Cromer. [5]

[1] A mastodon tooth was obtained by the late Col. Alexander from the Chillesford beds of Easton Cliff.

[2] If *Mastodon arvernensis* really has occurred at the base of the Coralline Crag, then the beds from which the remains of this animal have got into the Red and Fluvio-marine Crags may not improbably have been even anterior to the Coralline Crag.

[3] 'Hist. Antiq. and Geol. of Bacton,' Norwich, 1842.

[4] There is a freshwater Post-glacial bed at Mundesley (11) shown in section W, known as the insect bed, which might be confounded with the Pre-glacial beds of the Forest series, owing to the peculiar way it (like several other instances of Post-glacial beds) undercuts the Glacial beds, and is, so to speak, wedged into them.

[5] These localities are given from information furnished by the Rev. John Gunn to the author of the

The bed C is better exposed, and forms the base of the cliff at Woman Hythe (see Sections W and III). It consists of a dense black sandy clay, thickly packed with freshwater Mollusca, chiefly *Paludina contecta*, and is overlain by the Lower Glacial pebbly sands (6). These three beds A, B, and C, occur on the north Norfolk coast, while beds D and E are on the Suffolk coast, at the base of Kessingland Cliff (see Sections T and V). The base of this cliff is usually covered with talus, which obscures the section, but for a few years past the part shown in Section V has been very clear.

Bed D consists of a mottled clay, unstratified, its upper surface being full of small concretions and penetrated with roots. It extends from near the Lighthouse ravine almost to the southern extremity of Kessingland Cliff, and it has yielded Mr. W. M. Crowfoot, of Beccles, some mammalian remains.

Resting in a hollow of D, occurs bed E, which is a laminated clay, underlaid by prostrate trees, amongst which is a small colony of freshwater Mollusca. Most of the Mammalian remains obtained from Kessingland and Pakefield have, we believe, been found on the shore, doubtless washed out of beds D or E.

The relation of all these beds A, B, C, D, and E, to the Crag has been, and is likely to remain, a matter of uncertainty. As concerns A, B, and C, since there is, in our opinion, no marine bed exposed along the coast section between Eccles and Weybourn which is so old[1] as the Chillesford clay, by which to test their relative age, they may be coeval with the Upper Crag, or they may be posterior, or even anterior to it. The Chillesford clay may once have spread over that part of Norfolk, and been denuded prior to the formation of beds A, B, and C; in which case they are posterior. On the other hand, they may have accumulated on what was a land surface at the time when the Fluvio-marine Crag spread over East Norfolk and Suffolk, and the Red Crag over East Suffolk, as well as later, when these parts underwent the depression that gave rise to the Chillesford beds. Seeing how the Fluvio-marine and Red Crag deposits are confined to the more southern part of the area, and how much more thin the Chillesford clay is in the parts nearest to the Cromer coast, than it is further south, about Surlingham, Beccles, Easton Cliff, and Halesworth, and over the Coralline Crag of Aldboro', Sudbourn, &c., this latter alternative is far from improbable. Whichever way, however, we look at the question, the mammalian remains of beds A and B do, we think, more nearly represent the terrestrial fauna of the Upper Crag period than any others known to us; since, if

---

'Crag Mollusca,' in 1844. The indefatigable researches of Mr. Gunn in the Forest bed, and the costly collection made from it, and presented by him to the Norwich Museum, are so well known and highly appreciated, as to need no remark from us.

[1] We say this after repeated and most careful examination. What may be concealed below the beach between Weybourn and Mundesley at those places where the chalk does not rise to the beach line, or between Mundesley and Eccles where these occasional appearances of the chalk cease to occur, is another thing altogether.

not actually coeval with the Upper Crag beds themselves, they must have followed directly upon the elevation of the Chillesford clay into land.

With respect to beds D and E a different case presents itself. There occurs at the foot of the lighthouse ravine of Kessingland Cliff, a bed of micaceous clayey sand, which is marked in Section T as *5?*. So far as occasional clearances of the talus after storms has permitted Mr. Crowfoot (who has watched the section for some years) to see, the root-indented mottled clay D, with the freshwater bed E in its hollow, appears to lie up to this micaceous clayey sand, in the way shown in Section T. If this be so, and the bed *5?* be the Chillesford clay, then these beds, D and E, are posterior to the Crag. Even then, however, it by no means follows that D and E are coeval with A, B, and C, because, while the latter are overlain by the pebbly sand (6) and the other beds of the Lower Glacial series, the Middle Glacial (8) alone, as far as the talus allows us to see, rests upon D and E in the Kessingland section. As, therefore, the Lower Glacial beds are proved by outliers still remaining to the south (those of the pebbly sands being extensive) to have once overspread Kessingland, we cannot resist the conclusion that beds D and E, if coeval with A, B, and C, must have been covered by these Lower Glacial beds; and these, latter, to have been so *exactly* denuded prior to the Middle Glacial sands, as to have left the root-penetrated surface of the clay D intact, (as it may now be seen to exist, along a section of near a mile in extent) and yet have cleared away every trace of the Lower Glacial beds. All, therefore, that can safely be averred of beds A, B, and C is, that they are anterior to the Lower Glacial; and of beds D and E, that they are anterior to the Middle Glacial, and probably posterior to the Crag.

The Lower Glacial Series (Nos. 6 and 7).

The beds of this series (6, 6*a*, and 7) are shown in Sections C, D, E, F, G, H, J, K, L, M, N, O, R, S, U, and W. The lowest, the pebbly sands, 6,[1] consist of sands, mostly of a

---

[1] These sands were first brought to the notice of geologists by one of us under the name of "Bure Valley beds," and their range from the Bure Valley past Norwich to the neighbourhood of Southwold, as well as their inferiority to the contorted drift, and their superiority to the Chillesford clay, was shown by diagram and numerous vertical sections. This was in 1866 (see 'Quart. Jour. Geol. Soc.,' vol. xxii, p. 546). Their position, and that of the formations Nos. 1, $\overline{4}$, 5, 7, 8, 9, and 10, both above and below Norwich, will be found again represented in sections in the journal for the same year by the other of us, precisely as they are in that neighbourhood in the present sections (see vol. xxiii, p. 89). In 1868 the structure of Norfolk and Suffolk, illustrated by map and copious sections, was laid by us before the British Association in the same way in which it is represented in the present map and sections, with the exception that the identity of the pebbly sand of Weybourn which underlies the Cromer Till along the North Norfolk coast, with the Bure Valley beds, was expressly left open for further investigation; it being pointed out that the Weybourn sand and the overlying Cromer Till (with the base of which it is interbedded) occupied, relatively to the contorted drift, the same position as did the Bure Valley beds; so that they might either be the equivalent of these Bure Valley beds, or be a formation subsequent to them, and intervening between

deep orange colour, more or less interbedded with seams of rolled pebbles. The pebbles in some places so predominate as to form masses of shingle, while near their southern extremity, in the immediate neighbourhood of Halesworth (Section G) and of Henham (see northern end of Section C), they appear in the form of true beaches, as already mentioned at page vi. These sands and pebble beds contain shell fragments occasionally, but recognisable fossils only at Crostwick, Rackheath, Spixworth, Wroxham, and Belaugh inland, and along the coast at Weybourn, between that place and Runton, and about Trimmingham. They form the base of the whole Glacial series, and indicate the first setting in of the great Glacial subsidence. While their pebbles and shingle thus indicate shore conditions, their fauna is as truly Fluvio-marine as the beds $\overline{4}$ and $5'$; and while in some parts, as in Sections D, E (at Ingate), O (at Hartford Bridges), S, VIII, IX, X, XI, XII, and XV, they rest on and more or less indent the Chillesford clay; in others, as in Section G, they occupy its place, lying *in the beached form* up against foreshores of this clay.[1]

Their fauna has been investigated with some perseverance, but it is well worth further research. So far as yet known, it differs from that of the Fluvio-marine Crag and Chillesford beds in the disappearance, or in the increasing rarity of certain forms, rather than in the introduction of new ones. Of the three specially common species of *Tellina* characteristic of the Chillesford beds, *T. obliqua*, *T. lata*, and *T. prætenuis*, the latter is, in these

them and the contorted drift. (See abridgment of the paper in 'Geological Magazine,' vol. v, p. 452.) Having, soon after this, satisfied ourselves that they actually were such equivalent, we took the opportunity of laying a section disclosed by the Norwich Sewer Works before the Geological Society, in April, 1869, to assert this; and at page 446 of vol. xxv of their journal the succession of the beds about Norwich is once more shown in section; and the pebbly sands (Bure Valley beds), which in that section are shown as succeeding the Chillesford clay, are expressly stated to "expand northwards into the Weybourn sand and Boulder Till of the Cromer Cliff section, to be unconformable to the Crag and Chillesford beds, to be palæontologically distinct from them, and to be characterised by the first appearance in England of *Tellina Balthica*." In the 'Geological Magazine' for January, 1870 (pp. 19, 20, and 21), this position of the pebbly sands and Cromer Till was again pointed out very distinctly, and the character of the fauna which had been obtained by us from those sands at Belaugh, Rackheath, Weybourn, and Runton (Woman Hythe), explained. Lastly, in the same magazine for September, 1871, a woodcut of the coast section from Cromer to Woman Hythe is given exactly as in the present Section W. Subsequently to all this Mr. Prestwich, in the 'Quarterly Journal of the Geological Society' for November, 1871, brings forward this pebbly sand as something new, assigning to it the name of "Westleton Shingle"; and, apparently overlooking much of what had been thus brought forward by us, altogether misrepresents our views as thus matured, basing them apparently on the paper of 1866. Hence the above explanation. We have, under these circumstances, here refrained from assigning any other name to these sands than that of "pebbly" to denote their character; but if a local name be desirable, none can be so proper as that of "Bure Valley beds," both by reason of its priority, and of the fact that their greatest exposure, and their principal fossiliferous localities, occur in the valleys of the Bure and its tributaries.

[1] Some of the patches of Chillesford beds shown underlying the pebbly sands in Sections N and O are, of course, merely hypothetical, and are inserted under the mass of newer beds only to show the patchy way in which they remain wherever the chalk floor is exposed.

sands, extremely rare, and as we get no trace of it in later beds it was evidently then dying out.

There is one species, however, which comes in with these beds that is wholly unknown in any older ones, and is especially characteristic of every Glacial and Post Glacial deposit, viz. *Tellina Balthica* (*Psammobia solidula*). It has been objected that the occurrence of this shell at Weybourn is due to geographical causes, and that notwithstanding its presence there, some part of the Crag is comprised in the section at that place (No. IV); but apart from the fact that there is nothing at Weybourn answering physically to the Chillesford clay,[1] the Glacial, and non-Crag age, of this shell is proved by its absence from every exposure of sand (and they are many) that can be proved to belong to the Crag series by having the Chillesford clay over it; industriously as these beds have, many of them for half a century, been searched by collectors. Such Crag beds range up to within twelve miles of the Weybourn Cliff, occurring at Coltishall, Horstead, and Burgh, (see Sections IX, X, and XI) and in the neighbourhood of Coltishall and Horstead they are in the immediate contiguity of the pebbly sands with *Tellina Balthica*. Thus while, in every bed of this neighbourhood which can be proved to belong to the Crag by having the Chillesford clay over it, not a trace occurs of this shell, several fossiliferous sections of the Lower Glacial sands resting direct on the chalk are to be found in the immediate neighbourhood of such proved Crag beds; and these swarm with *Tellina Balthica*. Pits of the sands thus swarming with this shell occur at Belaugh (see Section N), only eight and eleven furlongs respectively from the well-known pit at Coltishall (No. X), and about three furlongs further from the equally well-known pit at Horstead (No. XI), in both of which the sands 5′, full of shells, are exposed with the Chillesford clay over them, but in which sands no trace of this shell, as before observed, has been detected. No impartial observer can, we think, doubt that such *Tellina Balthica* sands are those of the Lower Glacial formation, resting direct on the chalk in spaces from which the Chillesford beds have been previously denuded. It is only near the chalk floor that Mollusca usually occur in these beds, though where they overlie the freshwater bed C at Woman Hythe (Sections W and III) there occurs in them high up, where they are interbedded with the Till, a colony of *Mya truncata* double, and with siphonal ends uppermost, as they lived. We took from these sands at Weybourn part of a mammalian humerus.

Where these sands pass into shingle masses, we evidently approach the shore of the sea of their period, while where they are obliquely bedded in deep sections, we find its actual beach line. Tracing them from these parts towards the north Norfolk coast, and observing the way they are overlain by the contorted drift, where the Cromer Till is not present, as well as the the special characters of that drift, we find the Till (6a) to be merely the deeper water deposit of this early Glacial estuary, in which these shingle

---

[1] A few inches, or sometimes a foot, or little more, of gray clay over the large flints which rest on the chalk, and among which shells are interbedded, can, we think, never be mistaken for so well-defined a formation as the Chillesford.

d

masses accumulated; and that the Fauna of this Till, a glacier-fed deposit nearly destitute of Molluscan remains, is represented by the shells occurring in the sands of Crostwick, Rackheath, Belaugh, &c.   If we follow the coast section (W) we see that this stratified formation, the Till, is lenticularly developed; that is, it attenuates greatly towards either extremity of the Cliff section, being in its greatest thickness about Trimingham and Sidestrand.   Not only so, but we see throughout the Coast section, *that the pebbly sands are interbedded with the Till*, and that where this is thickest these pebbly sands begin to change their character, becoming very silty and interstratified with bands of ash-coloured Till, full of chalk débris, while the pebble seams are thin and intermittent.[1] The Till is found in well sinkings for a few miles inland of the North Norfolk coast, but it does not come to the surface, and it is not present where the pebbly sands come out along the valleys of the Bure and of its tributary rivulets (see Section N).

This clay, 6*a*, often strongly stratified, which was once supposed to be identical with the great Boulder clay (No. 9), thus turns out to be a mere estuary deposit, rapidly attenuating in every direction from its centre.   Part of the shores of this estuary we can define with approach to accuracy by the shingle masses and pebble beaches which lined them, while the abundance of rolled chalk lumps, and especially the great sheets of chalk several feet thick (VI of Section W) that are interstratified in the Till, prove that one side of this estuary must have been fed by a glacier, some part of which, at least, was in contact with the chalk district.

At the Hasboro' or Eastern extremity of the Coast section, there exists an unconformity between the Till and the overlying contorted drift, which is here uncontorted and finely stratified; but, so far as the churning up of the two formations at the other, or Western extremity of the Coast section allows of observation, the Till there seems to pass into the contorted drift without interruption.   Probably the unconformity of the Till at the Hasboro' extremity was due merely to some current, as we find the base of the contorted drift itself filling up, in the form of sand, hollows in the Till, and this sand unconformably overlain by the main stratified mass of that drift, as shown in Sections I and II.

The Contorted Drift.—The most extensive deposit of the Lower Glacial series is the contorted drift of Sir Charles Lyell (No. 7).   This bed along the Norfolk coast changes its aspect materially.   In the Western part of the section it becomes more chalky, by an intermixture of fine chalk débris and chalky silt with the red mud of which the Eastern half is mainly composed; and in this Western portion, masses, sometimes of enormous size, of marl, or reconstructed chalk (VII of Section W), are enveloped in it. A study of these masses shows clearly that the agency by which they were introduced was that to which the contortions were due, since masses of this material are usually

---

[1] They are here also much charged with lignite and peat débris, as is the Till itself in some places. This lignite and peat were doubtless swept off the old forest-covered surface by the land ice, then beginning to gather, and carried into this estuary.

present with the contortions; and as these masses cease eastwards, so do the contortions, the drift becoming, save for the presence of sandgalls, uncontorted; and about Bacton and Hasboro' putting on the stratified condition shown in Sections I and II, wherein bands of fine mud and sand alternate with chalky silt, and bands of clay more or less intermixed with chalk débris. Some of this interstratified material is scarcely distinguishable from the material of the great chalky clay No. 9; while more of it is identical with the marl of which the masses, No. VII, are composed. These masses, again, are identical with a formation that covers much of North-west Norfolk, and occurs also in the South-west, consisting of soft chalk finely ground up by the enveloping land ice sheet of the period, and spread out from its seaward termination into a deposit that frequently shows stratification (sometimes very fine), and in which great sandgalls, like those so abundant in the red mud portion of it, occur.[1] It is in the district where this reconstructed soft chalk of North-west Norfolk changes horizontally into the red mud or brick-earth that the included masses of marl, such as the Coast section presents, become frequent; being sometimes acres in extent, and worked for Marl pits, or for limekilns. The mass supplying one of these limekilns shows itself enveloped in the red mud in the Cliff section on the south side of Cromer Town, and another, west of Woman Hythe, which is some 300 yards long, has sunk down to the chalk, squeezing out the Till on either side. This reddish-brown mud is easily followed from the Cromer coast southward to Norwich in a continuous deposit; but the Marl masses in it cease a few miles north and north-west of that city. About Norwich it is worked as brick-earth in numerous sections, but, as the contortions and Marl masses are absent from it here,[2] it has been regarded by the Norfolk Geologists as "the Lower Boulder clay." There can be no question that in these contortions and Marl masses we have evidence of the grounding in the red mud of icebergs detached from the sea foot of the land glacier occupying the Chalk country, which were laden at their bottoms with masses of the same degraded chalk which was extruded from the glacier foot, and spread out under the sea over North-west Norfolk. South of Norwich the deposit becomes very intermittent, and often very thin, its thinness being apparently due principally to intra-glacial denudation.

[1] Even in this district it will suddenly change into a sandy stratified silt, or into compact yellow brick earth. Sections of either of these conditions of this drift occur with the clay No. 9 over it (beyond the limits of the map) near Guist, in the Wensum valley; that valley having been excavated out of the contorted drift, in the interval between this drift and the Middle Glacial No. 8, which, with the clay No. 9, occurs in the valley, and in some places *in the bottom of it*.

[2] It is contorted in a pit four furlongs north-north-west of Thorpe Asylum, but that, we think, is not a contortion of the period of this drift, but due to the intraglacial excavation of the valley here, which causes the later Glacial beds to plunge into the valley. The pit three furlongs north-west of Upton Church (near the eastern end of section O), shows in the most striking manner the contorted drift in its red mud form, contorted, and containing sandgalls, just as it is about Cromer, overlain by the great chalky clay No. 9. Section II represents a spot in Hasboro' Cliff, where a tongue-like piece of the contorted drift (finely stratified with chalky silt and clay with chalk débris) has been lifted during the accumulation of the overlying sand, 8, which has found its way under the piece without its becoming detached.

In tracing this drift from the East coast at Hopton, and from the numerous inland sections of the extreme North of Suffolk and East of Norfolk, towards the Cromer coast, we see it in a fine section at West Somerton, overlain by the sand, No. 8, which is again close at hand overlain by the clay, No. 9 (see Section M), and we then lose it by the intervention of several miles of marsh. On the cliff beginning to rise at Eccles the drift appears again in the identical form of reddish-brown brick-earth possessed by it at Somerton, but overlain by a thin bed of stony clay (No. XI), to be presently noticed. From this place westwards it can be seen to assume gradually the finely stratified appearance it possesses at Hasboro', and then to change still further west into its original red Brick-earth condition.

This deposit must once have spread far to the southward, for what appear to be outliers of it occur at Kesgrave (five miles south-west of Woodbridge), and at Blaxhall, half a mile east-north-east of the church (seven miles north-east of Woodbridge); while beyond the limits of the map we have found it in the south-west of Suffolk, near Boxford, and probably at Sudbury. The Blaxhall outlier contains marl masses similar to those in the North Norfolk coast section.

The breaking off of this deposit into outliers southward is evidently due to a great denudation of the Lower Glacial formation prior to the accumulation of the Middle Glacial sands, which occupy to a great extent troughs or valleys in the Lower Glacial beds; and it is quite possible that outliers of it may be concealed under the tablelands of Middle Glacial sand which separate the East Anglian valleys from each other in Section A. It is clear that the valley system of East Anglia had its inception in this denudation, though it would be beyond the scope of the present outline to show this further than appears from the sections accompanying the map.[1]

The Contorted Drift has yielded no fauna as yet worth mentioning; but *Tellina Balthica* and fragments of *Cardium edule, Cyprina islandica,* and *Mactra ovalis,* and of a *Mya,* are not unfrequent in it. We took from it at Elsing (twelve miles west-north-west of Norwich) the femur of a small mammal.

THE MIDDLE GLACIAL (No. 8).

This formation is principally composed of sand within the limits of the map, though gravel is more or less intermixed in places; but southwards, towards and over Essex, the formation consists mainly of gravel. These sands have their greatest thickness over the Red Crag region, and were long confounded with that formation, under the term "unproductive sands of the Crag." Although the sands over the Crag are treated by us as all

---

[1] We should explain, however, that it is to this intraglacial erosion that the chalky clay (No. 9) owes its abnormal position in the bottoms of valleys cut through the older Glacial beds—a position which, before we had discovered the true explanation of it, led us to suppose (as suggested in vol. xxiii of the 'Quarterly Journal of the Geological Society,' p. 89) that the clay in the valley bottoms was a different Boulder clay of subsequent origin to the clay No. 9.

belonging to the Middle Glacial formation, it is not impossible that the lower part of them, which is sometimes very red, may belong to the pebbly sand, No. 0; but no means occur to us by which that could be determined, the whole sand mass presenting no line of division, and being unfossiliferous alike. It is easy, however, to see that they have no connection with the Red Crag, for along the north side of the Stour and south side of the Orwell, this Crag occurs only in places under them, the sands there resting for the most part on the London clay. Moreover did any part of the sands covering the Crag belong to that formation, it would be to the Chillesford beds, which overlie the Red Crag to the north at Chillesford, and to the south at Walton. Over the greater part of the Red Crag area, however, no trace of the Chillesford clay appears in the numerous deep sections of sand with this Crag at their base, which there occur. The Red Crag has, evidently prior to the Middle Glacial, been denuded of some of its upper part, even over the country east of the Deben, since the Scrobicularia beds, which, by losing upwards their oblique stratification, indicate an increased depth of water which must have carried them over the Crags of Butley, Capel, Boyton, &c., do not occur south or west of Chillesford. Between the Deben and the Orwell this denudation has been greater, for at one point, north-west of Nacton, the Middle Glacial has for an interval of three miles no Crag under it; while between the Orwell and the Stour, and across the Stour towards Harwich and Walton, only small patches of Crag occur at intervals under the Middle Glacial, that formation for the most part resting direct on the London clay; and sometimes, as at Wrabness, only the phosphatic nodule bed has escaped. As the Chillesford beds, which must once have spread from Walton Cliff northwards over the Red Crag, have for the most part[1] gone with it, this denudation must have been posterior to them, and most probably was the same as that which intervened between the contorted drift and the Middle Glacial in Norfolk, as the two (apparent) outliers of the

RUGGED SURFACE OF OBLIQUE RED CRAG ENVELOPED IN MIDDLE GLACIAL SAND IN A PIT A QUARTER OF A MILE S.W. OF MELTON CHURCH, NEAR WOODBRIDGE.

The dark oblique bedded mass is the Crag.

---

[1] There may of course be patches of these beds over the Red Crag between the Stour and the Alde, which, for want of sections, are not shown on the map, and are lost in the expanse of Middle Glacial sand.

contorted drift at Blaxhall, and Kesgrave, lie just within the limits of the Red Crag area. The foregoing figure shows the manner in which these Middle Glacial sands along the Deben region rest on and envelope rugged surfaces of the Red Crag.

Over the greater part of their range these sands are unfossiliferous, and it is only in the neighbourhood of Yarmouth that they have yielded a fauna.[1] This fauna is a very interesting and important one, but it requires great patience to obtain, owing to the sparse occurrence and fragmentary condition of the specimens. The two principal places from which it has been procured, are Billockby, eight miles north-west of Yarmouth, and Hopton Cliff. At the latter place (see Section U) the sands are underlain by the contorted drift, No. 7, and overlain by the chalky clay, No. 9. At Billockby they are overlain by the clay No. 9, while No. 7 comes out along the lower ground. No question can thus arise as to the position of the sands from which this fauna has been obtained. The specimens come from a thin shelly seam at the top of the formation some four or five feet below the clay, No. 9. This fauna is specially interesting, not only as showing a much older aspect than that of any of the Glacial beds of Scotland, and even of Bridlington, but also in its approaching that of the Coralline Crag by the presence of such species as *Turritella incrassata*, *Nassa granulata*, *Chemnitzia internodula*, *Dentalium dentalis*, *Limopsis pygmœa*, *Cytherea rudis*, *Cardita scalaris*, *C. corbis*, *Woodia digitaria*, *Astarte Burtinii*, *A. Omalii*, and *Erycinella ovalis*, the three last of which are not known living, *Cardita scalaris* being a Pacific shell, and the rest Mediterranean and Atlantic species, not ranging so far north as Britain. On the other hand, its affinity with the Red and Fluvio-marine Crags and with the Chillesford bed is shown by the presence of *Cerithium tricinctum*, *Tellina obliqua*, and *Nucula Cobboldiæ*—species not known as living—as well as by that of several other Red Crag forms which are now living in the Mediterranean and range into British seas, but which are unknown further north.[2] In this respect the Middle Glacial fauna presents a contrast (which no difference of latitude will explain) not merely to the Post-glacial beds of Kelsea, March, Hunstanton, and the Nar valley, but also to the fossiliferous so-called Glacial beds of the Severn valley, of Wales, of the North-west of England, and of Scotland. Messrs. Crosskey and Robertson, to whom some of the sand was submitted, found on a cursory examination that the Foraminifera occurring in it were, like the shells, much worn, and that they presented an arctic character, varied by the presence of one or two Tertiary forms. The peculiarity of this fauna naturally prompts a suspicion that it may be derivative from the Crag, and it is necessary therefore to examine that question.

---

[1] We have observed fragments of shells in the Middle Glacial, however, at other places, viz. at Wisset, two miles north-west of Halesworth; at the Brick kiln on the North of Stowmarket (where the sand is overlain by a Post-glacial Brick clay) ; and at Helmingstone, six miles north of Ipswich.

[2] This Fauna, as well as that of the Lower Glacial, will be given by the author of the 'Crag Mollusca' in the tabular list to appear in the concluding part of his 'Supplement.' In the meantime see, as to the Middle Glacial Fauna (latest results), 'Geol. Mag.' vol. viii, p. 410; and as to the Lower Glacial Fauna, 'Quart. Journ. Geol. Soc.,' vol. xxvi, p. 92.

Premising that the general condition of the specimens justifies the assumption that if any of the species be derivative, the whole are so, since, the most remarkable species are quite as well preserved as such forms as *Purpura lapillus, Trophon antiquus,* and *Tellina Balthica* (whose genuineness would be doubted by none), the following reasons for the genuineness of the entire fauna offer themselves :

1. So far as the Crag fauna is known, it would require in order to furnish by derivation the Middle Glacial shells, the Coralline, Red, and Fluvio-marine Crags, as well as some other bed for four of them which do not occur in the Crag, viz. *Tellina Balthica, Venus fluctuosa, Loripes lactea,* and the not unfrequent *Trophon mediglacialis,* which is the characteristic shell of the formation. 2. Not a trace or fragment of most of the common strong shells of the Coralline and Red Crags has occurred. 3. The sinistral form of *Trophon antiquus,*[1] which is profuse throughout the Red and Fluvio-marine Crags, and frequent in the Chillesford and Lower Glacial sands, is absent ; while specimens of the dextral form, and especially fragments of the columella and mouth showing the dextral turn, are abundant. No derivation from the Red Crag, in which myriads of these strong sinistral Trophons occur, could have taken place without their fragments being present abundantly. 4. The specimens of *Pectunculus glycimeris* which make up so large a part of the Red Crag are large shells, whereas this species though abundant in the Middle Glacial is mostly in the condition of very small individuals and fry, the largest specimen or fragment not nearly equalling the average size of the Crag shells. Any derivation from the Red Crag would have brought an abundance of these large specimens, or of their fragments, into the Middle Glacial sand. 5. Some of the species, such as *Venus fasciata,* are, although greatly worn, and mostly consisting merely of the hinge portions of the shell, perhaps fifty times as abundant as in any known Crag bed.

These five reasons, to which others might be added, seem to justify our regarding the Middle Glacial fauna as contemporaneous and not derivative ; but although contemporaneous, it is evidently one which did not live on the spot where it is found, since not only are all the specimens, with the exception of the *Anomiæ,* more or less rolled, but the limited extent of the fossiliferous area suggests that it was only here that a current bringing the shelly sand from some other part of the sea bottom impinged. Interspersed with these rolled and worn Molluscan remains there occur in great numbers small and perfect valves, always single, of the tender papyraceous *Anomia ephippium.* These are all young shells from one eighth to one quarter of an inch in diameter ; and their occurrence among such a rolled accumulation suggests the idea that they adhered to floating bodies which were brought to the spot by the same current that swept the shelly sand along the bottom ; their tender valves having sunk as these floating bodies decayed, and become intermixed with the worn bottom-travelled Mollusca.

---

[1] See, as to the history in time of this shell, which is an indirect confirmation of the above argument, p. 19 of the ' Supplement.'

THE GREAT CHALKY CLAY, OR UPPER GLACIAL (No. 9).

This wide-spread unstratified deposit extends over the eastern and the east central counties, and is there wholly unfossiliferous. In the parts traversed by the sections it is mostly of no great thickness, owing apparently to denudation; but further from the coast, and away from the main valleys, its thickness is more considerable, amounting in the west of Suffolk to as much as 120 feet,[1] and in Cambridgeshire to still more. Where it does not descend, as it often does, in a plunge, but rests evenly on the Middle Glacial sand, as around Woodbridge and in Sections T and U, this clay not unfrequently passes down into that sand by a passage bed consisting of a few feet of the clay which becomes more sandy downwards, until it shades off into the subjacent sand. In other places it lies evenly upon the sand in a conformable manner without any such passage bed; and there can be little question that the succession was effected by tranquil deposition, uninterrupted by any change of conditions other than those which produced a new kind of sediment. This clay all may admit to be the result of the degradation of the Cretaceous, Oolitic, and Liassic districts by land ice—the *moraine profonde* of the great enveloping ice sheet of the Upper Glacial period; but no one can look at the even way in which it overlies the sand in Kessingland and Hopton Cliffs (Sections T and U) without being convinced that it could not have been *so placed* by the action of land ice, since this could not have failed to crumple up and distort those sands, just as the intruding bergs bringing the marl masses churned up the contorted drift. Section P shows what an ice plough of the period of the clay No. 9 would accomplish. The Crag and the overlying sand, 8, of this section appear on one side of the cutting at the East Suffolk Railway Junction; but in the short space of the width of the cutting they are cut off so entirely, that the opposite side presents nothing but a churn up of sand and London clay; while a piece of the Red Crag thus ploughed out has been left two miles off, at a high level, in a mass of gravel about the junction of the Middle Glacial sand with the chalky clay, and is exposed in a pit south of the "Sparrows nest." If this be the result of the grounding of an ice island, as we cannot doubt it was, how much of the sand of Hopton and Kessingland Cliffs would have been left under the pressure of the land-ice sheet itself? Nevertheless this clay, not only in these cliffs, but all over East Anglia, is quite as unstratified as is that of Scotland and the north of England, and it is everywhere, except on the Yorkshire coast, quite as destitute of organic remains.

The species given in the Crag Mollusca and its Supplement from the Upper Glacial formation are from the Bridlington bed, the age of which appears to us to be posterior to that of the great chalky clay of East Anglia (No. 9); the bed occurring in a continuation of

---

[1] This was the total thickness in the cutting and contractor's well beneath it at Horseheath on the borders of Cambridgeshire and Suffolk. The cutting and well at Old North Road Station of the Cambridge and Oxford Railway showed this clay there to be about 160 feet in thickness.

that clay, which is absent from the country south of Yorkshire and the north-east of Lincolnshire.

THE YORKSHIRE COAST SECTION FROM THE MOUTH OF THE HUMBER TO SPEETON (50 miles).[1]

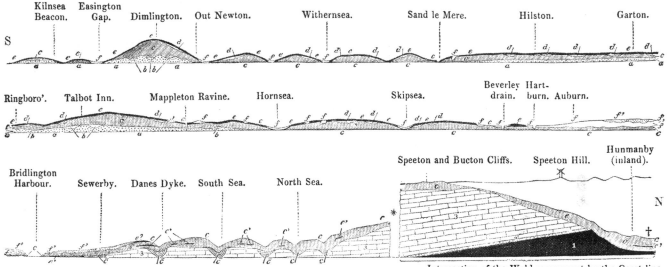

Intersection of the Wold escarpment by the Coast line.

Vertical scale, 500 feet to the inch. Horizontal scale of the part between Kilnsea and Bridlington, $2\frac{1}{2}$ miles to the inch; of the part between Bridlington and the break in the section, 2 miles to the inch; and of the Wold intersection, 2 inches to the mile. 1. Oolitic formations. 2. The Red Chalk. 3. The Chalk. *a.* Blue clay, principally composed of Oolitic and chalk *débris* (No. 9 of the Norfolk and Suffolk map and sections). *b.* Occasional thin beds of sand and gravel, and bands of clay intermediate between *a* and *c.* *c.* The purple clay with chalk *débris* in its lower part. This clay occurs as far south as Cleethorpes in Lincolnshire (where it is capped by *e*), but it then becomes lost under the great marsh of East Lincolnshire. *c′* Sand and gravel beds in *c*, from one of which at the beach line immediately north of Bridlington Harbour the Bridlington shells referred to in the 'Crag Mollusca' and its 'Supplement' have been obtained. $\bar{c}$. Moraines of rolled chalk lumps under *c*, which are probably terrestrial equivalents of the lowest portion of the Marine Glacial Clay, *c*, in the southern part of the section. *d.* The Hessle sand and gravel (Kelsea Hill Gravel). *e.* The Hessle clay with boulders. *f.* Sands and gravels posterior to *e*, and which at Hornsea contain fresh-water mollusca. *f′.* Still later gravel, principally composed of chalk fragments. The recent Cyclas marls omitted. The asterisk marks an interval omitted of four miles. † marks the source of the Hertford river, which, rising near the cliff, flows away from it inland. The lowest part of *c*, or that with the chalk *débris* in greatest abundance, dies out some way south of Bridlington, near which, its place becomes occupied by the chalk moraines, $\bar{c}$.

Along the South Yorkshire coast the unstratified chalky clay (No. 9) of East Anglia occurs in the lower part of the cliff, beneath which, judging from adjacent borings inland, it descends near the Humber end, about 100 feet, being in some of the borings underlain

[1] Taken from a paper by S. V. Wood, jun., and J. L. Rome on the "Glacial and Post-glacial structure of South Yorkshire and North Lincolnshire" in 'Quart. Journ. Geol. Soc.,' vol. xxiv, p. 146. The cut is reproduced here by the kind permission of the Council of the Geological Society.

by a few feet of sand. This thickness, added to what appears above the beach line, does not differ much from the maximum thickness of the clay in Suffolk. Over this clay, and separated from it in places by the beds *b*, there appears another clay, *c*, differing from it in colour (being of a purplish brown), but more particularly differing from it in its contents; for while in the clay, *a*, chalk débris is the principal, and débris of beds older than the chalk the subordinate ingredient, in the overlying purple clay, *c*, these proportions are reversed. Further, the purple clay, *c*, is not only less abundantly supplied with chalk, but this ingredient diminishes so much upwards, that where the uppermost part of the clay remains undenuded, in the highest cliffs, such as those of Dimlington and the "Talbot," it is wholly absent; showing the gradual release of the chalk country from those degrading agencies which supplied its débris so profusely to the earlier and under-lying East Anglian clay, *a*. North of Hornsea no trace of *a* occurs above the beach line, and where the chalk floor comes to the surface about a mile north of the Bridlington shell bed [1] (which is at the beach line immediately on the north side of the harbour), nothing like the chalky clay *a* appears over it; but instead, the purple clay *c* is seen resting direct on the chalk, underlain occasionally by some moraines of rolled chalk, *c̄*, that are probably coeval with the lowest part of *c* further south. The position of the Bridlington bed seems therefore to be about the middle part of the purple clay, *c*, where there is some chalk intermixture; and though posterior to the great chalky clay of East Anglia, the bed was evidently succeeded by all that part of the Upper Glacial period in which the rest of the purple clay, both with and without chalk, accumulated. Although the Bridlington fauna does not present nearly such Crag affinities as that of either the Lower or Middle Glacial, and is much more arctic than either, it is nevertheless distinguishable from that of every Scotch, north of England, and Welsh bed, by the presence of those special Upper Crag forms *Tellina obliqua*, and *Nucula Cobboldiæ*, shells not now living, and whose nearest living affinities occur in the Pacific; while all the species of the Scotch, Welsh, and north of England beds are to be found on one side or other of the Atlantic, or in the Arctic seas connected with it.[2]

THE PLATEAU GRAVEL (No. 10).

This gravel is everywhere unfossiliferous, and is composed almost entirely of flint. It is difficult in some cases to form an opinion whether it is of Glacial or Post-glacial age. Most of that which is shown in the sections under the number 10 is doubtless

---

[1] The bed was always so covered by the shingle as to be exposed only at rare intervals, and it is now, we believe, buried under the artificial works of an esplanade. Borings in the harbour showed it to be under-lain by twenty-eight feet of the clay *c*, under which were fifteen feet of gravel, and then the chalk.

[2] At Dimlington Cliff, a little below the beds *b*, Sir Charles Lyell and Mr. Thomas MᶜK. Hughes found a thin band of sand intercalated in *a* which contained Mollusca, some, as Mr. Hughes informs us, having their valves united. The species they found have not been communicated to us.

Post-glacial, but with respect to that which caps Mousehold Heath (see Section O), and that which caps the high land of Poringland and Strumpshaw (see Section L), it seems the same as the gravel which has an extensive spread in West Norfolk (beyond the limits of the map); since, like it, the gravel of these places, especially that of Monsehold, contains beds of very large flints more or less rolled. These gravels of West Norfolk set in almost along the same line as that about which the Middle Glacial sand ceases, *i. e.* along a line extending from Hingham in the south, to Wells in the north. This West Norfolk gravel is also composed almost entirely of large flints, which are mostly so rolled as to resemble cannon shot. These cannon-shot gravels sometimes contain masses of sand formed of chalk grains; and as they are never overlain by the chalky clay (9), but in a few instances have this clay, under them, it may be that, if not of Post-glacial age, they are a local modification of such clay due to the action of some powerful current over this part of Norfolk, which dissolved all the soluble part of the morainic material forming that clay, and rolled the flints into the cannon-shot form. The absence of these thick gravels over all the southern part of East Anglia is a peculiar feature, but some thick beds of gravel, which occur on the Chalk Wold of Yorkshire about Speeton and Bucton, seem to bear a relation to the purple clay (*c*) capping the Wold at those places; similar to that which the plateau gravels of Norfolk bear to the clay, No. 9. These Wold gravels, moreover, seem absent over the clay, *c*, where it occurs further south, viz. over Holderness.

THE POST-GLACIAL FORMATIONS (No. 11).

Several localities for shells from marine beds of this age are given in the 'Supplement to the Crag Mollusca,' viz., Kelsea Hill, Paull Cliff, March, Hunstanton, and the Nar Brickearth.

The Kelsea Hill bed is in Yorkshire, adjoining the railway from Hull to Withernsea, one mile east of the Burstwick station. It consists of sand and gravel, rich in individuals of shells, all more or less rolled, and is specially notable for an abundant admixture of the river shell, *Cyrena fluminalis*, with the marine Mollusca; pointing presumptively to the inference that the bed was accumulated after the glacial submergence had given place to emergence, so that a river flowed not far distant from this spot. The deposit was described many years ago in Prof. Phillips' 'Geology of Yorkshire,' and more particularly again by Mr. Prestwich in vol. xvii of the 'Quarterly Journal of the Geological Society,' wherein a more copious list of Mollusca than our researches have afforded is given by Mr. Gwyn Jeffreys. The whole of these consist of species still living, and with three or four exceptions that are Scandinavian they are British. The geological position of this deposit will be found fully examined in the paper from which the preceding section of the Yorkshire Coast is taken; and by the permission of the Council of the Geological Society the accompanying cut from that paper of the section afforded by Kelsea Hill in 1867 is given.

KELSEA HILL BALLAST PIT IN 1867 (Extreme height of section, 35 feet).

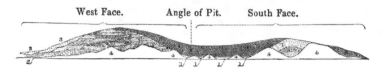

1. Sand and gravel with Marine shells and *Cyrena fluminalis* (*d* of the Yorkshire coast section). 2. The Hessle Clay (*e* of the coast section). 3. Newer Gravel (*f* of the coast section). N.B.—The Clay, No. 2 at the extreme left of the section, is probably only a subaërial wash down of No. 2, which is *in sitû* in the centre of the section. 4. Talus.

The Kelsea gravel is overlain by a clay containing some stones and boulders (mostly small), which, from its identity with that of the well-known Hessle section, has been termed "the Hessle clay." This clay is shown in the preceding coast section under the letter *e*, and there wraps the whole of the denuded edges of the purple clay to which the Bridlington bed belongs like a cloth, and is underlain irregularly by beds of sand and gravel called "the Hessle sand" (*d*) that are presumably identical with the gravel and sand of Kelsea Hill. This presumption accords with the molluscan contents of the gravel when contrasted with those of the Bridlington bed; for at Kelsea there is not only an absence of the specially arctic and American forms which characterise the Bridlington bed, but also of those characteristic crag forms *Tellina obliqua* and *Nucula Cobboldiæ*.

Paull Cliff is on the Humber; but it has been nearly destroyed in making the battery at that place. The sand and gravel of Paull, there can be little question, belongs to the Hessle sand (*d*), as it has yielded similar shells to the Kelsea gravel, Mr. Prestwich finding in it the *Cyrena;* and it rests on either *a* or *c*.

The Hunstanton gravel resembles in its palæontological aspects the Kelsea Hill bed in consisting entirely of living species, and none but those inhabiting British seas have yet been obtained by us from it.[1] It is not, however, overlain by anything answering to the Hessle clay, which caps the Kelsea gravel, that clay not having been traced further

---

[1] This gravel and that at March was first described by Mr. Seeley in the 'Quarterly Journal of the Geological Society,' vol. xxii, p. 470, but two shells that he names from Hunstanton, *Nassa reticosa* and *Tellina obliqua*, have not been found by others, and are believed to be *Nassa reticulata*, and the large, rounded form of *T. Balthica*, both of which occur there. The shell also, called by him *Astarte crebricostata*, from March, is most probably *A. borealis*, which occurs plentifully there, and which is a somewhat abnormal form of this shell, and approaching *A. crebricostata*. We have not yet been able to find *Tellina proxima* given by him from March, nor to recognise anything like the Glacial clay there, the clay with some minute fragments of chalk, which underlies and seems connected with the gravel, being, we consider, of Post-glacial age.

south than Steeping, in Lincolnshire, on the opposite side of the Wash, and about 20 miles west of Hunstanton ; neither does the *Cyrena* occur in it.[1]

The March gravel occurs around March railway station, and in pits in the town itself. It is in the midst of the Cambridgeshire Fen,[2] and there are no means of testing geologically its position, as it forms but small islands rising out of the recent peat and alluvium of the great level of the Fen. It is very rich in individuals of Mollusca, which are in good preservation, and its fauna, so far as yet known, resembles that of Kelsea and Hunstanton in consisting entirely of species now living, and which, with two exceptions, still inhabit British seas. In all these gravels, as well as in the Nar Brick-earth, *Ostrea edulis*, which is absent from all the East Anglian, and, indeed, from all the English glacial beds,[3] is abundant ; and there can be little doubt that the four are synchronous and belong to the earlier, or *Cyrena fluminalis* part of the Post-glacial period.

The fifth fossiliferous marine Post-glacial formation referred to, is that of the Brick-earth of the Nar Valley, in north-west Norfolk, which occurs at East Winch, Bilney, Pentney, East Walton, Tottenhill, and other places. Our knowledge of the Mollusca of this formation is wholly due to the late Mr. C. B. Rose.[4] Its fauna, like that of the gravels of Kelsea Hill, Hunstanton, and March, consists entirely of species still living, and which, with one exception, occur in British Seas. The shells of this formation are in fine preservation ; and there can be little doubt of the deposit being one of an estuary connected with that sea, which a Post-glacial depression, shown by the Hessle beds, caused to overflow the lower elevations of the eastern side of England.

The clay XI, shown in Sections I and II as capping Hasboro' Cliff, is a deposit of not more than four or five feet in thickness and is unfossiliferous. It seems destitute of chalk, but is full of small stones, chiefly flint ; and rising from the sea level at Eccles Cliff, it dies out at an elevation of about fifty feet along the coast between Bacton and Mundesley. It is not improbable that it may be synchronous with the Hessle clay, although, at present, we are without the means for a satisfactory comparison.[5]

---

[1] *Cyrena fluminalis*, however, occurs in the fluviatile deposit of Barnewell, in Cambridgeshire, which we regard as coeval with the March and Hunstanton marine beds.

[2] Similar fossiliferous gravel occurs at other places in the Fen, such as Whittlesey, Wimblington, and elsewhere.

[3] In some earlier papers by us this shell was given from the Middle Glacial, but it was a mistake.

[4] See 'Geol. Mag.,' vol. ii, p. 8, and the prior papers there recited. A list of the Mollusca is there given by Mr. Rose, to which should be added *Nassa pygmœa*.

[5] This clay obscures the surface of the low country for some miles south-east of Hasboro' so as to render the mapping of that part uncertain. Whether it has been mistaken on the Hasboro' Cliff for the Great Chalky clay, No. 9, we know not ; but we do not recognise this latter clay either at Hasboro' or anywhere else along the North Norfolk coast section, though it occurs on the East Norfolk coast from Caistor to Winterton. On the top of Dunwich Cliff (Section R) there occurs a Post-glacial bed, from three to seven feet in thickness, consisting of a loam which in places contains some fragments of the clay, No. 9. This may, perhaps, be the same as the capping clay, XI, of Hasboro'. It is shown in Section R under the same

The rest of the Post-glacial formations shown in the map and sections under the number 11, are with some exceptions, (among which must be included beds wherein *Cyrena fluminalis* occurs, such as those at Stutton on the Stour estuary, and at Gedgrave near Orford, regarded by us as belonging to the older part of this period,) probably newer than these marine gravels, and belong to the later part of the Post-glacial period. In the preceding Yorkshire Coast section the earlier Post-glacial series, the Hessle sand and clay (*d* and *e*), are excavated or removed to give place to numerous later beds (*f*) of sand and gravel, which are of considerable thickness, and some of which contain freshwater Mollusca. It is clear from this coast section, therefore, that extensive beds, especially river gravels, accumulated over the north-east of England after the land had emerged from the Hessle clay re-depression; and to these we consider the principal part of the East Anglian Post-glacial Valley beds, which are shown in the map, under the number 11, belong.[1]

The recent alluvium, shown in the map under the figure (that of a crow flying) which has been adopted in their maps by the Geological Survey to distinguish these deposits, is mainly due to that considerable depression anterior to historical times, which buried so much forest ground all round the English coasts. This last depression brought the sea water into valleys which during the preceding (later Post-glacial) period were dry and forest-covered; and filling them, has given rise to the Broads of East Norfolk. The same depression has produced the wide flats of alluvium which fill so much of the valleys of the Waveney, Ant, and Yare, in East Norfolk, by silting up the lower parts of these valleys with modern estuarine mud, which was found at Yarmouth, in a well boring, to be 170 feet in thickness.[2]

The rest of the recent deposits consist of shingle, such as the great bank which shuts in the Alde from the sea at Orford, or of blown sand, which at Lowestoft has buried the sea cliff, and with some deserted foreshore has produced new land called the Denes.

S. V. Wood, Jun.

F. W. Harmer.

February, 1872.

N.B.—The lithographic map having been reduced to one fourth the original scale, from a survey made on the one inch Ordnance sheets, those who may have occasion to use it for field purposes will find it convenient to employ these sheets, (which are of very

shading and number as the Post-glacial gravel (10), but it, in fact, passes over that gravel where the cliff is highest (under the Ruins), and where a small portion of the clay No. 9 remains *in situ*, but the two patches, numbered 10, capping the southern half of the section, consist entirely of this loam.

[1] Sections of the Crag and Glacial beds along valley sides often show a Post-Glacial gravel over them, as, *e. g.*, Sections IX, X, XI, and XIV. It would, however, be scarcely possible to map all these patches, and if done their representation would obscure that of the older beds.

[2] See Prestwich in ' Quart. Journ. Geol. Soc.,' vol. xvi, p. 449.

small cost) and to measure off on them from the lithographic map all distances multiplied four times. In this way, by measuring from the parish church of any place, the site of which is shown by a cross, or from any bend of a stream, a greater approach to accuracy will be obtained, as well as the advantage of the contour surface furnished by the Ordnance sheet. The names of a few places (Norwich among them), where the shading is close and intricate, have been omitted, or they would have obscured the Geology. As, however, the site of the churches of all these is shown by the crosses, they can be readily identified on the Ordnance sheet. Wherever the lines of section traverse pits affording sections which exhibit the beds clearly, they are indicated by indentations in the surface line, and their position is written over them. Thorpe and Sizewell Cliff is run down and grassed. From Section T, northward to Lowestoft, the cliff is much obscured by talus, but traces of a boss of the pebbles, No. 6, appear at one part. From Lowestoft to near the southern end of Section U, the cliff is obscured by being buried in blown sand. The cliff by Caistor, Ormesby, Scratby, and Winterton, is similarly obscured, but in places, at the top, No. 9 is exposed. None of these portions of the East Anglian Cliff are, therefore, represented in the sections, but the map shows of what the uppermost part of those portions of the cliff consists. Many names of places have been wrongly spelt by the lithographer in the map, but they can easily be recognised.

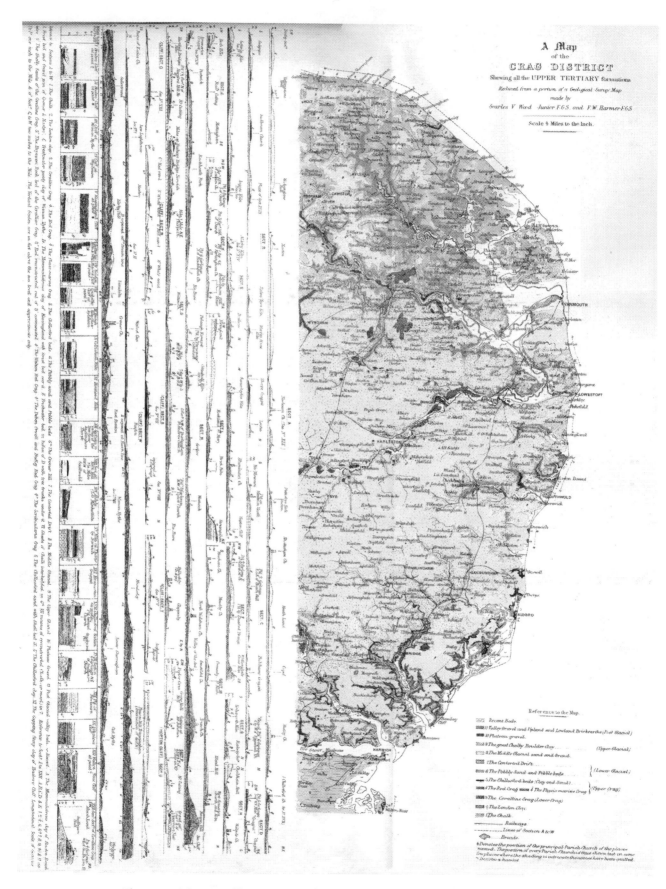

The material originally positioned here is too large for reproduction in this reissue. A PDF can be downloaded from the web address given on page iv of this book, by clicking on 'Resources Available'.

SUPPLEMENT

TO THE

MOLLUSCA FROM THE CRAG;

BEING

DESCRIPTIONS OF ADDITIONAL SPECIES,

AND

REMARKS ON SPECIES PREVIOUSLY DESCRIBED.

═══════════

CEPHALOPODA.

In the first part of the 'Crag Mollusca' it was observed that no remains of an animal belonging to this class had been detected in any section of the Upper Tertiaries, or what was there called the periods of the Crag deposits. I am equally unable now to introduce the name of any Cephalopod which may be presumed to have lived during the time of the Crag or any of the succeeding periods, although I have searched zealously in the hope of obtaining the terminal portion of the bone of the " Cuttle Fish." This bone is in some places left in great numbers on our own shores, and is an organic remain we might expect to find, but as yet I have not seen a vestige of such a fossil in the Upper Tertiaries of the East of England. The mucro of *Belosepia*, presumed to be similar to that of *Sepia*, is by no means rare in the sandy beds of Bracklesham or of Grignon.

1

GASTEROPODA.

PULMONATA.

HELIX SUTTONENSIS, *S. Wood*.   Tab. I, fig. 2, *a—c*.

*Spec. Char. H. Testá apertè umbilicatá, orbiculatá, supra convexiusculá, costulato-trans-
versè striatá, subtùs convexá, lævigatá ; anfractibus septem subcarinatis, versus peri-
pheriam inconspicue subangulatis ; aperturá angustato-lunulatá ; peristomate reflexo ?
umbilico parvo.*

Diameter, ¼ of an inch.

Locality.   Coralline Crag, Sutton.

A single specimen rewarded my researches in 1867, and this is the first instance, so far as I know, of a land animal having been observed in the truly marine
deposit of the Coralline Crag. This deposit I have always imagined to have been quite out
of the reach of fresh-water streams, and the only way I can account for the presence of this
Helix at Sutton is that it was probably carried out to sea upon a piece of drift wood,
which, in its decay, left it among the marine Mollusca.

This is an elegant little shell, and symmetrical in its form. It has seven volutions,
the first of which is perfectly smooth, but the others are beautifully ornamented with
thickened and rounded, but not imbricated, ridges, or thickened lines of increase ; about
eighty on the outer volution, and the suture or juncture between the whorls is deep and
very distinct, the edge of the succeeding volution being slightly elevated. The under
surface of the shell is quite smooth, the umbilicus small, and broadly funnel-shaped.

On comparing my fossil with existing species I find it most nearly related to *Helix
calathus* and *H. bifrons*, but to neither of them can it be strictly referred ; the spire is less
elevated than in the first, but more so than in the second, and there is a difference in the
ornamental ridges of both ; in one they are fewer and larger, and in the other smaller.
In my shell the umbilicus is more funnel-shaped, and as my specimen was probably a
full grown individual, judging from a slightly reflected margin to the aperture, it is not so
large as either of the recent species ; it bears rather a closer resemblance to *bifrons*, but a
specimen of that species of corresponding size to the Crag shell has only six volutions.   I
have therefore given to it the above name to commemorate a locality that has yielded me
so many Crag species.

The near relationship of the shell to Madeira species is not without its significance on
the subject of the climate of the Coralline Crag period in Britain, for though a suite of
land Mollusca would be necessary before any just inference could be drawn on this point,
yet this solitary form harmonises with what I and Messrs. Forbes and Hanley (in opposition to the author of the ' Brit. Conch.') hold to be the general facies of the Cor. Crag
Fauna, viz. that it is more southern than British.

HELIX RYSA, ' Crag Moll.,' vol. i, p. 4, Tab. I, fig. 1, must for the present retain that name. I have not been able to find an existing species to which it can be referred. The specimen I figured was then unique. Mr. Canham has obtained from the " diggers " at Waldringfield a second individual, which, he says, came out of the Red Crag at that locality.

PUPA MUSCORUM, *Müller.* Supplement, Tab. I, fig. 7, *a, b.*

Localities. Red Crag, Butley. Fluvio-marine Crag, Bramerton. Post-glacial, Clacton and Stutton.

A specimen of this species has recently been found in the Red Crag at Butley by Mr. A. Bell, which I have had figured, and another by Mr. Harmer at Bramerton; each of these shows a tooth in the aperture. My specimens from Clacton and Stutton are nearly all endentulous. I have given to this shell the above name, having previously used it in my former ' Catalogue.' The confusion respecting the true *muscorum* of Linné still exists, and perhaps may never be cleared up, but the *muscorum* of Müller appears to be admitted, and this name is entitled to precedence before that of *marginata*, Mont.

LIMNÆA PINGELII? *Möller.* Supplement, Tab. IV, fig. 4.

LYMNÆA PINGELII (LYMNOPHYSA PINGELII, *Beck*) in Möller, Ind. Moll. Groenl., p. 5.

" *Testá ovato-elongata ; spira conica, acutiuscula ; anfr.* 5 ; *sutura profundiori ; apertura dimidio testæ longitudinis breviori ; rima umbilicali angustiori. L.* 6, 5'''.*"—Möller.

Locality. Red Crag, Butley.

This specimen was found in the Red Crag at Butley by Mr. A. Bell, and I have assigned it, though with doubt, to the above-named species ; it is shorter and more inflated than any of our British species. This specimen is in the cabinet of Mr. Reed, of York. Fig. 8, *b,* Tab. I, ' Crag Moll.,' vol. i, may, perhaps, be referred to the same species, and fig. 8, *a,* of the same plate may be what Möller described as *L. Holböllii*, which is more elongated, and it has a larger umbilicus. I have several specimens of this from Butley. These appear to represent the existing northern forms of this genus.

MELAMPUS FUSIFORMIS, *S. Wood.* Crag Moll., vol. i, p 12, Tab. I, fig. 14 (as *Conovulus myosotis*), and Supplement, Tab. I, fig. 1.

MELAMPUS FUSIFORMIS, *A. Bell.* Ann. and Mag. Nat. Hist., 1870.

Axis, ½ an inch.

Localities. Red Crag, Sutton. Fluvio-marine Crag, Thorpe, in Suffolk, and Bramerton.

The specimen figured in Supplement, Tab. I, was found by Mr. A. Bell at Thorpe, in Suffolk, and in the 'Crag Moll.,' vol. i, Tab. I, figs. 14 and 15, are represented some specimens from the Red Crag of Sutton, which are there called *Conovulus myosotis* (with a doubt). These, I think, may be united with the shell figured in Supplement, Tab. I, although they have no denticles on the inside of the outer lip. A similar, but smaller, specimen has been found at Thorpe, in Suffolk, by Mr. E. Cavell, in which the outer lip has also indications of denticles. This shell is much larger and more fusiform than the existing *myosotis*, has a more pointed base, and is a thicker shell.

Mr. Bell, in the 'Geol. Mag.,' vol. vi, p. 41, gives *Limnæa peregra* and *L. truncatula* as species from the Red Crag of Butley. I have also found *Planorbis complanatus* and *Pl. spirorbis* at the same locality.

These two or three fresh-water shells thus occurring in the Red Crag do not appear to me to indicate estuarine conditions, as they may have been introduced into the Red Crag sea down those rills of fresh water that we see on every beach between tide marks, coming from the land and meandering over the shore to the brink of the waves. The principal part of the Red Crag, including the Crag at Butley, from which these shells were obtained, being a beach or foreshore deposit, *i. e.* one formed between high and low water marks, land and fresh-water shells so carried down would become incorporated with the purely marine deposit, and thus impart a different aspect to a formation from that of a fluvio-marine one, which is produced by the gradual intermingling in an estuary of a fresh-water river with the salt water of the sea. It is clear from the position of the Red Crag at Butley relatively to the Coralline Rock bank against which it abuts, that the shore was immediately contiguous to the places from which these fresh-water shells were derived. Even those parts of the Red Crag which appear to have been formed actually under water, such as that under which the phosphatic nodules are worked, were in the immediate contiguity of the foreshore deposits with which they are associated, and may easily have received introductions of land and fresh-water shells in a similar way.

PECTINIBRANCHIATA.

MARINE.

OVULA LEATHESII, *Sowerby*. Crag Moll., vol. i, p. 14, Tab. II, fig. 1, *a, b*.

Localities. Cor. Crag, Sutton and near Orford. Red Crag, Walton Naze and Butley.
Two varieties of this Crag fossil were figured as above referred to, both of which were considered there as referable to *Bulla spelta*, Linné.

At fig. 23, Tab. VII, of this Supplement I have given a front view of the short variety, the back of which was represented in 'Crag Moll.,' Tab. II, fig. 1, *b*, from the Red Crag.

This somewhat resembles *O. Adriatica*, Sow., but I believe it is only a short variety of *O. spelta*, which I will here call var. *brevior*. Specimens from the Cor. Crag in Mr. Cavell's collection exhibit also the same differences. Mr. Bell gives the name of *O. Adriatica* ('Ann. and Mag. Nat. Hist.,' May, 1871, p. 9) as a Crag species, which I imagine to be the short variety. This thickened margin, the presumed indication of the adult, may be seen on specimens of various sizes, like the variations in *Cypræa Europæa*. I have obtained the shell from Butley, where it is smaller than in the Walton bed. I am informed by Mr. Charlesworth that the specimen, upon the authority of which this species was introduced by Dr. Woodward into his Norwich Crag list, is most probably spurious as a Norwich Crag shell.

Ovulum obtusum from China, Sow., 'Thesaur. Conch.,' vol. ii, p. 474, pl. c, figs. 22, 23, can scarcely be distinguished from our Crag shells.

CYPRÆA EUROPÆA, *Mont.* Crag Moll., vol. i, p. 17, Tab. II, fig. 6, and Supplement, Tab. V, fig. 24.

Localities. Cor. Crag, Sutton, and near Orford. Red Crag passim. Middle Glacial, Billockby.

The figure in Supplement, Tab. V, represents a very globose form from the Red Crag of Sutton, which I believe is merely a variety of *C. Europæa*, although it is more spherical and less elongated than the generality of specimens, and it has rather more numerous ridges. In Mr. Bell's paper on "Some new or little-known Shells of the Crag" ('Ann. and Mag. Nat. Hist.,' September, 1870) is the name of *Cypræa Dertonensis*, Mich., but I do not know the shell he alludes to. If it be the present Crag shell I cannot acquiesce in the reference. The shell figured and described by Michelotti ('Desc. des Foss. Terr. Mioc. de l'Ital.,' p. 331, pl. xiv, fig. 10) appears to me to be a different species. I have not seen Mr. Bell's specimen.

Fragments of *C. Europæa* have occurred in the Middle Glacial sand of Hopton Cliff. I am informed by Mr. Charlesworth that the specimen upon the authority of which *Cypræa Europæa* was introduced into Dr. Woodward's Norwich Crag list is probably spurious as a Norwich Crag shell.

ROSTELLARIA LUCIDA? *J. Sowerby.* Supplement, Tab. II, fig. 14.

Locality. Red Crag, Sutton.

In my 'Catalogue,' as also in the 'Crag Mollusca,' vol. i, p. 24, mention is made of a shell found in the Red Crag at Sutton, to which I gave the name of *R. plurimacosta*, and then stated it as greatly resembling the London Clay species (*R. lucida*). The specimens found by me are all more or less in a mutilated condition, and I regret to say that my knowledge is not improved by the sight of any better than I then possessed. I am induced to have it figured because the specimens have the lithological character of the Red Crag shells ; and although it is a species probably derived from an older deposit than the one in which it was found, it has not the appearance of the *known* derivative fossils from the older tertiaries, and I think it is just possible to have lived in the Coralline Crag sea and been derived from that Crag. My specimens differ slightly from the Highgate fossil ; the costæ upon the Crag shell are rather closer, thicker, and more obtuse, and the volutions not quite so convex, with the intermediate striæ closer and not so fine, but the Crag specimens will not admit of fair comparison. Our shell seems to agree rather better with the figure and description given by M. Deshayes (' An. sans Vert. du Bass. de Par.,' t. iii, p. 461, pl. 92, figs. 4—7), which is more slender, and the costæ a little larger, with the striæ closer ; the long rostrum there shown has not been preserved in any of our British fossils. The figure given by Sowerby of the Bracklesham shell (Dixon, ' Geol. of Sussex,' p. 187, tab. v, fig. 21) much resembles the Crag specimens.

ANCILLARIA GLANDIFORMIS ? *Lam.* Supplement, Tab. V, fig. 7.

Locality. Red Crag, Waldringfield.

Another shell, most probably derived from some anterior deposit, has been obligingly presented to me by Mr. Charlesworth, who obtained it from the nodule pits in the Red Crag at Waldringfield. It is in a very mutilated condition, but I think it may be referred to *Ancillaria glandiformis*, Lamarck, ' Ann. du Mus.,' tab. xvi, p. 305. It seems to correspond with fig. 7, *a, b*, tab. vi, ' Foss. Moll. des Wien. Beck,' vol. i, where Dr. Hörnes has figured several varieties of the species. My specimen is in a similar condition to most of the Red Crag shells in respect to lithological character.

VOLUTA NODOSA ? *J. Sowerby.* Supplement, Tab. V, fig. 6, *a, b*.

VOLUTA NODOSA, *J. Sow.* Min. Conch., tab. 399, fig. 2.

Locality. Red Crag, Waldringfield.

A single specimen, which I have referred as above, has been obtained by the Rev. Mr. Canham from the nodule diggers in the Red Crag at Waldringfield ; it is in all probability a specimen derived from some anterior formation, perhaps from the same bed which has supplied to the Crag the specimens of *Rostellaria lucida*.

Our specimen has undergone considerable water action, as most of the exterior ornament is obliterated. Mr. Edwards speaks of this older Tertiary species being very variable, and has represented several different forms (' Eocene Moll.,' p. 141, Tab. XIX, fig. 1) ; he gives it as a species from the London Clay at Highgate, and also from the Bracklesham beds.

Voluta luctatrix, *Solander.* Supplement, Tab. VI, fig. 14.

Locality. Red Crag, Waldringfield.

Another extraneous specimen has been obtained by Mr. Canham from the same place, which I have referred to the young state of *Voluta luctatrix.* Our shell corresponds with fig. 3, *d, e,* Tab. XIX, of 'Eocene Mollusca;' it is rather less in size, but I have no doubt of its identity; the specimen has undergone a good deal of rough treatment by its removal into the bed of the Red Crag. This species (*luctatrix*) is very abundant at Barton, where specimens of all ages and sizes may be found, from $4\frac{1}{2}$ inches down to $\frac{3}{8}$ths of an inch, and the young species vary much in shape, being comparatively more elongated than when full grown. Perfect specimens have a small apex or pullus, and two or three of the upper volutions are smooth, or free from ornament of any kind, like the pullus of true *Voluta,* with its very obtuse apex. Swainson proposed for these Eocene fossils with a small apex the term *Volutilites,* but this is a name of hybrid composition, and does not appear to be generally adopted. This Eocene form of Volute has nearly died out. One shell of this character has been obtained from the Aguilhas Bank, and named *V. abyssicola,* which is probably a true descendant of ne of our Eocene species.

Voluta Lamberti, *J. Sowerby.* Crag. Moll., vol. i, p. 20, Tab. II, fig. 3.

Localities. Cor. Crag passim. Red Crag passim. Fluvio-marine Crag, Yarn Hill, near Southwold.

In the 'Crag Mollusca' I regarded the presence of this shell in the Red Crag as due to derivation from the Coralline, but there can, I think, be now no reason for doubting that the shell was a denizen of the Red Crag sea, because it was lately found by Mr. Charlesworth and Mr. V. Colchester in association with *Pecten princeps* in the pit at Yarn Hill, near Potter's Bridge, Southwold, which belongs either to the Fluvio-marine Crag or the Chillesford bed (though to which it is hard to say). There can, therefore, be no doubt that it lived through the Red and Fluvio-marine Crag period. There is no known living shell with which it can in my opinion be identified.

Mitra ebenus? var. uniplicatus, *S. Wood.* Supplement, Tab. III, fig. 6.

Locality. Coralline Crag, Orford. Red Crag, Waldringfield (*A. Bell*).

When I first obtained the above represented specimen, I had it figured under an impression that it was a distinct species (or even genus), having but one fold or ridge upon tne columella, but I have since seen two or three more specimens in Mr. Bell's possession resembling it in every respect, with the exception of the columella, which in his specimens had three, and in one instance four folds, denoting it to be a true *Mitra*. I have, therefore, retained for it the above name. It much resembles, and probably may be the same as, *Voluta pyramidella* of Brocchi, p. 318, Tab. IV, fig. 5 ; my shell has an obtuse apex, the volutions much flattened, a very slight shoulder, and a deep suture. Mr. Bell gives *M. ebenus* as a Red Crag shell from Waldringfield, but I have not seen the specimen.

MITRA FUSIFORMIS, *Brocchi*. Supplement, Tab. V, fig. 3, *a, b*.

VOLUTA FUSIFORMIS, *Broc.* Conch. Foss. Subapen., vol. ii, p. 315, 1814.
MITRA — *Grat.* Conch. Foss. du Bas. de l'Ad., t. 37, figs. 6, 7.
— — *Bellardi.* Mon. della Mitr. Foss. del Piem., p. 5, t. 1, figs. 6—10.
— — *Hörnes.* Foss. Vien. Bas., p. 98, t. 10, figs. 4—7.

Spec. Char. M. " *Testá fusiformi-elongata, lævi ; anfractibus convexiusculis, postice subangulatis ; apertura elongata ; columella recta 4—6 plicata ; spira elata.*" (Bellardi.)
Length, 2 inches, nearly."
Locality. Red Crag, Waldringfield.

A single specimen of a shell, which I have with very little doubt referred to a common Continental fossil species, has been obtained from the nodule workings by the Rev. Mr. Canham. This specimen has, like many of its associates in the Red Crag at Waldringfield, undergone some rough treatment. Most of its outer coating has been removed. It is a very elongated specimen, but I believe where it is abundant the same form may be observed. Our specimen has only four folds upon the columella; the first is the most prominent, diminishing towards the base, but these plaits or folds are said to vary in number from four to six. This is probably a derived specimen.

In vol. ii, 'Moll. Sic.,' Phillippi refers *M. fusiformis* to *M. zonata*, Swainson and Risso. Weinkauff gives this name to Marryat. This latter name seems to have been imposed on the existing shell from the coloured band on the exterior. I have retained Brocchi's name, which seems to be generally adopted for this fossil. It is a variable species, and Bellardi assigns as synonyms *M. plicatella*, Lam., *M. pyramidella*, and *M. incognita*, Grateloup.

TEREBRA CANALIS, *S. Wood*. Supplement, Tab. IV, fig. 1.

Length, 1⅛th inch.
Localities. Coralline Crag near Orford. Red Crag, Waldringfield (*Bell* and *Canham*).

When describing this shell in the 'Crag Moll.' I had only a few fragments to assist in the determination of the species. Since then I have obtained some better specimens, and have here given an additional figure.

This may possibly hereafter be referred to *T. fuscata*, Broc., which is said to be very variable, but it differs considerably from the typical form of that species. The Crag shell has a longer and more inflected canal, with a more obtuse apex, and the spiral band is narrower and less distinctly marked, with the longitudinal lines less prominent. Specimens of *fuscata* from the Vienna and Bordeaux beds attain to the length of three inches and upwards. The Crag shell seems somewhat to resemble a variety figured by Dr. Speyer, 'Ober. Oligoc.,' &c., p. 13, tab. i, figs. 7, 8, *a, b*. M. Nyst considers it as a dextral variety of *T. inversa* (Catal. in 'Bull. de l'Acad. Roy. des Sci., Lettr., et des Beaux Arts de Belgique,' chap. xvii, p. 420). I have for the present retained my original name. The shell referred to by Mr. Bell in 'Ann. and Mag. Nat. Hist.' for May, 1871, under the name *T. exilis*, Bell, I have seen, and regard as a variety only of *canalis*.

COLUMBELLA? HOLBÖLLII, *Möller*. Supplement, Tab. VI, fig. 21.

FUSUS HOLBÖLLII, *Möll*. Ind. Moll. Groenl., p. 15

" *Testa fusiformi, elongata, alba, lævi, epidermide cornea, fusco-lutea, solidiore obtecta; anfr. 9 sensim crescentibus, planulatis; spira acuminata. Long.* 2, 4′″."—Möll.

Locality. Upper Glacial, Bridlington.

A specimen of this species is in the British Museum among the Bridlington fossils, and it is enumerated by the late Dr. S. P. Woodward in his list of Bridlington fossils in the 'Geol. Mag.,' vol. i, p. 53. This species is also found fossil in the Belfast deposit, where I believe it is by no means uncommon. It is found living in the Spitzbergen and Greenland seas.

COLUMBELLA? SULCATA, *J. Sowerby*. Crag Moll., vol. i, p. 23, Tab. II, fig. 2. Supplement, Tab. II, fig. 16.

Localities. Cor. Crag, Sutton. Red Crag, Walton and Sutton. Fluvio-marine Crag, Bramerton? Middle Glacial, Hopton?

The figure in Supplement, Tab. II, represents a specimen found by G. Gibson, Esq., of Saffron Walden, in the Red Crag at Walton-on-the-Naze. It is nearly double the size of any specimen of this species previously obtained by myself.

In the 'Crag Moll.,' vol. i, p. 23, the character given to this species is " apex acute." This is an error, for although in some of the specimens the spire is much elevated and elongately tapering, the apex is always more or less obtuse or mammillated. The species of this genus from Turin and Astigiana are represented by Bellardi as being very acutely pointed. This species has been obtained from the Coralline Crag by Mr. Bell, as well as a fragment of it by myself.

2

It is a very aberrant form of *Columbella*. The young state of the Crag shell much resembles *Buc. minus*, Phil.; it has then a sharp and plain outer lip and a longer canal (see 'Crag Moll.,' Tab. II, fig. 2, *d*). A fragment, consisting of the outer lip with its denticulations, and a little of the exterior of the shell on which the striated markings are visible, obtained from the Middle Glacial of Hopton, seems referable to this species. It is given in Dr. Woodward's Norwich Crag list as a Bramerton shell, but I have not seen it myself from there.

PYRULA RETICULATA, *Lamarck*.   Crag Moll., vol. i, p. 42, Tab. II, fig. 12.

Localities.   Cor. Crag, Ramsholt.   Red Crag, Waldringfield, Walton Naze.

In the Appendix, p. 311, vol. ii, I stated that I thought the cast of *Pyrula* figured (Tab. XXXI, fig. 6) was the same as the one previously figured from the Coralline Crag, and the specific name *reticulata* was in consequence proposed to be altered. I now think they belong to two different species, and I here restore to the Cor. Crag shell the name *reticulata* originally given to it, and the Sandstone cast, which is of a different age from that of the Crag, may retain the name of *acclinis* until it can be better determined. This Cor. Crag shell has been referred to *P. condita*, Brongn., by M. Nyst, and to *P. cancellata*, Grateloup, by Mr. A. Bell, and to *P. subintermedia*, Bronn, by Mr. Jeffreys. Two or three specimens of what appear to be *P. reticulata* of the Cor. Crag have been obtained from the Red Crag. Mr. Bell gives it from Waldringfield. It is probably derivative in the Red Crag. Hörnes gives eight synonyms to *P. reticulata*.

CASSIS SABURON, *Bruguière*.   Supplement, Tab. VI, fig. 2, *a, b*.

> LE SABURON, *Adanson*.   Senegal, p. 112, pl. vii, fig. 8, 1758.
> CASSIDEA SABURON, *Brug*.   Ency., p. 420.

Locality.   Red Crag, Waldringfield.

This has been obtained from the diggers in the Red Crag at Waldringfield by Mr. Canham. It is probably an extraneous fossil, and derived from some anterior formation. The shell has undergone a good deal of water action, and I cannot perceive a trace of striation upon the surface; still, it so appears to correspond in all other respects with the species to which I have referred it, that I imagine the striæ have been rubbed off, or the outer surface has decorticated away, as it is quite smooth; it is also a little disfigured by the loss of a portion of its canal, but this was probably done in the lifetime of the animal and clumsily repaired.

This is a living species, with an extensive geographical range, being found on the coast of Spain, Portugal, and Algiers. It is a fossil of the Bordeaux beds and the Vienna basin, and M. Nyst gives it from the "Crag gris" of Belgium, and it may possibly have lived in the Coralline Crag Sea.

CASSIDARIA BICATENATA, *J. Sow.* Crag Moll., vol. i, p. 27, Tab. IV, fig. 5.

Localities. Cor. Crag, Ramsholt and near Orford. Red Crag, Sutton, Bawdsey, Felixstowe, and Waldringfield. Fluvio-marine Crag, near Norwich?

Var. ECATENATA. Supplement, Tab. VI, fig. 1, *a*.

Locality. Cor. Crag, near Orford.

The figure represents a specimen of *Cassidaria* from Mr. Cavell's cabinet, with a more elongated form than what I have had previously figured as *bicatenata*, and it is free from the double chain-like ornament upon the upper part of the volution. It approaches very near to *C. Tyrrhena*, Chemn., but it is, I believe, specifically distinct. It is from the Coralline Crag at Orford. I have also found the general form of *bicatenata* with its two rows of nodules or catenæ in the Cor. Crag at Ramsholt. Our present specimen differs also in the ornamentation, not having the small intermediate line between the ridges.

Fig. 1, *b*, Tab. VI, represents a small specimen in Mr. Canham's cabinet of *Cassidaria bicatenata* from the Red Crag at Waldringfield, which I have had figured to show that the thickened margin to the outer lip was in this case formed before the shell had attained to half of its usual size. The present specimen having had its aperture and anterior termination enveloped, may perhaps have had its natural growth checked and so became adult, though of so small a size; but this does not necessarily follow, because the thickening of the lip in this group of Mollusca is by no means an indication of the adult condition, for in the recent forms of *Cassis* this thickening of the lip may sometimes be seen to have occurred two or three times in the growth of the animal. This species is included in the list of shells from the neighbourhood of Norwich obtained by Sir C. Lyell in 1838, and identified for him by the late Mr. G. Sowerby and myself ('Mag. Nat. Hist.,' n. s., vol. iii, p. 329). I can therefore scarcely doubt its being a genuine Norwich Crag shell, though it does not appear to have been found of late years in the neighbourhood of Norwich.

NASSA GRANIFERA, *Dujardin.* Supplement, Tab. VI, fig. 11.

> BUCCINUM GRANIFERUM, *Dujard.* Geol. Tr. France, vol. ii, pt. 2, pl. 20, figs. 11, 12.
> NASSA GRANIFERA, *A. Bell.* Ann. and Mag. Nat. Hist., May, 1871.

Axis, $\frac{5}{16}$ths of an inch, nearly.

Localities. Coralline Crag, Gedgrave (*A. Bell*), Sutton (*S. Wood*). Red Crag, Sutton (*S. Wood*).

The specimen figured was lent to me by Mr. Bell, who obtained it from a dealer at Orford. I have a second, but less perfect, specimen from the Cor. Crag of Sutton. I have also found one in the Red Crag of Sutton, which I consider may be referred to the same species. In the 'Crag Moll.,' vol. i, p. 30, this was considered as specifically different from *N. granulata*, and the differences are there pointed out.

N. granulata, 'Nyst. Foss. Belg.,' p. 575, pl. xliii, fig. 11, appears to belong to this species.

NASSA PYGMÆA, *Lam.*   Crag Moll., Appendix, p. 315, Tab. XXXI, fig. 5; Supplement, Tab. VI, fig. 6.

Localities.   Cor. Crag, Sutton?   Red Crag, Butley (*Bell*)?   Post-glacial, Nar Brick-earth (*Rose*).

The condition of the specimen fig. 5 of Tab. XXXI renders it doubtful whether this species has occurred in the Cor. Crag. It is given by Mr. Bell ('Ann. and Mag. Nat. Hist.,' May, 1871) as from the Red Crag of Butley, but I have not seen the specimen. The shell figured in Supplement, Tab. VI, fig. 6, is one of a suite in excellent preservation obtained by Mr. Rose from the Nar Brick-earth, and on the tablet was the name of *Nassa pygmæa*, by Mr. Jeffreys.

NASSA GRANULATA, *J. Sowerby.*   Crag Moll., vol. i, p. 29, Tab. III, fig. 3.

Localities.   Cor. Crag passim.   Red Crag passim.   Middle Glacial, Billockby and Hopton.

Several specimens of this species, most of them very imperfect, but with their distinctive characters well preserved, have been obtained from the Middle Glacial sands at Billockby and at Hopton Cliff.

NASSA INCRASSATA, *Müller.*   Crag Moll., vol. i, p. 29, Tab. III, fig. 4.

Localities.   Cor. Crag, Sutton.   Red Crag, Sutton and Butley.   Fluvio-marine Crag, Bramerton.   Middle Glacial, Billockby.   Post-glacial, Nar Valley.

I have seen this shell from the Fluvio-marine Crag of the Bramerton Pit. The body-whorl and mouth of this shell have also been obtained from the Middle Glacial sands of Billockby. Several specimens in fine preservation have also been obtained by Mr. Rose from the brick-earth of the Nar Valley.

NASSA DENSECOSTATA, *A. Bell.* Supplement, Tab. VI, fig. 8.

NASSA DENSECOSTATA, *A. Bell.* Ann. and Mag. Nat. Hist., May, 1871.

Locality. Cor. Crag, near Orford.

A single specimen, not perfect, has been put into my hands for figuring. This came, Mr. A. Bell informs me, from the Coralline Crag, near Orford. Its nearest resemblance is to *N. limata*, Ch. (*B. prismaticum*, Broc.), but is much more slender and more numerously costated, as pointed out by Mr. Bell. I have seen only this specimen, and its specific distinction will require some further evidence for its confirmation.

NASSA PROPINQUA, *J. Sowerby.* Crag Moll., vol. i, p. 30, Tab. III, fig. 2.

Localities. Red Crag, Walton Naze, and Sutton; Chillesford bed, Easton Bavent.

This shell Mr. Bell gives ('Ann. and Mag. Nat. Hist.,' September, 1870) from the Chillesford shell-bed of Easton Bavent Cliff, but I have not seen the specimen. He says it is the same as *N. trivittata*, Say., *Buc. trivittatum*, Gould (1st ed., 'Inv. Massach.,' p. 309, fig. 211, and in 2nd ed., p. 364, fig. 632), neither of which figures well represent the Crag shell, but the recent species is said to be very variable. This name (*N. propinqua*) was given to me in MS. by S. P. Woodward as from the Cor. Crag (in Mr. Wigham's collection), but I have not been able to find the specimen. Sowerby's name has, I believe, priority over that of *trivittata*.

NASSA PULCHELLA, *Andrzejowski.* Supplement, Tab. VI, fig. 7.

NASSA PULCHELLA, *Andrz.* Coq. Foss. Volh. Bull. Mosc., vi, p. 438, t. 2, fig. 2, 1833.
BUCCINUM RETICULATUM, *Dubois.* Coq. Foss. Wolh-Pod., p. 27, pl. 1, figs. 28, 29.
— — *Hörnes.* Vienna Foss., p. 151, t. 12, fig. 18, 1856.

Spec. Char. *N. Testá mediocre, ovato-conicà, longitudinaliter plicatá, striis transversis distantibus decussatá, subgranulosá; anfractibus planiusculis; aperturá magná ovatá, labro intus, dentato.*

Axis, $\frac{1}{2}$ an inch.

Localities. Coralline Crag, near Orford. Red Crag, Waldringfield (*Bell*).

The specimen figured was obtained by Mr. Bell from the Cor. Crag, near Orford, and it was put into my hands with the name of *N. pulchella*, and it is the shell described under that name in the 'Ann. and Mag. Nat. Hist.,' May, 1871, as *N. pulchella*, A. Bell, n. s., but I believe it to be referable to the Russian fossil.   Mr. Miller, of Ipswich, has obtained from the same locality a second specimen, and Mr. Harmer has very recently sent me another.   It somewhat resembles a dwarf variety of *N. reticulata,* but it is, I believe, a full-grown shell and specifically distinct, as described by Mr. Bell.   The accidental coincidence in the adoption of the name is curious.   Mr. Bell also gives it (loc. cit.) from the Red Crag of Waldringfield.

NASSA PUSILLINA, *S. Wood.*   Supplement, Tab. II, fig. 7.

Spec. Char.   N. Testá parvá, elongato-conoideá, longitudinaliter costatá ; costis 5—6, spiraliter striatá ; striis paucis, magnis, elevatis ; anfractibus planiusculis aperturá ovatá ; labro extus varicoso, intus dentato.

Axis, ⅜ths of an inch.

Localities.   Fluvio-marine Crag, Bramerton.   Red Crag, Butley.   Middle Glacial, Billockby.

In Dr. S. P. Woodward's list of Norwich Crag fossils this is inserted as " *Nassa* ——, sp. (slender pointed), Norwich ; examples in all collections ;" from which remark, I presume, it is not rare near Norwich.   The varices are rounded, with a considerable space between each ; the suture is well defined, and the ribs slightly oblique, but the artist has made them rather too much inclined.   It is very unlike all the other Crag *Nassas,* coming nearest to *N. consociata,* but quite distinct ; it is identical in all respects with a shell in the Museum of the Geological Society presented by the late Jas. Smith, of Jordan Hill, but which is without a name, marked " Raised Beach, Gibraltar."   It may, therefore, probably some day be found living in the Mediterranean area.   It is by no means rare at Billockby.

It is the shell referred to as *N. pusio* in the paper by S. V. Wood, jun., and F. W. Harmer, in ' Brit. Assoc. Reports ' for 1870, but I find the name *pusio* has been previously occupied.

NASSA RETICULATA, *Linné.*   Supplement, Tab. VI, fig. 5.

BUCCINUM RETICULATUM, *Linn.*   Syst. Nat., edit. xii, p. 1205.

Locality.   Post-glacial, Kelsey Hill and Hunstanton.

This shell is very common at Kelsey Hill, but I do not know it from any Glacial or Pre-glacial formation in Britain.   The specimen figured was found by my son at Kelsey. It has a great range in the recent state.   The shell figured under this name from Bordeaux is, I believe, distinct.   A fragment of *N. reticulata,* obtained from the beds of the Severn Valley, was sent me for inspection by Mr. G. Maw.

NASSA RETICOSA, *J. Sowerby*. Crag Moll., vol. i, p. 33, Tab. III, fig. 10 ; Supplement, Tab. IV, fig. 3, and Tab. VII, fig. 15.

Localities. Red Crag passim ; Chillesford bed, Easton Bavent (*Bell*).

This, as before shown, is a very variable shell, and I have here introduced two more forms which I think may be referred to the same species. Fig. 3, Tab. IV, of the present Supplement represents a small specimen obtained by Mr. Bell from the Red Crag of Butley, which has a perfectly cancellated exterior without any thickened ribs or varices, which might be called var. *simplex*, and fig. 15, Tab. VII, is a shell I found in the Red Crag of Sutton which has an angulated shoulder to the volution, and belongs, I believe, to this species, and I will call it var. *scalarina*.

Mr. Bell informs me that he has obtained *N. reticosa* from the Chillesford bed of Easton Bavent Cliff, but I have not seen the specimen. Some fragments of spires from the Middle Glacial sand of Billockby seem to belong to the *costata* variety of this shell, but they are too doubtful to justify an identification.

NASSA CONGLOBATA, *Broc.* ('Crag. Moll.,' vol. i, p. 32, Tab. III, fig. 9), is possibly *Buccinum* (*Desmoulea*) *abbreviatum*, Chemn., a shell living on the coast of Senegal, but the shell is so rare in the Crag that I am not certain the fossil and recent shells are absolutely identical.

NASSA MONENSIS, Forbes, and N. PLIOCENA, Strickland, are still uncertain species. These two shells, originally described by Strickland, are not to be found. Mr. Etheridge, of the Geol. Survey, kindly endeavoured with much trouble to find them, but without success. My own shell (Tab. III, fig. 5, ' Crag Moll.') is possibly a var. of *reticosa*. I have not been able to find another like it.

NASSA LABIOSA, *J. Sow*. Crag. Moll., vol. i, p. 28, Tab. VII, fig. 22.

This shell was by E. Forbes referred to *Buc. semistriatum*, Broc. (' Mem. Geol. Surv.,' 1846, p. 428), and lately by Mr. Jeffreys, 'Geol. Journ.,' vol. xxvii, p. 144. I still think the two shells specifically distinct for the reasons stated at the above reference.

Buccinum Dalei, *J. Sowerby.* Crag. Moll., vol. i, p. 34, Tab. III, fig. 10. Supplement, Tab. II, fig. 9.

Localities. Cor. Crag, Ramsholt, Sutton, and near Orford. Red Crag passim. Fluvio-marine Crag, Bramerton; Chillesford bed, Easton Bavent.

The figure in Supplement represents a specimen of *Buccinum Dalei* found in the Red Crag at Walton-on-the-Naze by the Rev. T. Wiltshire, who has obligingly presented it to me. This specimen has the volutions in a reversed direction, that is to say from right to left. The late Dr. S. P. Woodward told me in 1864 that he had also found the fragment of a specimen of this species in the Coralline Crag with a sinistral volution, and as this shell had not been previously known in that reversed condition, I thought it deserving of a special representation. Mr. Robert Bell has very recently showed me a similar specimen from the Red Crag of Waldringfield. The circumstance that these specimens should have been discovered within a short period would seem rather to indicate a slight tendency in this species to vary its mode of volution, and perhaps if a few individuals of this form congregated together a progeny possessing a sinistral volution might have been produced.

The form of this species resembles in its general contour that of *Buc. undatum*, but it has a more distinct plait or tooth at the base of the columella, like that of *Nassa*, and Mr. Hancock pointed out that the animal had a different kind of " lingual ribbon " from *B. undatum*. In consequence of this character, and of possessing a different form of operculum, Dr. W. Stimpson, in an elaborate paper published in the 'Canadian Naturalist,' 1865, vol. ii (wherein he describes fifteen recent species in the genus *Buccinum*), has, at p. 366, proposed for it the generic name of *Liomesus* " with *Buc. Dalei* as the type." Mr. Jeffreys (' Brit. Conch.,' vol. iv, p. 297, 1867) has given to this shell the name of *Buccinopsis*.[1] Dr. Gray in 1859 proposed a genus under the name of *Cominella*, to receive species resembling *Buccinum*, having an operculum like that of the *Murices* and *Fusi*, in which the nucleus is terminal at the inner base of the mouth, increasing by semi-elliptical layers.

If the form of the operculum be sufficient of itself to consitute a generic character, I think our species will have to be referred to *Cominella*, should that be of prior establishment. With this uncertainty, and being unable to ascertain the date of priority for these different names, I have left our Crag species in its original position of *Buccinum*. A specimen of *B. Dalei* is in the Norwich Museum from the Fluvio-marine Crag, and

---

[1] This has no generic connection with *Buccinanops*, a word of similar meaning proposed by D'Orbigny, 1839, of which Herrmannsen says " Etym. vocabulum hybridum non admittendum ;" neither is it generically related to two Eocene species figured by Deshayes with the name *Buccinopsis* (' An. sans Vert. du Bas de Par.,' t. xi, pl. xciii, figs. 21—23 and 29—32), afterwards described as *Truncaria*, Adams.

Mr. A. Bell gives this species from the Chillesford Bed of Easton Bavent Cliff, in 'Ann. and Mag. Nat. Hist.' for Sept., 1870.

BUCCINUM PSEUDO-DALEI, *S. Wood.* Supplement, Tab. V, fig. 4; Tab. VI, fig. 9.

Locality. Cor. Crag, near Orford.

Tab. V, fig. 4, represents a specimen lately obtained by Mr. A. Bell from the Coralline Crag near Orford, which, though resembling *B. Dalei*, departs from that shell sufficiently to entitle it, I think, to a distinct specific name, and I propose to call it *pseudo-Dalei*. The exterior of the specimen is not quite perfect, but it appears to have been covered with fine striæ, smaller and finer than I have seen upon any specimens of *B. Dalei*. The form of the aperture is also different, being more expanded at the base, and the columella is more twisted. The apex of this specimen is obscured. There is also a general angularity of aspect presented by the shell, in which it contrasts with *B. Dalei*.

Tab. VII, fig. 9, represents the fragment of a shell now in the British Museum found in the Coralline Crag at Orford by Henry Woodward, Esq.; it is marked as the apex of *B. Dalei*, which I believe it is, or even more probably of the above *pseudo-Dalei*. It seems from its depression and from the early expansion of the volutions to have belonged to the present species, which in the perfect shell has unfortunately this part hidden. Several fragments of *B. Dalei* in my cabinet from the Cor. Crag of Sutton have the first two or three volutions filled with calcareous matter.

BUCCINUM GLACIALE, *Linné.* Supplement, Tab. II, fig. 1.

BUCCINUM GLACIALE, *Linn.*	Syst. Nat., 12th ed., No. 474, p. 1204.
— — *Chemn.*	Conch. Cab., vol. x, p. 180, t. 152, figs. 1446, 1447.
TRITONIUM — *Fabr.*	Faun. Groenl., No. 397, 1780.

Length, 2 inches.
Locality. Red Crag, Sutton? and Walton Naze.

The figure of this species here given is from a recent shell in the British Museum, but I have seen a perfect specimen that was, I believe, obtained by the late Mr. Edward Acton from some of the nodule diggers in the parish of Sutton, undoubtedly belonging to this species. This I should have preferred to figure, but I was not able to obtain it for that purpose. There can be no doubt that it came from the Crag, and I have myself found a fragment of what I believe belongs to this species at Walton-on-the-Naze.

BUCCINUM UNDATUM, *Linné*.   Crag Moll., vol. i, p. 35, Tab. III, fig. 12.   Supplement,
Tab. II.

Localities.   Red Crag, Sutton.   Upper Glacial, Bridlington?   Post-glacial, Kelseahill
gravel (*Jeffreys*), and Nar brick-earth (*Rose*).

Localities of var. *tenerum*.—Red Crag passim.   Fluvio-marine Crag, Bramerton and
Thorpe;   Chillesford bed, Bramerton and Horstead.   Middle Glacial, Billockby and
Hopton.   Post-glacial, The March gravel.

In the 'Crag Mollusca' I have figured three different forms under the specific name
of *undatum*, and I have here introduced three more.

Fig. 2, Tab. II, Supplement, appears to resemble what has been called *B.
Grœnlandicum*.   Fig. 3 of the same plate is an extreme variety as to its ornamentation,
which I will call *Buc. undatum*, var. *clathratum*.   *Buc. tenerum* of 'Min. Conch.,' t. 486,
figs. 3, 4, which the late Dr. S. P. Woodward in his list of Norwich Crag shells refers to
cyaneum? may possibly be distinct.   Fragments of *tenerum* (principally of the columella)
occur in the Middle Glacial sands of Billockby and Hopton Cliff, while perfect specimens
of it exactly resembling those of the Red Crag and of all ages abound in the Post-glacial
Fen gravel of March.   The constant features maintained by *tenerum* in the Crag, when
found to recur in so modern a formation as the Fen gravel, impress me with the belief
that this is a distinct species.   *Buc. undatum* ('Crag Moll.,' Tab. III, fig. 12, *c*) is very
rare in the Crag, while the form *tenerum* swarms in it.   In the numerous specimens
sent to me from March by Mr. Harmer I have not seen the true form of *B. undatum*,
and as this Post-glacial gravel of March presents a more Arctic character than the
present British seas I am disposed to believe *B. tenerum* may be *B. cyaneum*.

Supplement, Tab. II, fig. 5, represents a specimen from the Red Crag of Butley which
has much perplexed me.   There is an angularity in the upper part of the volution, below
which it is contracted.   The upper part of the volution is slightly striated, and there are
some striæ on the base or lower part of the body-whorl, with very faint indications of
undulations upon the spire, like those of *undatum*, with reflected imbrications upon the
columella.   The spire is much depressed, but it looks like a distortion, and I have
considered it as such for the present.   I will call it *B. undatum*, var. *ovulum*.   The axis
of the shell is $\frac{9}{16}$ths of an inch.

PURPURA LAPILLUS, *Linné*.   Crag. Moll., vol. i, p. 36, Tab. IV, fig. 6.

Localities.   Red Crag passim.   Fluvio-marine Crag, Bramerton and Thorpe.   Lower
Glacial, Belaugh, Rackheath, and Weybourne.   Middle Glacial, Billockby and Hopton.
Post-glacial, Kelsey Hill and March.

The variety *crispata* occurs in the Chillesford bed at most of its localities and in the
Lower and Middle Glacial sands.   In the Middle Glacial the form *incrassata* is in

association with *crispata*. I am disposed to think that *incrassata* is entitled to specific value; it much resembles *Fusus decemcostatus*, Gould. The common living British form, *crispata*, occurs in the fen gravels, and is said to be abundant at Kelsea Hill. *P. incrassata* is given in Woodward's list as from Thorpe, but rare.

TROPHON[1] ANTIQUUS, var. STRIATUS. Crag. Moll., vol. i, p. 44, Tab. V, fig. 1, *c, d*.

Localities. Red Crag passim, except Walton and Bentley. Fluvio-marine Crag passim; Chil. bed, Horstead, Coltishall, Aldeby, and Easton Bavent. Lower Glacial, Belaugh, Rackheath, and Weybourne. Middle Glacial, Hopton and Billockby.

TROPHON ANTIQUUS, var. STRIATUS CONTRARIUS. Id. Tab., fig. 1, *d, e, f, g, i, j*.

Localities. Red Crag passim. Fluvio-marine Crag passim; Chillesford bed, Horstead, Coltishall, Aldeby, and Easton Bavent. Lower Glacial, Belaugh, Rackheath, and Weybourne. Middle Glacial, Billockby?

TROPHON ANTIQUUS, var. CARINATUS. Id. Tab., fig. 1, *a, b*.

Localities. Red Crag, Sutton and Butley. Fluvio-marine Crag, Bramerton? Upper Glacial, Bridlington (*Woodward*).

TROPHON ANTIQUUS, var. CARINATUS CONTRARIUS. Id. Tab., fig. 1, *k*. Supplement, Tab. I, fig. 10, *a, b, c*.

Localities. Red Crag, Newbourn. Fluvio-marine Crag, Bramerton? Upper Glacial, Bridlington.

Some few years since, I found at Newbourn a specimen of this sinistral shell, which exhibits three ridges of carinæ upon the upper volution, and these are continued over the body-whorl, a form of sculpture I had not before seen upon any Crag specimens from Suffolk or Essex, and I have also had figured, by the obliging permission of the Committee of the Norwich Museum, a specimen of this reversed form which has the same carinæ, even more prominent (fig. 10, *a*); this latter shell was presented to the Museum

[1] *Neptunea*, Bolten, 1798, has been proposed by Messrs. H. and A. Adams as a generic name for this shell. *Tritonium*, O. Fabr., was adopted by Loven, and this has precedence; but it is difficult now to say what species was intended as the type of that genus. The name of *Trophon** has been previously given to my Crag shells, and as the differences between these are merely artificial or conventional, I have here retained the one I employed in the 'Crag Mollusca.'

* *Trophon* appears to be masculine, being, according to Mr. Jeffreys, a contraction of *Trophonius*.

by the late Col. Alexander, but, unfortunately, it has no special locality attached. It is undoubtedly a Crag shell, and from its appearance it looks like one of the Fluvio-marine specimens of Bramerton. The sinistral form of *T. antiquus*, occasionally found at the present day in the British seas, is simply striated, and not carinated, corresponding in that respect with the shell so abundant at Walton-on-the-Naze.

In the 'Crag Mollusca,' vol. i, p. 45, I have expressed an opinion that this left-handed striated whelk was, in British seas, probably the original form, in opposition to the general statement of conchologists that it is merely a variety, in consequence of the difference displayed from the common right-handed shell of the present day.

The great majority of shelled univalved Mollusca have the volutions turned in a dextral direction, that is, from left to right, but whether the original inflexion was given to the right or to the left we do not know, or why they should have taken the one in preference to the other. Among the Cephalopoda, the oldest known form is the straight one, as in *Orthoceras*. The bend from this seems to have been first in a vertical direction, such as *Phragmoceras* or *Toxoceras*. The deviation from that vertical direction was, I conjecture, due to the partial atrophy of the organs on one side, from a slightly altered position of the heart, until the highly oblique growth of the *Turrilite* was reached.

Fusus sinistrorsus, Lam., is now an inhabitant of the Mediterranean Sea, and it is also a fossil in the newer Tertiaries of Sicily, and this may be a descendant of the older form of the Walton Crag sea. I can perceive no difference sufficient to constitute the Mediterranean shell a different species from the Crag fossil. It is, therefore, somewhat remarkable that in *Trophon antiquus* this sinistral form should be the only one found in the Crag of Belgium, appearing there in the middle and upper beds, both dextral and sinistral forms being unknown in the lower, as they are also in the Cor. Crag of this country, thus apparently showing that the dextral form of this shell was of more modern origin than the sinistral, and that it had not appeared during the earlier part of the Red Crag. The left-handed "Almond Whelk" is the only form of this variable species which is found in the Red Crag of Walton-on-the-Naze (the whole fauna of which locality is, in my opinion, clearly older than that of any other part of the Red Crag); for while I have seen thousands of the sinistral shell from this locality, I have never met with one of the dextral form there, or seen a specimen of it in the possession of any collector from this place. In the rest of the Red Crag the dextral and sinistral forms of the striated shell seem present in about equal proportions, and the same thing occurs in the Fluvio-marine Crag, in the Chillesford bed passim, and in the Lower Glacial sands of Belaugh, Rackheath, and Weybourne. In the Middle Glacial sands, however, the only trace of the sinistral form that has occurred is the pullus of some sinistral *Trophon*, which is probably *contrarius*; while several perfect young specimens, and one full grown, as well as numerous fragments of the columella of the dextral shell, have occurred. It would thus seem that the life of this species, so far as the seas of Britain and Belgium reveal it, exhibits the curious feature of having begun exclusively left-handed, then to have varied by the birth of dextral individuals, and so progressed through the various subsequent formations until, by the decrease of the sinistral and increase of the dextral individuals, the species has attained in these seas to its present condition, wherein millions of dextral individuals occur to one sinistral. The late Edward Forbes mentions the occurrence in the Irish Drift of the sinistral form, and if it be the case that *Nucula Cobboldiæ* occurs in that Drift in association with it, as has been said, that occurrence would be in accordance with the antiquity of both these shells in the seas of Britain.

TROPHON BERNICIENSIS? *King.* Supplement, Tab. I, fig. 8, *a, b.*

FUSUS BERNICIENSIS, *King.* Ann. and Mag. Nat. Hist., vol. xviii, p. 246.

Locality. Fluvio-marine Crag, near Norwich.

The above (fig. *b*) represents a specimen which was sent to me by the late Dr. S. P. Woodward for examination ; it was accompanied by the following note :—" I have compared my new *Fusus* from the Norwich Crag with the figure of *T. Spitzbergensis,* and find them agree very well in the character of the spiral striæ ; but the recent shell has a more contracted canal." The very young condition of the " Norwich Crag " shell is much like *Spitzbergensis,* but the resemblance is less so when it is full grown. I think it more resembles the British shell *T. Berniciensis,* King, although with this it has not a perfect identity. Another specimen (fig. 8 *a*) has been more recently obtained from the same neighbourhood which appears to belong to the same species. This was found by Mr. John King, of St. Andrew's, Norwich, who has obligingly permitted me to have it figured. I have given to it provisionally the above name with a doubt, but if from future discoveries it should prove to be a new species I would suggest that it be called *T. Woodwardii.*

TROPHON NORVEGICUS, *Chemn.* Supplement, Tab. V, fig. 14.

In the ' Crag Moll.,' vol. ii, Tab. XXXI, fig. 1, is represented a specimen of what I then believed to be the species here referred to, and I have now another very nearly perfect from the cabinet of Mr. Canham, who has obligingly permitted me the use of it. It came from the nodule pits of the Red Crag at Waldringfield. This shows a more elongated form than the living shell, with a comparatively smaller aperture and more recurved canal. The outer lip is not quite perfect, but if it were, it would rather help to diminish the smaller proportions of the aperture.

TROPHON VENTRICOSUS? *Gray.*　Supplement, Tab. III, fig. 4.

> FUSUS VENTRICOSUS, *Gray.*　In Beechey's Voyage, p. 117.
> — 　— 　*Gould.*　Inv. Massachusetts, p. 285, fig. 200.

Locality.　Upper Glacial, Bridlington.

The specimen figured was sent to me by Mr. Leckenby ; it is not perfect, and unfit for fair description, but I have doubtfully referred it as above.　Dr. S. P. Woodward introduced this name in his list of Bridlington species, depending for so doing upon the above-mentioned specimen, as I have done.

TROPHON TURTONI, *Bean.*　Crag Moll., Appendix, p. 312, Tab. XXXI, fig. 2.　Supplement, Tab. I, fig. 11, *a, b.*

Locality.　Red Crag, Butley and Waldringfield.

In the 'Crag Moll.,' vol. ii, p. 312, I was able to indicate the presence of this species in the Red Crag, but my specimen there illustrated (Tab. XXXI, fig. 2) was so fragmentary that it could not be depended upon, and although the specimen I have had figured is not quite perfect it is sufficiently so to justify me in referring it to the species above named.　Since my specimen was engraved I have seen, in the collection of the Rev. Mr. Canham, an individual of this species obtained from the nodule pit in the Red Crag at Waldringfield, rather larger than my own, with the aperture more perfect, as also another by Mr. Bell, from Butley ; one of these I should have preferred to have had figured had I been previously aware of their existence.

TROPHON ELEGANS, *Charlesworth.*　Crag Moll., vol. i, p. 46, Tab. V, fig. 2.

In Mr. Canham's rich collection of Crag fossils is a fine specimen of what has been figured under the name of *Trophon elegans*, Charlesworth.　The locality was not then known, and it is somewhat singular that this second individual should also have been picked up on the beach at Felixstow.　Fig. 6, *a, b,* Tab. II, represents a shell found in the Red Crag at Butley by Mr. A. Bell, which, when first shown to me, I thought was a new species, but as it may possibly be the immature form of *T. elegans* I hesitate to give it a new name.

Trophon Sabini, *Hancock*. Supplement, Tab. II, fig. 15.

Locality. Upper Glacial, Bridlington.

A species is introduced into Dr. Woodward's list of Bridlington fossils under this name, but to which no authority is attached. I presume the shell belonging to Mr. Leckenby figured here was the one intended. It is given as a distinct species by Dr. Woodward, and I have followed him. This is the shell, I suppose, referred to by E. Forbes in ' Mem. Geol. Surv.,' p. 426, No. 119, 1846, from the Irish Drift, as well as from Bridlington, but he has given to his fossil the name of *T. Sabini*, Gray. *Buc. Sabinii*, Gray, is another species. The present shell seems more slender than *T. Islandicus*, with rather a deeper suture and a smaller apex, and it is, I imagine, the same as a shell figured by Gould, ' Invert. Massach.,' p. 284, fig. 199, called *F. Islandicus*, var. *pygmæus*.

Trophon altus, *S. Wood*. Crag. Moll., vol. i, p. 47, Tab. VI, fig. 13; and Supplement, Tab. II, fig. 17, *a, b*.

Localities. Red Crag, Butley. Fluvio-marine Crag, Bramerton.

I have here given representations of two specimens which, I believe, belong to the fossil I previously figured and described, with the above name, in ' Crag. Moll.' One is an elongated variety, with obsolete costæ, which I found in the Red Crag at Butley; the other is a shorter and more inflated shell, without ribs or striæ, found by Mr. A. Bell also at Butley. The nearest approach to this species, as pointed out by Mr. A. Bell, is a specimen in the British Museum from Newfoundland, named *F. cretaceus*, Reeve, and Mr. Bell has sent to me for examination some fossil specimens he has received from Dr. Dawson with the locality of " Rivière du Loup," which appear to be identical with the Crag shell, differing from *F. cretaceus*, in having an obtuse or mammillated apex, and attached to them is the name of *Buccinofusus Kroyeri*. This Canadian fossil is, I think, the same as the Crag shell, and it appears to present the same difference from *cretaceus* of Reeve (the one having an obtuse apex, while the other is pointed) as is considered specifically to distinguish *F. Islandicus* from *F. propinquus*. I think the Canadian fossil is not the *Kroyeri* of Möller.

This Crag fossil was originally called *Murex pullus* by S. Woodward, and I would have adopted the specific name, but it is neither the *pullus* of Linné nor the *pullus* of Pennant, and, in order to avoid confusion, I have thought it best to give it a new name.

TROPHON PROPINQUUS, *Alder*.   Crag Moll., Appendix, p. 313, Tab. XXXI, fig. 3.
Supplement, Tab. II, fig. 15, *a*, *b*; Supplement, Tab. II, fig. 15.

Localities.   Cor. Crag, near Orford.   Red Crag, Sutton and Butley.   Upper Glacial, Bridlington.

In the 'British Conchology,' vol. iv, pp. 333—341, are described four species under the respective names of *Fusus Islandicus*, *F. gracilis*, *F. propinquus*, and *F. buccinatus* (or *Jeffreysianus*, Fisch.); not one of these is there admitted by the author to have been an inhabitant of the Crag sea, either of the Coralline or of the Red, and he says (p. 336) " I do not consider the Crag specimens which have been referred to this species (*gracilis*) by Searles Wood, Woodward, and Nyst, identical with the above.   These last agree with the North American form, which is smaller and more tumid and has a short spire.   If such should prove to be distinct it might be called *curtus*."[1]

Having expressed my dissent to Mr. Jeffreys, he obligingly sent me some of his recent specimens for examination, but this has not altered my previously formed opinion.   I still consider *gracilis*, *propinquus*, and *buccinatus* (or *Jeffreysianus*) as Crag species.   The shell called *Islandicus* has a mammillated apex, and is probably distinct.   This latter I have not yet seen from the Crag, either the Red or Coralline.

Supplement, Tab. II, fig. 15, *a*, represents a specimen with a very straight canal, obtained by Mr. A. Bell from the Red Crag, Butley, and this I consider merely as an abnormal form of *F. propinquus*; and fig. 15, *b*, is that of a distorted variety of the same species found by myself in the Red Crag of Sutton.   Fig. 21 of Supplement, Tab. VII, is a reversed form obtained by Mr. Robert Bell at Waldringfield.

TROPHON LECKENBYI, *S. Wood*.   Supplement, Tab. VII, fig. 1.

Locality.   Upper Glacial, Bridlington.

The specimen figured has been in my possession for several years, but is, unfortunately, imperfect, and must, I think, have been given me by Mr. Leckenby, of Scarborough, to whom we are much indebted for obtaining the authentic fauna of the Bridlington bed.[2]

Though resembling *gracilis*, *Islandicus*, *propinquus*, and *Jeffreysianus*, it differs

---

[1] *Fusus curtus*, James Smith, a Clyde fossil ('Trans. Geol. Soc.,' 2nd ser., vol. vi, p. 156, No. 26), is probably *Mangelia Trevelyana*, Forb. and Hanl., and *F. curtus* of James Sowerby is a London Clay shell from Highgate, M. C. T. 199, fig. 5, and is quite distinct.

[2] I am sorry to say that spurious shells have been put on the scientific market as from Bridlington. Amongst them one that Mr. Leckenby detected, and sent me to look at, was an Eocene shell from a bed whose fossils are of similar colour to those from Bridlington.

from them all in two respects—one in that it is much less tapering, and the other in the shallowness of the suture. It is strongly marked with spiral striæ. Mr. Jeffreys sent me for comparison a shell obtained by him in the Porcupine dredgings, and which he intended to call *turgidulus*, which resembles our shell, but is much thinner and less strongly striated than ours. I have named the present shell after Mr. Leckenby.

TROPHON ACTONI, *S. Wood*. Supplement, Tab. II, fig. 13.

Locality. Red Crag, Butley.

In this figure I have represented a specimen found by myself at Butley, which I am unable to refer to any known species. Its principal distinction is a slight shoulder to the volution or obtuse angularity at the upper part. The outer lip is a little sinuated, like many northern species of this genus.

If this should prove (by the discovery of better specimens) a new species, I propose to call it *Tr. Actoni*, in commemoration of the late Edward Acton, surgeon, of Grundisburgh, a zealous collector of Crag fossils, and a liberal distributor of his specimens where he thought they would contribute to disseminate information.

A specimen very recently sent to me for examination by Mr. James Reeve, from the Fluvio-marine Crag of Bramerton, seems to belong to the same species, but the specimen, like my own, is imperfect, and unfit for correct determination.

TROPHON SARSII, *Jeffreys*. Supplement, Tab. I, fig, 9.

Length, $1\frac{5}{8}$ths inch.
Localities. Red Crag, Waldringfield (*Bell*), and Butley.

A specimen represented as above referred to was obtained by myself in 1868, from Butley, and since the figure was engraved Mr. A. Bell has shown me a similar specimen from the nodule pit at Waldringfield.

In the paper called 'Nature,' December 9th, 1869, Mr. Jeffreys has, in his report on the deep-sea dredgings, given to a shell there obtained, which he says is the same as my Crag fossil, the name of *Fusus Sarsii*, in compliment to the late Prof. Sars, who had obtained the same shell living near the Laffoden Isles; I have, therefore, adopted the name for my Crag shell.

TROPHON CRATICULATUS, *Fabricius*. Supplement, Tab. III, fig. 1, *a, b*.

Locality. Upper Glacial, Bridlington.
Mr. Leckenby has obligingly sent to me for description a fossil from Bridlington which

4

he tells me has been determined by Mr. Jeffreys to belong to the above-named species; the label accompanying the specimen says, " once marked *Gunneri* by S. P. W." In the Appendix to the ' Crag Mollusca,' p. 313, Tab. XXXI, fig. 4, is described a shell from the neighbourhood of Wexford which had been previously considered by E. Forbes (' Mem. Geol. Surv.,' 1846, p. 425) as *Trophon Fabricii*, Möller, with the name of *craticulatus*, Fab., given by Möller as a synonym. The shell I have here figured is probably the same species, but it is different in some of its characters, and I have in consequence had it represented as a variety. The exterior is ribbed and decussated by two or more raised spiral ridges, but it has not the upper part of the volution projecting and fimbriated. I believe in the assignment, and have adopted the above name as being the older one, on the authority of Möller. This name must, therefore, be given to the figure in the ' App. to Crag Moll.' *Fusus craticulatus*, Broc., is a very different shell.

TROPHON BAMFFIUS, *Donovan*. Supplement, Tab. III, fig. 2.

MUREX BAMFFIUS, *Don.* Brit. Shells, vol. v, p. 169, fig. 1.

Localities. Post Glacial, March and Kelsea Hill.

The figure of this shell was taken from a specimen from the Clyde beds, and was introduced in order to show the difference from the next species (figs. 10 *a* and *b*), but since the plate was engraved I have seen a suite of specimens from the post-glacial gravel of March ; so that this species is an East Anglian fossil. The March specimens are rather larger than any that I have seen from the Clyde beds, and approach slightly nearer to *scalariformis*. A specimen of *Bamffius* was sent me for examination by Mr. G. Maw, from the Severn Valley beds. I have not met with this species from the Crag or any East Anglian glacial bed. It is given under the name *clathratus* by Mr. Jeffreys from Kelsea Hill.

TROPHON SCALARIFORMIS, *Gould.* Crag Moll., vol. i, p. 48, Tab. VI, fig. 7. Supplement, Tab. III, fig. 10 *a, b.*

Localities. Red Crag, Sutton, Bawdsey, Butley. Fluvio-marine Crag, Bramerton. Middle Glacial, Billockby and Hopton. Upper Glacial, Bridlington. Post-glacial, Kelsea Hill.

The figure is taken from a Bridlington shell belonging to Mr. Leckenby, which seems to have been the authority for the occurrence of *Bamffius* at Bridlington. The shell, however, seems to me not to belong to that species, but to be the young of the much larger Crag shell, *scalariformis* (' Crag Moll.,' Tab. VI, fig. 7), which (identical in size and in all other respects with the Crag shell) is common at Bridlington.

The costæ of *scalariformis* are much fewer than those of *Bamffius*, and the shell is of more than double the linear dimensions of *Bamffius*. *Scalariformis* occurs in the Fluvio-marine Crag of Bramerton, but is very rare in it, and a specimen has been obtained from the Middle Glacial sand of Billockby, and another from that of Hopton, both of which fell to pieces, but their fragments exhibit clearly all the characters which distinguish this shell from *Bamffius*. I have not met with it from any of the localities of the Chillesford bed or from the Lower Glacial. It is given by Mr. Jeffreys as from Kelsea Hill.

TROPHON BARVICENSIS, *Johnson.* Supplement, Tab. VI, fig. 20.

> FUSUS BARVICENSIS, *Johnson.* Edinb. Phil. Journ., vol. xiii, p. 221.
> TROPHON — *Forb. et Hanl.* Brit. Moll., vol. iii, p. 442, pl. cxi, figs. 5, 6.

Localities. Red Crag, Walton, and Shottisham (*Bell*), Waldringfield.

Mr. Canham has sent to me a specimen he has obtained from the Red Crag at Waldringfield, which I have referred and represented as above. The fossil has been rubbed and its more prominent portions worn down, but it seems to correspond in all other respects. The periodical reflections of the outer lip called ribs or costæ are about ten in number upon the last remaining whorl; these are decussated by four or five spiral lines or ridges, and it has an angular or projecting shoulder a little below the suture. The specimen measures half an inch in length. Mr. Bell gives the species as from Walton and Shottisham ('Ann. and Mag. Nat. Hist.,' May, 1871), but I have not seen the specimens.

TROPHON GUNNERI, *Lovén.* Supplement, Tab. III, fig. 18 *a, b.*

> TRITONIUM GUNNERI, *Lovén.* Ind. Moll. Scand., p. 12, No. 84.

Localities. Upper Glacial, Bridlington. Post Glacial, Kelsea Hill.

My figure represents a specimen in the British Museum from Bridlington, and the late Dr. S. P. Woodward has introduced the name into his list of Norwich Crag shells, but with a mark of doubt. I have not, however, seen it from the Crag of either Suffolk or Norfolk. This much resembles *scalariformis*, but the fimbriæ are furnished with projecting fronds. Mr. Jeffreys has ('Quart. Journ. Geol. Soc.,' vol. xvii, p. 450) identified the shell as a Kelsea Hill species.

TROPHON PAULULUS, Crag Moll., vol. i, Tab. VI, fig. 6, is, according to Mr. Jeffreys, the young of his *Defrancia teres*, 'Brit. Conch.,' vol. v, p. 219. In this I think he is correct. The peculiar ornamentation of the upper volutions is spoken of in the 'Brit. Moll.,' as also in 'Brit. Conch.,' but it has never been represented.

TROPHON MURICATUS. Crag Moll., vol. i, Tab. VI, fig. 5.

A perfect specimen has occurred in the Middle Glacial sand of Billockby, but I have not met with it from the Fluvio-marine Crag, the Chillesford bed, or from the Lower Glacial sands.

TROPHON MEDIGLACIALIS, *S. Wood.* Supplement, Tab. VII, fig. 12, *a, b.*

Spec. Char. *T. Testá elongato-fusiformi, anfractibus rotundatis, longitudinaliter costatá, costis (8—10) elevatis obtusis; spiraliter lineatá, lineis paucis elevatis; aperturá ovatá; caudá elongatá.*

Locality. Middle Glacial, Billockby, and Hopton.

Length, half an inch nearly.

About a dozen specimens of this shell have occurred in the Middle Glacial sand of Billockby, and five in that of Hopton, and as they do not vary greatly in size, I infer that the specimen figured is a full-grown shell, or nearly so. Like almost all the fossils from this formation, the specimens are in a more or less injured condition; but the one figured has suffered but little, as the spiral striæ are preserved on it. The upper part of the whorls do not show any spiral striation, and this appears, not only in the specimen figured, but on such others as retain the external markings. As all the specmens, however, are more or less worn, this absence of striata on the upper part of the whorl may possibly be due to erosion. The costæ appear to be nearly equal in number on all the volutions. None of the specimens indicate any less tapering form than that figured, while some are slightly more tapering and slender.

This species being unknown to me from any other formation than the Middle Glacial sand of East Anglia, while it is somewhat numerous there, it appears to be characteristic of that formation; I have therefore assigned to it the specific name of *mediglacialis.*

TROPHON ? BILLOCKBIENSIS, *S. Wood.* Supplement, Tab. VII, fig. 13.

Locality. Middle Glacial, Billockby.

Length, ¼ of an inch.

A unique specimen in good preservation from Billockby, shown in fig. 13, is the foundation for this species, but whether it be a young specimen or a full-grown shell, there are no means of judging. It differs from *mediglacialis* in its less tapering form, the length of the shell being hardly twice its breadth; while in *mediglacialis* it is very nearly three times. It also differs from all the specimens of *mediglacialis* in the spiral striation, which is equally distributed over the whorls. I have never seen any but full-grown forms or nearly so of *Purpura tetragona*, and it is not impossible that the shell in question may be the young of that species. In this uncertainty I have provisionally given it the above name.

FUSUS CRISPUS? *Borson.* Supplement, Tab. 11, fig. 10.

FUSUS CRISPUS, *Borson* (fide *Mich.*). Oritt. Piemont., p. 317, No. 17.
— — *Michelotti.* Desc. des Foss. Mioc. de l'Ital., September, p. 272, t. ix, figs. 17, 18.

Spec. Char. "*F. Testa elongato-fusoidea, solida ; anfractibus convexis, longitudinaliter costatis ; costis crassis, rotundatis, transversim plicatis, plicis super costas lamellosis, in interstitiis filiformibus, apertura subovata, canali elongatiusculo, aperto, cylindraceo ; labro intus profunde sulcato ; columella lævigata.*"—Mich.

Locality. Red Crag, Sutton.

The specimen figured was obtained by Mr. Whincopp from the workmen at the nodule excavations in the Red Crag, and he has kindly permitted me to figure it, and though much worn it retains some of the outer coating with its ornaments. I have also obtained from the Red Crag at Sutton a specimen which appears to belong to the same species, but in a more worn condition. This is probably a derived species.

FUSUS ABRASUS, *S. Wood.* Supplement, Tab. II, fig. 8.

Locality. Red Crag, near Woodbridge.

This represents another specimen from the collection of Mr. Whincopp. It appears to be a fossil extraneous to the Red Crag, and it has been much altered, and the outer coating apparently removed. It has somewhat the form and ornaments of *F. rugosus* or *F. costiferus,* but the ribs incline too much to the right, and it is too elongated. It is a very much abraded shell, and as I am unable to refer it to any species known to myself, I have given to it provisionally the above name.

FUSUS IMPERSPICUUS (*T. imperspicuum*). Crag Moll., vol. i, p. 50, Tab. VI, fig. 12.

The late Dr. S. P. Woodward (*in Lit.*) suggested this might be *F. latericeus.* I have carefully again compared my shell with that species, and I think they are specifically different. I have given another view of the Crag shell, showing the opening, Supplement, Tab. II, fig. 4.

Fusus Forbesii, Strickland. *F. cinereus*, Say., *Buc. plicosum*, Gould, is mentioned at p. 314, 'Append. Crag Moll.' I have not seen this as a fossil from the Upper Tertiaries of the *East* of England.

TRITON HEPTAGONUS. Crag Moll., vol. i, p. 41, Tab. IV, fig. 8.

The specimen figured, as above referred to, is, I regret to say, the only one I have seen. Mr. Jeffreys, in ' Brit. Conch.,' vol. v, p. 218, refers this Crag shell to *T. cutaceus*. I do not coincide in that opinion ; at the same time it does not strictly accord with the Subapennine fossil; neither can I find a published species with which it can be identified. I propose to call it *Triton connectens*.

MUREX CORALLINUS, *Scacchi*. Supplement, Tab. II, fig. 12 *a, b*.

> MUREX CORALLINUS, *Scac.* Faun. del. Nap., p. 11, fig. 15.
> — INCONSPICUA, *G. B. Sow.* Conch. Illust. Murex, fig. 81.
> — BADIUS, *Reeve.* Conch. Icon. Murex, pl. 32, fig. 159 ?
> — ACICULATUS, *Jeffreys.* Brit. Conch., vol. iv, p. 310.
> FUSUS LAVATUS, *Phil.* En Moll. Sic., vol. i, p. 203.
> — CORALLINUS, *Id.* En Moll. Sic., vol. ii, p. 178, t. 25, fig. 29.

Spec. Char. M. Testá elongato-fusiformi; anfractibus 6—7, rotundatis; longitudinaliter plicatá; spiraliter striatá, striis vel lineis elevatis scabris, subæqualibus, labro incrassato intus plicato; caudá brevi, fistulosá.

Length, ⅝ths of an inch.

Locality. Cor. Crag, Gedgrave.

A single specimen (the only one known to me) was obtained by the late Dr. S. P. Woodward, and Mr. Horace Woodward has obligingly permitted me to have it figured. The only difference that I can detect between the recent shell and our fossil is that the latter is a trifle the longer, and, although it appears to be a full-grown specimen, the canal is not closed, as in the recent species.

MUREX CANHAMI, *S. Wood.* Supplement, Tab. VII, fig. 14.

Spec. Char. Testá oblongo-ovatá, fusiformi, valde striatá; anfractibus superne depressis, carinatis, scalariæformibus, septem fariam varicosis, varicibus lamellosis, in angulum protractis; superne squamato crenulatis; aperturá ovatá; canali brevi.

Length, $\frac{9}{16}$ths of an inch.

Locality. Red Crag, Waldringfield.

The specimen figured has been obligingly sent to me by Mr. Canham. It is in good preservation, deeply coloured with Red Crag, and it appears to me not to be a derivative. I have a fragment from the Coralline Crag near Orford that may probably be the same species.

The shell to which it appears to approach the nearest (judging from figure and description) is *Murex Haidingeri*, Hörnes, 'Vienna Foss.,' vol. i, p. 228, tab. xxiii, fig. 12; but I think it is distinct. It differs from *M. tortuosus*, the well-known Crag species, in having all its ridges frondiculated, and I have named it after the discoverer, the Rev. H. Canham.

MUREX ERINACEUS, *Linné*. Supplement, Tab. II, fig. 11.

MUREX DECUSSATUS, *Broc*. Conch. Foss. Subapen., p. 391, pl. vii, fig. 11.

Locality. Red Crag, Harwich? Butley. Fluvio-marine Crag, Bramerton? Post Glacial, Kelsea Hill.

At page 39, vol. i, of the 'Crag Mollusca,' is introduced a notice of this species as having been found in the Fluvio-marine Crag at Bramerton. This was sent to Sir Charles Lyell for examination, in whose possession I saw it. We were both of opinion that it was a genuine Crag shell, but a short time previous to the publication of my first volume it was unfortunately lost, and I was unable to have it figured. I have, therefore, now, in order to complete the Crag Mollusca, given the representation of a recent specimen, and since the Plate has been engraved, Mr. A. Bell has found a fragment of this species in the Red Crag at Butley.

In a paper by the late Mr. Webster, in the 'Trans. of the Geol. Soc.,' vol. ii, p. 220, 1814, is a List of Shells from Harwich (which, I presume, were intended as Crag species), and in this is the name of *Murex erinaceus*; but where these specimens are I cannot ascertain. There are two or three in that List it would be desirable to examine (viz. *Trochus alligatus*, *Venus gallina*, and *Pecten infirmatus*), to learn what shells were intended to be determined by those names.

Mr. Jeffreys identifies a fragment of *Murex erinaceus* from the Kelsea Hill Gravel.

PLEUROTOMA.

The genus *Pleurotoma*, as originally instituted by Lamarck, was intended for those fusiform shells possessing a slit in the outer lip for the excurrent canal. There is, however, a large group of this kind of shell in which the outlet varies from a deep and narrow incision, such as in *Pl. Babylonia*, through a broad and sloping sinus, until it vanishes in the genus *Fusus*. These have been divided into many genera, such as *Mangilia, Bela, Defrancia, Clavatula, Rhaphitoma*, and several more. The impossibility of drawing a satisfactory line between these genera seems to have induced M. Hörnes, in his work 'On the Vienna Fossils,' to group them all under the generic name of *Pleurotoma*, and in this Supplement I have followed his example.

PLEUROTOMA CORONATA? *Bellardi.* Supplement, Tab. VI, fig. 4, *a, b.*

> PLEUROTOMA CORONATA, *Bellardi.* Mon. Pleurot. Foss., p. 47, t. iii, fig. 5, 1847.

Locality. Red Crag, Waldringfield.

A single specimen from the nodule pit in the Red Crag at Waldringfield has been put into my hands for examination by the Rev. H. Canham. This shell has a smaller and shorter aperture than is shown in *Bellardi's* figure of this species; but the Crag specimen is unique, and I have given to it the above name provisionally. There is a distinct ridge on the lower part of the volution just above the suture; the knobs are on a somewhat angular whorl, and the sinus is not very distinct. It is probably a derived specimen.

PLEUROTOMA INTORTA, *Brocchi.* Crag Moll., vol. i, Tab. VI, fig. 4.

Although some more specimens have been turned out of the nodule pits, this species may, I think, be still classed among the extraneous fossils of the Red Crag. I have not yet seen it from the Coralline Crag. It much resembles *Pl. callosa*, Kiener, from the Gaboon River.

PLEUROTOMA INTERRUPTA, *Brocchi.* Supplement, Tab. V, fig. 1.

> MUREX INTERRUPTUS, *Broc.* Conch. Foss. Subapen, t. ix, fig. 21, 1814.
> PLEUROTOMA INTERRUPTA, *Bellardi.* Mon. Pleurot. Foss., p. 31, t. ii, fig. 16, 1845.
> — TURRIS, *Lam.* Hist. des An. sans Vert., vol. vii, p. 97.

Locality. Red Crag, Waldringfield.

This is another shell for the notice of which I am indebted to Mr. Canham. It is from the Red Crag at Waldringfield. It is probably a derived specimen, and has been much rubbed and water-worn, but there is enough of it left, I think, to justify the reference. It is the only specimen I have seen.

PLEUROTOMA TURRIFERA, *Nyst.* Crag Moll., vol. i, p. 53, Tab. VI, fig. 1 (as *P. turricula,* Brocch.).

Localities. As in 'Crag Moll.'

This, so far as I know, is confined to the Red Crag. The specimens I found were not restricted to the "Coprolite" bed, but were obtained from various parts of the Red Crag. They are all more or less rubbed and worn, and are probably derived specimens. I have seen nothing more perfect than those previously figured. This is quite distinct from *Murex turricula,* Mont., and as this latter has precedence in date, our present shell must have its name changed. M. Nyst proposed to call the present shell *Pleurotoma turrifera,* and I have adopted his proposition.

PLEUROTOMA NODIFERA, ? *Lamarck.* Supplement, Tab. VII, fig. 6.

PLEUROTOMA NODIFERA, *Lam.* Hist. des An. sans Vert., 2nd edit., t. ix, p. 353.

Locality. Red Crag, Waldringfield.

The specimen figured is from the nodule workings at Waldringfield, and is in a mutilated condition, its true character being obscured. It is probably an extraneous fossil, but it has the colour of the true shell of the Red Crag. The nearest species I can compare it with is the above, but it is very doubtful, and I am unwilling in its present state to consider it a new species. The specimen has one row of prominent nodules (fourteen in the last remaining volution), between which and the suture is a shallow and curving sinus, as shown by lines of growth. *Pl. nodosa,* Bellardi, has a similar row of nodules, but the form of that shell is different. The specimen is from the cabinet of Mr. Canham.

PLEUROTOMA INERMIS, *Partsch.* Crag Moll., vol. i, p. 55, Tab. VII, fig. 1, as *P. porrecta.* Supplement, Tab. III, fig. 2 *a, b.*

PLEUROTOMA INERMIS, *Partsch,* 1842, fide *Hörnes.*
— — *Hörnes.* Foss. Moll., Vienna, vol. i, p. 349, t. xxxviii, fig. 10, 1856.
— NIVALE ? *Lovén.* Ind. Moll. Scand., p. 14, 1846.

PLEUROTOMA GASTALDII, *Sismonda.* Syn. Method. Pied. Foss., p. 33, 1847.
   —   —   *Bellardi.* Mon. Pleurot. Foss. Pied., p. 44, t. xi, fig. 19, 1848.
   —   NIVALIS? *Jeffreys.* Brit. Moll., vol. iv, p. 388, pl. xci, fig. 4, 1867.

Localities. Cor. Crag, near Orford, and Ramsholt.

I have given two fresh views of this species of specimens from the Cor. Crag, near Orford, as they are variable in ornament; one is nearly smooth and the other shows a row of prominent nodules in the middle, formed by the thickened margin of the back of the sinus. The specimen previously figured, 'Crag Moll.,' vol. i, Tab. VII, fig. 1, is of an intermediate character.

M. Bellardi (p. 44, as above) has given the date of 1842 to the name of *Gastaldii.* I am not able to determine the right to priority. My copy of 'Syn. Meth. Pied. Foss.,' referred to, is dated 1847.

PLEUROTOMA TARENTINI,? *Philippi.* Supplement, Tab. III, fig. 5.

PLEUROTOMA TARENTINI, *Phil.* En. Moll. Sic., vol. ii, p. 175, t. xxvi, fig. 26.

Locality. Cor. Crag, near Orford and Sutton.

The specimen figured was sent to me by the late Dr. S. P. Woodward some years since, as having been found in the Coralline Crag, near Orford, and with it was the name of *Pleurotoma Renieri*, Phil. Some fragments of the same species have been long in my cabinet, but I had imagined them to be the young state of *Pl. turricula*, Broc. (*P. turrifera*). Mr. A. Bell has sent to me, since my figure was engraved, a specimen in better condition, accompanied by the name of *Pl. Tarentini*, while I have myself obtained another, and as I think it corresponds with that species rather than with *Pl. Renieri*, I have so referred it, although probably the two species may be united. It is ornamented with spiral ridges, the two upper ones being larger and further apart than the others; these ridges are continued over the base of the volution, but the upper part is not carinated. The sinus appears to be broad and shallow, curving towards the suture; the apex of the shell is obtuse, and the two first volutions smooth.

PLEUROTOMA MODIOLA, *Jan.* Crag Moll., vol. i, p. 54, Tab. VI, fig. 2 (as *P. carinata*).

FUSUS MODIOLUS, *Jan.* Catal., p. 10, No. 17, 1832, fide *Hörnes.*
PLEUROTOMA MODIOLA, *Bellardi.* Mon. Pleurot. Foss. Pied., p. 68, t. iii, fig. 9, 1841.
   —   —   *Hörnes.* Foss. Moll., Vienna, vol. i, p. 366, t. xxxix, fig. 12, 1856.

Localities. As in Crag Moll.

My Crag shell was referred to *Pl. carinata*, Phil., which, being subsequent in date to the above, I have here adopted the older one. I have not been able to find a specimen since the one previously figured. Dr. Hörnes represents the lip as deeply sinuated above the keel.

PLEUROTOMA CRISPATA, *Jan.* Supplement, Tab. VI, fig. 13.

PLEUROTOMA CRISPATA, *Jan.* Catal., p. 9, No. 25, fide *Bellardi.*
— — *Bellardi.* Mon. Pleur. Foss. Pied., p. 69, Tab. IV, fig. 2, var.
A. papillosa.
— — *Hörnes.* Foss Moll., Vien., vol. i, p. 367, t. xxxix, fig. 13.

Locality. Cor. Crag, Ramsholt.

Spec. Char. Pl. " *Testa turrita, spira elata, anfractibus convexis medio carinatis, postice concavis, longitudinaliter arcuatim striatis, antice convexis, transversim grosse striatis subcostatis ; striis æquidistantibus in primis 1 vel 2 in ultimo perplurimis usque ad canalem regulariter decressentibus, carina simplici ; sutures marginatis, margini simplici, filiformi ; canali brevi, contorto, apertura, ovata."*—Bellardi.

This shell was found by myself some years ago in the Coralline Crag, at Ramsholt, and as it has a sub-nodulous keel upon the upper whorls, it was (with other forms which I have now separated) regarded as a variety of the shell figured under the name of *P. semicolon* in 'Crag Moll.,' vol. i, p. 54. I now think it is specifically distinct, and have referred it to Bellardi's var. *carina nodulosa* of *P. crispata,* Jan. The sinus, as shown by the lines of growth, is broad, and curving from the keel to the suture, below which is a small ridge. The apex of my shell is obtuse and smooth, and the aperture is rather shorter comparatively than the species to which it is referred.

PLEUROTOMA ICENORUM, *S. Wood.* Crag Moll., vol. i, p. 54, Tab. VI, fig. 3 *a* (as *Pl. semicolon*).

Locality. Coralline Crag, Orford and Sutton.

In the 'Crag Moll.' this shell was doubtfully referred as var. *a* of *Pl. semicolon,* a Lower Tertiary species. Since then I have seen and found more and better specimens, which satisfy me that the Crag shell is quite distinct from *semicolon*; a view shared by Mr. Edwards in his 'Eocene Mollusca Univalves' (p. 244). I should have been inclined, from Basterot's figure of *Pl. denticula,* to refer the Crag shell to his species of that name; but another widely spread Eocene shell is described by Mr. Edwards under the name *Pl. denticula* ('Eocene Moll. Univalves,' p. 286), and considered by him to be identical with Basterot's species. As the Crag shell is clearly distinct from the shell thus identified by Mr. Edwards with *Pl. denticula,* Bast., there seems no alternative than to assign it another name as a new species, which I have accordingly done as *Pl. Icenorum.* It appears to be intermediate between *denticula* and *galerita.*

PLEUROTOMA BIPUNCTULA, *S. Wood.* Crag Moll., vol. i, p. 54, Tab. VI, fig. 3 *b* (as *P. semicolon*).

This shell, which I also doubtfully referred as var. *b* of *Pl. semicolon*, J. Sow, is, I am now satisfied, distinct from that species. It differs from the preceding *Pl. Icenorum* in several points, viz. in its ornament, in its more fusiform shape, and in the position of the sinus of the outer lip, which in this is more central, as indicated by the knobs left on the volution. It somewhat resembles *Pl. bicatena*, Grat., but I believe it is different, and it differs from the Eocene *Pl. denticula*.

Mr. Bell, in the ' Ann. and Mag. Nat. Hist.,' 1871, gives the name of *Pleur. pannum* to my shell. It possibly may be so, but I have not been able to see a specimen of that Bordeaux species. The description given by Basterot is insufficient, so that I could not venture to refer my shell in reliance on that, and the figure of this species by Bellardi, 'Piedm. Foss.,' Tab. II, fig. 2, is so inadequate, that it does not represent the position of the sinus, neither does his description (p. 27) supply the deficiency. I have therefore given to it a new name.

PLEUROTOMA LINEARIS. Crag Moll., vol. i, p. 56, Tab. VII, fig. 2.

Localities. Cor. Crag, Sutton. Red Crag, Sutton and Butley. Chillesford Bed, Aldeby (*Crowfoot* and *Dowson*). Middle Glacial, Billockby.

The red crag form of this shell figured in the 'Crag Mollusca' as the recent *linearis*, possesses the same number of ribs and the same sculpture as on the recent shell, but a specimen from the Middle Glacial sand of Billockby, in a state of preservation unusually perfect for that formation, departs from the recent form in respect that it possesses fourteen ribs on the body whorl and twelve on the next, or two beyond the respective numbers on these whorls possessed by recent shells. I have not, however, thought it necessary to figure the Middle Glacial shell, or to assign it as a new species, which, however, it may possibly be, as it approaches *perpulchra* in the character of its ribs and of its striation, though not identical with that species.

PLEUROTOMA ELEGANTULA, *A. Bell.* Supplement, Tab. III, fig. 8.

PLEUROTOMA ELEGANTULA, *A. Bell.* Ann. and Mag. Nat. Hist., p. 8, May, 1871.

Spec. Char. Pl. " *Shell stoutly fusiform; whorls* 7—9 *convex, ornamented with close-set ribs* 10—12 *on the second whorl* (*penultimate ?*) *; suture deep; mouth and canal open; pillar lip reflected; notch sinuated rather deeply.*"—A. Bell.

Length, $\frac{7}{10}$ths of an inch.

Localities.   Cor. Crag, Gedgrave, and Ramsholt.

The specimen figured was found by myself at Ramsholt, and since the engraving was made Mr. A. Bell has shown me some similar and rather better specimens which he has obtained from a dealer at Orford.   In the 'Ann. and Mag. Nat. Hist.' for May, 1871, Mr. Bell has described three species under the respective names of *Pleurotoma elegantula, Pl. volvula,* and *Pl. notata.*   These specimens I have seen, and although there is some slight variation among them, there did not appear to me sufficient differences for specific variation.   I have here adopted one of the names he has published, but I think from what I saw the others can be considered only as varieties.   A large series of these is desirable to determine their correct claim to specific isolation.   Our present shell has 7—8 volutions, the two first are smooth and the third has numerous riblets, while the rest of the shell has prominent obtuse costæ, with fine spiral striæ, which are visible only in places.

PLEUROTOMA CRASSA, *A. Bell.*   Supplement, Tab. VII, fig. 10.

CONOPLEURA CRASSA, *A. Bell.*   Ann. and Mag. Nat. Hist., p. 8, May, 1871.

Spec. Char.   " *Shell thick, shortly conical, smooth, polished; spire occupying about half the length of the shell; apex pointed; whorls 8—10 slightly convex at bottom, constricted towards the top; suture slight, forming a channel on the top of the whorl; ribs stout, but hardly raised above the surface; mouth short, open; canal short and broad; pillar lip straight, reflected with the callus, massed into a pad at the top, which forms one side of the labial notch; notch very large, broad, and deep; outer lip spreading.*"—A. Bell.

Length, $\frac{7}{16}$ths of an inch.

Locality.   Coralline Crag, Gedgrave.

Two specimens of this species have been obtained by Mr. Bell and sent to me with the MS. name of *Conopleura crassa* attached.   This specific name I have therefore retained for the present, but the generic one, I think, is not required.   This shell appears to be closely allied, and perhaps when more specimens are examined it may be referred to *sigmoidea,* Bronn, figured and described by Bellardi, 'Mon. della Pleur. Foss.,' p. 109, as *Raphitoma, sigmoidea,* Tab. IV, fig. 29 ; that shell, however, is described as " *Anfractibus ventricosis,*" and finely striated.   The Crag specimens appear to be free from striæ (judging from the two specimens I have seen) with rather flattened volutions.   The apex is small, but obtuse, and the two first volutions are without ridges.

PLEUROTOMA ATTENUATA, *Montagu*.   Supplement, Tab. III, fig. 7.

> MUREX ATTENUATUS, *Mont.*   Test. Brit., p. 266, pl. ix, fig. 6, 1803.
> PLEUROTOMA ATTENUATA, *Jeffreys*.   Brit. Conch., vol. iv, p. 377, pl. xc, fig. 2.

Locality.   Coralline Crag, Orford and Sutton.

A few specimens in my Collection appear to correspond with Montague's figure and expressive description, as well as with those of Forbes, and Hanley, and Mr. Jeffreys, although my specimens, being fossil, do not exhibit the spirally-coloured lines of the living shell.   The outer lip of my specimens is not quite perfect, but the lines of growth show an elegant curve with a moderately sized sinuation from the projecting portion up to the suture.   Fig. 6 *a*, Tab. VII, of 'Crag Moll.,' may, I think, be a form of this species; and so far as I have seen, I doubt if the shell called *Pl. gracilior*, A. Bell, 'Ann. and Mag. Nat. Hist.,' May, 1871, can be specifically separated from it.

PLEUROTOMA SEPTANGULARIS, *Montagu*.   Supplement, Tab. VI, fig. 16.

> MUREX SEPTANGULARIS, *Mont.*   Test. Brit., p. 268, Tab. IX, fig. 5.
> PLEUROTOMA HEPTAGONA, *Scacchi.*   Notiz., p. 42, t. i, fig. 9.
> —   SEPTANGULARE, *Phil.*   En. Moll. Sic.,' vol. ii, p. 169.

Locality.   Post-glacial, Nar Brickearth, Pentney (*Rose*).

Two specimens belonging undoubtedly to this well-known species are in Mr. Rose's cabinet, and I give them without hesitation; moreover, they are attested· in Mr. Jeffreys' handwriting.   Mr. Rose has kindly permitted me the use of them for illustration.

This is one of the existing British species which I have not seen or heard of from any Tertiary deposit in this country older than the Post-glacial.

PLEUROTOMA ELEGANTIOR, *S. Wood*.   Supplement, Tab. III, fig. 15.

Spec. Char.   Pl. Testa fusiformi, tereti; anfractibus 7, juxta suturam costulata obsolete angulatis, longitudinaliter subtilissime striatis, transversim costulatis; spira acuminata.

Length, ½ an inch.
Locality.   Upper Glacial, Bridlington.

The specimen figured was, among some Bridlington fossils, obligingly sent to me for examination by Mr. Leckenby.   It comes near to, but does not seem identical with, *M. elegans* of Möller.   The costæ are prominent, terminating at the upper angle of the volution, between which and the suture is the sinuation curving backwards. Considering it different, I have given to it the above specific name.

PLEUROTOMA SCALARIS? *Möller.* Supplement, Tab. III, fig. 12

DEFRANCIA SCALARIS, *Möll.* Ind. Moll. Groenl., p. 12, 1842.

Locality. Upper Glacial, Bridlington.

This was also sent to me by Mr. Leckenby, and with the specimen is a label on which is written, "considered by J. G. J. to be a large example of *Mang. pyramidalis.*" It seems to correspond with the description given by Möller of his species *scalaris*, and I have ventured, with a doubt, to refer it accordingly.

Mangelia cinerea, Möller, is given in Woodward's List as a species from Bridlington, which I have not been able to find. The nearest approach to *cinerea* is a shell represented in 'Crag Moll.,' vol. i, p. 64, Tab. VII, fig. 15, as *Clavatula plicifera.*

Defrancia cinerea, Möll., in British Museum, is probably an elongated var. of *scalaris.* These forms all depart from *turricula* in the absence of the prominent shoulder of that species.

PLEUROTOMA DOWSONI, *S. Wood.* Supplement, Tab. III, figs. 13 and 14.

Spec. Char. Pl. Testa ovato-fusiformi, clathrata; anfractibus convexis, tumidiusculis, juxta suturam angulatis, longitudinaliter lineis eminentibus cinctis, transversim plicatis; spira breviore.

Length, $\frac{9}{10}$ths.

Localities. Chillesford bed, Aldeby. Middle Glacial, Billockby, and Hopton? Upper Glacial, Bridlington.

One specimen was sent to me as from Bridlington, fig. 14, by Mr. Leckenby, and another, fig. 13, from Aldeby, was found by Messrs. Crowfoot and Dowson. They approach the recent *exarata* of Möller, but are shorter and much more tumid. These two specimens are in excellent preservation. About a dozen specimens of a shell from the Middle Glacial of Hopton and Billockby appear to belong to this species, but they are too much mutilated to enable me to refer them to it without a note of interrogation. They agree in the tumid form, and in so much of the sculpture as is preserved, their ribs being fewer than those of *proxima.* I have named the shell after Mr. E. T. Dowson, of Geldeston, to whose industry, in association with Mr. W. M. Crowfoot, of Beccles, we are indebted for so complete a fauna of the Aldeby deposit.

PLEUROTOMA ROBUSTA, *S. Wood.* Supplement, Tab. III, fig. 16.

Spec. Char. Pl. Testa ovato-fusiformi ; anfractibus 6, juxta suturam subangulatis, longitudinaliter costulata, obsoletè clathrata.

Length, ¾ths inch.
Locality. Upper Glacial, Bridlington.

From a specimen in my own possession, given to me some years ago by Mr. Leckenby. This is the largest species of that section of this genus, to which the generic name of *Mangelia* has been applied, that I have met with. The shell differs altogether from *turricula* in its more robust form, in its greater size, and in the indistinctness of the shoulder angle.

PLEUROTOMA TURRICULA, *Mont.* Crag Moll., vol. i, p. 62, Tab. VII, fig. 13 ; Supplement, Tab. VII, fig. 8.

Localities. Red Crag, Walton, Sutton, and Butley. Fluvio-marine Crag, Bramerton. Chillesford Bed, Aldeby. Middle Glacial, Hopton. Upper Glacial, Bridlington (*Woodward*). Post-glacial, March and Kelsea Hill.

Several specimens of this well-marked and common British shell have occurred in the Middle Glacial Sand of Billockby and Hopton, and I have a good suite of them from the March gravel. It is rare in the Fluvio-marine Crag, and has occurred in the Chillesford bed at Aldeby, and it is given as common at Kelsea Hill by Mr. Jeffreys.

The specimen in Supplement, Tab. VII, fig. 8, is from the Middle Glacial Sand of Hopton Cliff. It differs from the typical form of *turricula* (of which several individuals have been obtained from the same sand) in the relatively greater proportion of the body whorl, which is fully two thirds the length of the whole shell. As this, however, seems to me only such a modification as might be expected to occur to a chance individual, I have only ventured to regard it as a variety of *Pl. turricula.*

PLEUROTOMA ASSIMILIS, *S. Wood.* Supplement, Tab. VI, fig. 18.

Localities. Red Crag, Butley. Middle Glacial, Billockby and Hopton.
Spec. Char. Pl. Testa ovato-fusiformi, obsoletè clathrata ; longitudinaliter costulata anfractibus sex, juxta suturam rotundato-angulatis ; spira acuminata.

Length, 9/16ths of an inch.

The specimen figured is from the Middle Glacial Sand at Billockby, and another has occurred at Hopton. I have also found in the Red Crag of Butley a specimen of it quite perfect, and of similar size to that figured, which, so far as I know, is unique in the Crag.

I have given the name *assimilis* from its close similarity, in all other respects than size, to the large 'Bridlington shell *robusta*. The uniformity in size between the Crag specimen and the two Middle Glacial ones, renders it probable that the specimen figured is full grown; I have, therefore, considered this species as distinct from *robusta*, and assigned to it a separate name. It is, however, possible that, as some existing shells assume gigantic proportions under certain arctic conditions, the shell thus appearing in the Red Crag before the Glacial conditions had actually fallen upon Britain, and occurring in similar size in the Middle Glacial Sand by virtue of those causes, whatever they may be, to which the presence of so many southern species in these sands is due, became inflated by the truly arctic conditions under which the Bridlington shells lived, to the gigantic dimensions possessed by the species I have figured under the name of *robusta*. The shell differs from *Trevelyana* in the greater length of the body whorl relatively to the spire, and in having a less prominence of shoulder: also in its greater size.

PLEUROTOMA HYSTRIX, *Jan.* Supplement, Tab. VI, fig. 3, *a, b*.

> PLEUROTOMA HISTRIX, *Jan.* Catal., p. 10, No. 59, 1832, fide *Bellardi*.
> RHAPHITOMA HISTRIX, *Bellardi.* Mon. delle Pleur. Foss., p. 85, t. iv, fig. 14, 1847.
> DEFRANCIA HISTRIX, *A. Bell.* Ann. and Mag. Nat. Hist., September, 1870.

Spec. Char. " *Testa subfusiformi, elongatá, angusta, costis longitudinalibus et transversalibus exilissimis, lamellosis clathrata; in earum intersecatione papillis acutis, erectis hirsuta; anfractibus planiusculis, elongatis, posticé lævibus; spira elata, apertura ovato-elongata; labro intus sulcato; canali longinsculo.*"—Bellardi.

Length, $\frac{13}{16}$ths of an inch.

Localities. Cor. Crag, Sutton (*Bell*). Red Crag, Walton-on-the-Naze.

The specimen figured was obtained by myself from Walton, from which place another has since been obtained by Mr. Bell. It much resembles *Fusus cancellatus*, J. Sow., *Clavatula cancellata*, 'Crag Moll.,' but the present shell is more elongated, and there is a broad sinus adjoining the suture; this leaves a blank or naked depressed space at the top of the volution, or at least shows only lines of growth of the sinuated aperture. The shell is elegantly cancellated, and the outer lip is somewhat thickened and denticulated on the inside. This is not *Murex hystrix*, Linn. Mr. Bell gives the species as from the Cor. Crag. ('Ann. and Mag. Nat. Hist.,' May, 1871), but I have not seen the specimen.

PLEUROTOMA TENUISTRIATA, *A. Bell.* Crag Moll., vol. i, p. 62, Tab. VII, fig. 12 (as
Clavatula lævigata).

> PLEUROTOMA TENUISTRIATA, *A. Bell* Ann. and Mag. Nat. Hist. for May, 1871.

Locality. As in 'Crag Moll.'

6

In the 'Crag Mollusca' I referred this shell doubtfully to *Pleurotoma lævigatum* of Philippi, but the specimen there figured was not quite perfect. Mr. Bell has lately found one or two in better preservation, to which he has given the name of *Pl. tenuistriata*. As I am satisfied that this shell is distinct from *P. lævigata*, I have given it here under Mr. Bell's name.

PLEUROTOMA LÆVIGATA, *Phil.* Supplement, Tab. VI, fig. 15.

Localities. Red Crag, Walton and Butley.

The above figure represents a specimen of what I believe to be the true form of *lævigata*, found by myself in the Red Crag of Butley. Mr. Bell gives it also from Walton Naze.

PLEUROTOMA CYLINDRACEA ? *Möller.*

> DEFRANCIA CYLINDRACEA, *Möller.* Ind. Moll. Grœnl., p. 13, 1842.
> MANGELIA CYLINDRACEA, *Woodward.* Geol. Mag., vol. i, p. 53, 1864.

Locality. Upper Glacial, Bridlington?

This is given in Dr. Woodward's Bridlington list with a query, but I cannot find a shell among the Bridlington fossils that will correspond with Möller's description.

PLEUROTOMA SENILIS, *S. Wood.* Supplement, Tab. V, fig. 5.

Locality. Red Crag, Waldringfield.

The specimen figured was obtained by Mr. Canham from the Nodule pit at Waldringfield. It is much rubbed and worn, and the striæ are nearly obliterated. The costæ on the body whorl are, however, sinuated, from which I imagine it possessed a sinuated outer lip. Not being able to give it a correct diagnosis, I have called it provisionally *Pl. senilis*. It is probably a derived specimen.

PLEUROTOMA HISPIDULA, *Jan.* Supplement, Tab. III, fig. 3.

> RHAPHITOMA HISPIDULA, *Jan.*, fide *Bellardi.* Mon. Foss. Pied., p. 92, t. iv, fig. 17.
> PLEUROTOMA DECUSSATUM, *Phil.* En. Moll. Sic., vol. ii, p. 174, t. xxvi, fig. 23.
> — — *A. Bell.* Ann. and Mag. Nat. Hist., May, 1871.

Spec. Char. "*Testa fusiformi, plicis tenuibus longitudinalibus, lineisque transversis elevatis distantibus reticulata; anfractibus æque et parum convexis; apertura oblonga, spiram æquante.*"—Philippi.

Length, ¾ths of an inch.

Locality. Coralline Crag, near Orford.

The specimen figured was found by myself, and Mr. Bell has very recently obtained several more from the same place. There is a considerable difference among the specimens, some being more elongated than others; the shorter are strongly ribbed on the body whorl, while on the longer the ribs become nearly obsolete. On the last volution the whorls are very slightly shouldered, and the sinus is broad and shallow. *Clavatula concinnata,* 'Crag. Moll.,' vol. i, p. 61, Tab. VII, fig. 11, *a, b,* are probably varieties of this species. This is quite distinct from *Pl. decussata,* Lam., as also from *Pl. decussata* of Couthouy. *Rhaphitoma plicatella,* Jan., is closely allied, but seems to have fewer ribs.

PLEUROTOMA PYRAMIDALIS, *Ström.* Supplement, Tab. III, fig. 9, *a, b,* Tab. VII, fig. 22.

Localities. Red Crag, Butley. Fluvio-marine Crag, Thorpe, in Suffolk (*Bell*). Upper Glacial, Bridlington. Post-glacial, March and Kelsea Hill.

The specimen of this well-known northern shell represented in fig. 9 of Tab. III is one found by Mr. Bell in the Red Crag of Butley; and that in the fig. 22 of Tab. VII is one of a suite from March found by Mr. Harmer. They differ slightly, but may both, I think, be referred to this species. The Crag shell has the body whorl smooth, but there are indications of longitudinal riblets on the upper volutions. I cannot see any spiral striæ, but these may be obliterated by attrition. Some of the March specimens are perfectly smooth without appearing to have undergone much wear, but in others faint traces of riblets on the lower whorl are apparent. A specimen from Bridlington is in the British Museum, and it is given from Kelsea Hill by Mr. Jeffreys. The species is given from Thorpe in Suffolk by Mr. Bell in 'Ann. and Mag. of Nat. Hist.' for September, 1870.

PLEUROTOMA BICARINATA? *Couth.* Supplement, Tab. VI, fig. 17.

Locality. Red Crag, Butley.

The specimen figured was found by myself at Butley. It very much resembles *P. violacea* of Meig. and Ad., but seems still closer to *bicarinata,* Couth., as there are two very distinct carinæ on the whorls.

Many of the forms of *Pleurotomæ,* both those in the Crag and those living in British and Northern seas, run so much into each other, and are, withal, so inconstant in their characters, even among a group of individuals of apparently the same species, that I feel the greatest difficulty in assigning specific names, and I have therefore placed a note of interrogation against the name of this species.

PLEUROTOMA EQUALIS, *S. Wood.* Supplement, Tab. III, fig. 17.

Spec. Char. Pl. Testa elongato-fusiformi lævi, anfractibus 6, convexiusculis; obsolete angulatis; suturis distinctis, sub-depressis, spira brevi, apice obtuso, apertura dimidiam æquante; canali longiuscula, labro simplici acuto.

Locality. Red Crag, Butley.

Length, half an inch.

The specimen figured is the only one that I have seen, and was obtained by myself from Butley.

It is perfectly smooth and polished; it resembles in form *Pl. violacea,* Migh. and Ad., but my shell has no sculpture, and it appears too perfect to have been abraded; moreover, that shell is stated to measure $\frac{3}{10}$ths only of an inch in length, whereas my shell is at least $\frac{5}{10}$ths of an inch. I therefore presume it to be distinct, and have given to it a new name.

PLEUROTOMA RUFA, *Mont.* Supplement, Tab. VII, fig. 17.

Localities. Red Crag, Butley? Fluvio-marine Crag, Thorpe? Post-glacial, March.

The specimen figured of this well-known British shell is one of two found by Mr. Harmer in the March Gravel. The species is given by Mr. Bell from the Red Crag of Butley ('Ann. and Mag. Nat. Hist.,' September, 1870), and in Dr. Woodward's 'Norwich Crag List' (in 'White's Directory'), from Thorpe, but I have not seen the specimens for either of those occurrences.

PLEUROTOMA QUADRICINCTA, *S. Wood.* Supplement, Tab. VII, fig. 11.

Spec. Char. Pl. Testa turrita, sub-fusiformi, costata, costis sub-erectis, anfractibus 6—7, convexis transversim stratis, striis paucis; apertura ovata; labro intus lævi.

Locality. Red Crag, Butley.

Length, $\frac{5}{8}$ths of an inch.

The above represents one of two specimens in good preservation which I found in the Red Crag at Butley. The shell somewhat resembles *Murex harpula,* Broc., but in the description of that species it is distinctly stated as " interstitiis lævigatis." The ornamentation upon the Crag shell is peculiar; there are four strong, large, spiral lines upon the lower half of the volution (especially visible between the ribs), and the upper portion is covered with very fine spiral striæ. My shell has seven volutions, including the apical one, which is smooth; the costæ are prominent, 10—12, and the striæ are carried over the ribs when the surface has not been rubbed and worn.

PLEUROTOMA NEBULOSA, *S. Wood.* Crag Moll., vol. i, p. 60, Tab. VII, fig. 10, as
Clavatula nebula.

Localities. Red Crag, Sutton and Butley.

Mr. Jeffreys, ' Brit. Conch.,' vol. iv, p. 386, observes that this Crag shell does not agree with the living species *nebula* of Montague. In this I am disposed to agree with him, and, not being able to identify it with any other known shell, I have assigned to it the above name.

PLEUROTOMA NEBULA, *Mont.* Supplement, Tab. VII, fig. 7.

Locality. Coralline Crag, Sutton.

The above figure represents a specimen in my cabinet from the Coralline Crag of Sutton, of what, I believe, may be truly referred to the existing species of this name, agreeing with the Mediterranean variety of it.

Fig. 9, Tab. VI, of this Supplement represents the fragment of a shell which Mr. Canham had obtained from the Red Crag of Waldringfield. I felt induced to have it figured in order to call attention to its existence, and to leave nothing unnoticed up to the present time that has been found in the Upper Tertiaries of the East of England. This fragment has been named and published in the ' Ann. and Mag. Nat. Hist.,' 1871, by Mr. Bell, under the name of *Pl. violacea* var. *gigantea* — *Pl. arctica*, Adams. There were two specimens, but both unfortunately in the same mutilated condition, with nothing but the last volution remaining, and this is entirely destitute of sculpture, showing only that it possessed a sinuation at the upper part of the outer lip. The reference of these fragments to *violacea* by Mr. Bell does not quite meet with my approval, as the shell to which they belonged must have exceeded an inch in length, while the longitudinal dimensions of *violacea*, as given by the American authors, are only $\frac{3}{10}$ths of an inch ; and there is no ornament on our present shell to assist in its determination. There is a fragment of a shell in my cabinet from the Cor. Crag (which much resembles our Red Crag shell), and this has distant and obsolete costæ, and covered with fine spiral striæ, which might have been upon the Red Crag shell. As Mr. Bell has given to this the name of *arctica*, I have not thought it necessary to alter it, still I think it might be called *ambigua*. I doubt its specific connection with *violacea*.

PLEUROTOMA RUGULOSA, *Philippi*.

> PLEUROTOMA RUGULOSA, *Phil.*　En. Moll. Sic., vol. ii, p. 169, t. xxvi, fig. 8.
> —　　　　—　　*Jeffreys.*　Brit. Conch., vol. iv, p. 381, pl. xc, fig. 4.

Locality.　Coralline Crag, Sutton.

Very recently I have found two or three specimens which I believe may be fairly referred to the above-named species.　They were obtained too late to be represented in my plate.

PLEUROTOMA STRIOLATA, *Philippi*.

> PLEUROTOMA STRIOLATUM (SCACCHI), *Phil.*　En. Moll. Sic., vol. ii, p. 168, Tab. xxvi,
> fig. 7.

Locality.　Coralline Crag, near Orford.

Since my plates were engraved for this work, Mr. Robert Bell has shown me a specimen from near Orford with the above name, which I believe is correctly referred, and more recently I have obtained a similar one from the same locality.

CANCELLARIA CONTORTA? *Basterot*.　Supplement, Tab. VI, fig. 19.

> CANCELLARIA CONTORTA, *Bast.*　Foss. Env. de Bord., p. 47, pl. ii, fig. 3, 1825.
> —　　　　—　　*Hörnes.*　Vienna Foss., vol. i, p. 311, t. xxxiv, figs. 7, 8.
> —　　　　—　　*Bellardi.*　Foss. Tert. de Pied., p. 29, t. iii, figs. 7—10.

Locality.　Coralline Crag, Gedgrave.

Two specimens of this species have been obtained by Mr. A. Bell and sent to me with the above name.　They differ slightly from the general form of *contorta* in being rather more elongated, corresponding with what Dr. Hörnes has considered a variety (fig. 8 of his plate), which has comparatively a smaller and shorter aperture, that is less than the length of the spire.　It has about sixteen longitudinal ridges, crossed by spiral striæ, alternating large and small, and the outer lip is strongly denticulated.　It is an elegantly formed shell, and not at all appropriately named.　Having had no other means of identifying this than figures, I have given the reference with doubt.

CANCELLARIA GRACILENTA, *S. Wood.*　Supplement, Tab. III, fig. 23.

Spec. Char.　*C. Testa minuta ovato-acuta, extremitate utraque acuta anfractibus convexis longitudinaliter creberrime costulatis, transversim vel spiraliter striatis, suturis profundis; apertura ovata; labro acuto, columella triplicata.*

Length, ⅜ths of an inch.

Localities. Coralline Crag, Sutton, and near Orford.

When describing *C. costellifera* ('Crag Moll.,' vol. i, p. 66) I considered the above represented specimen merely as a variety of that shell; but I now believe it to be distinct. Mr. Bell has lately sent to me a specimen from near Orford, which may be referred to it; and with this was the name of *C. Bonellii*, var. *Dertonensis;* but I think my shell differs sufficiently to be specifically removed from that species; it is more elongated, has finer sculpture and a deeper suture, with more convex volutions; the upper two of which are free from ornament.

CANCELLARIA SUBANGULOSA, *S. Wood.*   Crag Moll., vol. i, p. 66, Tab. VII, fig. 20.
Supplement, Tab. III, fig. 27.

ADMETE REEDII, *A. Bell.*   Ann. and Mag. Nat. Hist., September, 1870.

Localities.   Cor. Crag, Sutton, and near Orford.

Mr. Bell placed in my hands the specimen figured in Tab. III, fig. 27, of this Supplement, and he has since described it under the name of *Admete Reedii*, as above, but except in point of size I cannot distinguish between it and a numerous suite of specimens in my possession of *subangulosa*, all small, but varying much in size, which, like this larger shell *Reedii*, possess no fold on the columella. This fold is shown on the specimen figured in the 'Crag Mollusca,' Tab. VII, fig. 20, but it does not appear to be a constant character. Specimens of *subangulosa* were given by me to Dr. Koenen, who recognised it as a German Oligocene species ('Unter Oligoc. Fauna' von Helmstädt, S. 473), and Dr. O. Speyer has, in his 'Conchology of the Cassel Tertiaries' (Cassel, 1867), figured several forms which he groups together as this shell. In some the shoulder is almost obsolete, as it is in *gracilenta*, and the folds on the columella faint and double, and others where the fold is very distinct and treble.

It would seem, therefore, that this is a very variable shell, and that the folds on the columella are not a reliable character in it. Mr. Bell's large shell *Reedii* shows six whorls, while those specimens of mine, that are identical with it in all respects save size, show but four. This strengthens my view that my *subangulosa* is only the young of *Reedii;* but why the full-grown shell should be so rare I am unable to suggest. One of Dr. Speyer's figures shows six whorls.

CANCELLARIA BELLARDI? *Mich.*   Supplement, Tab. III, fig. 25.

CANCELLARIA BELLARDII.   Descrip. des Foss. Misc. de l'Italie Septent., p. 225.
—   EVULSA, *Sow.*   Bellardi, Descript. d. Cancell. foss. du Piemont, p. 25.

Locality. Red Crag, Sutton.

At page 67, ' Crag Moll.,' vol. i, two specimens are spoken of as having been found in the Red Crag of Sutton, and considered as worn individuals of *Cancellaria læviuscula*, Sow. One of these is represented in the above figure. It resembles a shell obligingly sent to me by M. Bosquet with the name *C. Bellardi* attached, and I have referred my specimen to it. It is doubtless a derivative in the Red Crag.

CANCELLARIA BONELLII ? *Bellardi*. Supplement, Tab. III, fig. 26.

CANCELLARIA BONELLII, *Bellardi*. Desc. d. Cancell. foss. Piemont, p. 24.

Locality. Red Crag, Sutton.

The other of these specimens is represented in the above figure, and I have referred it with doubt to Bellardi's species *Bonellii*. This also is doubtless a derivative form in the Red Crag.

CANCELLARIA ? CHARLESWORTHII, *S. Wood*. Supplement, Tab. III, fig. 22, *a, b*.

Locality. Red Crag, Waldringfield.

The shell represented in this figure was given to me by Mr. Charlesworth, who obtained it from the diggers at Waldringfield. It is not in a condition for fair comparison ; a portion of the outer shell remains, and this shows a few large costæ somewhat resembling *C crassicosta*, Bellardi (Pl. IX, figs. 7, 8), or to *C. inermis*, Hörnes (Tab. XXXIV, fig. 10), but to neither can it, I think, be referred. I have called it provisionally *C. Charlesworthii*. This also is doubtless a derivative form in the Red Crag.

CANCELLARIA CANCELLATA ? *Linné*. Supplement, Tab. III, fig. 24.

VOLUTA CANCELLATA, *Linné*. Sys. Nat., 12th ed., p. 91.

Localities. Cor. Crag, Ramsholt, Sutton. Living, Mediterranean.

In my Catalogue, 1842, is inserted the name of *Cancellaria granulata ;* an imperfect specimen only was found by myself, and am I sorry to say that I have seen nothing better. This specimen, now in the British Museum, is represented in the above figure, and I believe it to belong to *C. cancellata*, Linné. Another, but smaller, fragment I have since found in the Cor. Crag of Sutton. In consequence of the imperfect state of the specimens, I have placed a note of interrogation against the name.

CANCELLARIA SPINULOSA? *Broc.*   Supplement, Tab. VI, fig. 10.

VOLUTA SPINULOSA, *Broc.*   Conch. foss. sub-app., tab. iii, fig. 15.

Locality.   Cor. Crag, Sutton.

The specimen figured was found by myself, and though not in perfect condition, appears to belong to Brocchi's species *spinulosa*.   It is, however, destitute of the spinous volutions of that species as represented by Brocchi; but these may have been rubbed off. I have, however, considered it a variety under the name of *subspinulosa*.   The form of my shell much resembles an older tertiary genus called *Mesostoma*, by Deshayes, but that has no fold upon the colummella; still I think the two are nearly related.

APORRHAIS PESPELICANI, *Linn.*   Crag Moll., vol. i, p. 25, Tab. II, fig. 4 *a b.*

Localities.   Cor. Crag., Ramsholt, Gedgrave, and Sutton.   Red Crag, Sutton, Newbourn, Brightwell, Bawdsey.   Fluvio-marine Crag, Thorpe in Suffolk (Bell).   Post-glacial, March Gravel, Nar Brick-earth (Rose).

The shell figured, as above referred to, was from the Red Crag at Brightwell, and it corresponds precisely with the common form of the British shell of that name.   This so-called *pespelicani* has been lately found in considerable numbers in the Cor. Crag at Gedgrave; but the specimens there found differ somewhat in being more elongated and more delicately marked than those from the Red Crag, representing an intermediate form between it and *A. McAndreæ*, Jeffreys (*A. pescarbonis*, Forb and Hanl).

Our Coralline Crag variety is a slender shell, and the nodules smaller than upon the common British form (in some specimens as many as twenty on the last volution), and in that respect it closely approaches *Chenopus Serresianus* (Phil. " En. Moll. Sic.," vol. ii, p. 185, Tab. XXVI, fig. *b.*), which probably is the same species.

The Crag shell appears to me to be the connecting link between *Ap. pespelicani*, and *Rostellaria pescarbonis*, Brongniart.   Several of these finely marked specimens have lately come into my possession, but after my plates were finished, or I would have had one represented.

I do not know the *Aporrhais* in any newer formation than the Red and Fluvio-marine Crag of Suffolk, until we come to Post-glacial beds of the Nar Valley, where, I believe, it is not rare; the specimens from there being of the large and coarse form, like those of the Red Crag.   A fragment has been found in the March Gravel by Mr. Harmer, and Mr. Bell gives it from Thorpe, Suffolk.

7

CERITHIUM ? ABERRANS, *S. Wood.*   Supplement, Tab. III, fig. 20.

Length, ½ an inch.

Locality.   Coralline Crag, near Orford.

The figure above referred to is the representation of a specimen I picked out of a tray of shells which Mr. Henry Woodward obligingly showed to me; and who said that they had been found by his late brother, Dr. S. P. Woodward, and, he believed, all in the Coralline Crag.   There was no special locality attached to any one of the specimens, but they had the aspect of the shells from Orford; and as the late Dr. Woodward had collected from the Cor. Crag only in that neighbourhood, there can be little doubt of its being from that locality.   Neither do I think that there is any doubt as to this shell being a genuine fossil of the Coralline Crag.   It appears to be destitute of nodules or thickenings in the lines of growth, which is the general character of the genus *Cerithium.* In the form of the aperture it is like a shell called *Bittium filosum*, from Neeah Bay, but that is more elongated.   I have given to my shell provisionally the above name.

N.B.—Since the engraving was made (some years since), and before the figure could be compared, I regret to say the fossil was lost or mislaid, and has not since been found. This is unfortunate, as it was put into the hands of the engraver before the shell had undergone a thorough examination, trusting to future opportunity for comparison.

CERITHIUM PERPULCHRUM, *S. Wood.*   Crag Moll., vol. i, p. 72, Tab. VIII, fig. 10.

This may perhaps be referred to *C. mamillatum*, Riso.   See Phil. " En. Moll. Sic.," vol. i, p. 194, Tab. XI, figs. 11, 12.

M. Nyst gives from the Belgian Crag, *Cerithium trilineatum*, Phil. var. *inversum.*   I have not seen this variety from the English Crag.

CERITHIUM RETICULATUM, *Da Costa.*   Supplement, Tab. V, fig. 22.

STROMBIFORMIS RETICULATUS, *Da Costa.*   Brit. Conch., p. 117, pl. viii, fig. 13, 1778.

Locality.   Red Crag, Walton-Naze (Bell)?   Post Glacial, Nar Brick-earth.

Several specimens of this species have been found by Mr. Rose in the Nar Brick-earth at West Bilney, and he has permitted me to have one of them figured.

It was one of the Nar specimens, I am informed, that constituted the authority under which the species was introduced into Dr. Woodward's list of Norwich Crag shells.   This species, though somewhat resembling the Red Crag, *C. variculosum* (' Crag Moll.,' Tab. VIII, fig. 3), is, I consider, quite distinct: the form of the whorls separately, as well as that which they give to the complete shell being very different. Mr. A. Bell gives *C. reticulatum* from the Red Crag of Walton ('Ann. and Mag.,' September, 1870), but I have not seen the specimen, and imagine that it may probably be the *variculosum* of that locality.

CERITHIUM TRICINCTUM, *Broc.* Crag Moll., p. 69, Tab. VIII, fig. 1, *a* and *b*. Supplement, Tab. III, fig. 19.

Locality. Cor. Crag, Sutton, and near Orford. Red Crag *passim*. Fluvio-marine Crag, Bramerton. Chillesford Bed, Bramerton and Horstead. Middle Glacial, Hopton.

The above figure, 19, was made from a fragment of this species, which I had obtained from the Cor. Crag of Sutton, and it was inserted in order to justify its admission among the shells of that formation. Since the plate was engraved, however, I am glad to say that I have obtained three other specimens from the Coralline Crag of the neighbourhood of Orford (Gomer pit), one of which is nearly perfect, and would be exactly represented by the old figure 1*a*, of Tab. VIII, of the 'Crag Mollusca.' This species is common in the Red Crag and in the Fluvio-marine Crag at Bramerton, as well as in the Chillesford Bed at that place. It occurs also, but rarely, in what I consider to be its Fluvio-marine representative—the Crag of Horstead, but I have not seen it from the other localities of that bed. I have not met with it from the Lower Glacial sands; but one specimen, which fig. 19, of Tab. III, would well represent, has occurred in the Middle Glacial sand of Hopton. I do not know it living.

Mr. Charlesworth has also given me a specimen, too mutilated for figuring of a species of this genus, but the characters are not sufficiently distinct for specific determination. In this the volutions are more close and numerous than those of *C. tricinctum*, as were remarked to me by Mr. Charlesworth; it is, I believe, distinct. He obtained it from the nodule bed at Waldringfield, and is probably a derivative in the Red Crag.

Fig. 21, of Tab. III, of this 'Supplement,' represents an imperfect specimen obtained by myself from the Post Glacial Freshwater Deposit at Grays. Mutilated specimens of *Melania inquinata*, and another species of *Melania*, have got into this deposit at Grays, from the Eocene Woolwich sands, and the specimen produced may therefore be of similar origin, as it resembles *Cerithium (Potamides) intermedium*, of Sowerby, from those sands ('Min. Cor.,' Tab. CXLVII, fig. 3.) I was induced to figure its form, from the difference between the character and composition of this specimen, and that of these older tertiary derivatives; the *Melaniæ* being strong shells, much abraded, while the specimen figured was so fragile that it fell to pieces, leaving only the fragment figured, and disclosed that it was filled with the material of the Grays Deposit, and not that of the Eocene one, which at first induced me to suppose that it might have been a living denizen of the waters of the Grays deposit. I have figured it simply as a shell found in the Grays Bed, without venturing to express a positive opinion about it.

CERITHIOPSIS LACTEA ? *Möller*.   Supplement, Tab. IV, fig. 16.

> TURRITELLA LACTEA, *Möll.*   Ind. Moll. Groenl., p. 9, 1842.
> —   RETICULATA, *Mighels* and *Adams*.   Bost. Journ., iv, p. 50, pl. iv, fig. 19.
> —   —   *Binney*.   Gould's Inv. Massach., 2nd ed., p. 318, fig. 586.

Locality.   Coralline Crag, Sutton.

I have found a single specimen only, which is here referred to the above-named species, although with some doubt.   The aperture of my shell is not quite perfect; the figure is represented as rather too conical.

CERITHIOPSIS TUBERCULARIS, *Mont*.   Crag Moll., vol i, Tab. VIII, fig. 5 (as *Cerithium*).

Localities.   Cor. Crag, Sutton, and near Orford.   Red Crag, Shottisham (Bell). Fluvio-marine Crag, Bramerton.   Middle Glacial, Billockby.

Specimens of this shell are abundant in the Coralline Crag at Sutton, exhibiting great variation, as I have shown in 'Crag Moll.,' Tab. VIII, some being much elongated, while others are short and tumid; but I have not met with a specimen having only two rows of tubercles, like that which has been called *C. Clarkei* ('Brit. Moll.,' vol. iii, p. 368, Tab. CIII, fig. 6).   This appears to be merely the absence of the middle row, a character common in *C. perversum*, where most of the upper volutions have only two.   *C. Barleei* I do not know. One imperfect specimen of *C. tubercularis* has been found by Mr. Reeve, and fragments by others, in the Fluvio-marine Crag of Bramerton; and an imperfect specimen has occurred in the Middle Glacial of Billockby.   It is given from the Red Crag, Shottisham, by Mr. Bell ('Ann. and Mag. Nat. Hist.,' May, 1871).   It has not been yet met with in the Chillesford Bed anywhere, or in the Lower Glacial Sands, to my knowledge.

TURRITELLA INCRASSATA, *J. Sow*.   Crag Moll., vol. i, p. 75, Tab. IX, fig. 7.

Localities.   Coralline, Red, and Fluvio-marine Crags *passim*, Chillesford Bed, Bramerton.   Middle Glacial, Billockby and Hopton.

The above name was employed by me from a belief that it was prior to the one used by Brocchi (*T. triplicata*), which I still believe is the case.   There is a date upon the plate in 'Min. Conch.' of April 1, 1814.   This species is very abundant in the Cor. Crag., and is very variable in ornamentation.   The apex is very seldom preserved, except in very young individuals which have a slender form, tapering up to a point which is smooth and obtuse.

Though common in the Red and Fluvio-marine Crags, it becomes rare in the Chillesford bed, and I have not met with it in the Lower Glacial Sands. It is an extremely abundant shell in the Middle Glacial Sands, especially at Billockby, at which place I have never met with the allied form *terebra*. The form in these sands is the var. *triplicata*, fig. 7*a*, and 7*c* of Tab. IX of 'Crag Moll.'

TURRITELLA TEREBRA, *Linné*. Crag Moll., vol. i, p. 74, Tab. IX, fig. 9.

Localities. Fluvio-marine Crag, Bramerton. Chillesford Bed *passim*. Middle Glacial, Clippesby, and Hopton. Upper Glacial, Bridlington. Post-glacial, March and Kelsey Hill, Nar Brick-earth (Rose).

This species is common in the Fluvio-marine Crag of Bramerton, but I have not met with it in the Cor. Crag, nor with a clear example of it in the Red Crag. It occurs, but is not very common, in the Chillesford bed at all its localities ; while at Clippesby (only a mile from Billockby where we find the form *incrassata* alone), *terebra* exclusively occurs. Compared with *incrassata* it is rare in the Middle Glacial, but specimens are not unfrequent. I have not seen it from the Lower Glacial. It is common in the Post-glacial gravel of March, and occurs at Kelsey Hill. Specimens of it sent to me by Mr. Maw from the Severn Valley beds, as well as those from March, shew a tendency in occasional individuals to approach the form *incrassata*, but there is no difficulty in distinguishing a group of the one form from a group of the other. Mr. Rose obtained it from the Nar Brick-earth, at West Bilney, and East Winch.

TURRITELLA EROSA, *Couthouy*. Crag Moll., vol. 1, p. 76, Tab. IX, fig. 8.

TURRITELLA EROSA, *Couthouy*. Boston Jour. Nat. Hist., 2, p. 103, Tab. III, fig. 1.
— CLATHRATULA, *S. Wood*. Crag Moll., vol. i, p. 76, Tab. IX, fig. 8.

Locality. Upper Glacial, Bridlington.

The shell, from Bridlington, was in the 'Crag Mollusca' assigned by me as a new species (*clathratula*). I am now satisfied that it is identical with Couthouy's North American shell *erosa* (*polaris* of Beck), which is given by Möller as a Greenland shell. The statement at p. 83, of vol. iv, of 'Brit. Conch.,' that this shell has been found by me in the Cor. Crag of Sutton is an error. I know it from no older East Anglian bed than that of Bridlington.

TURRITELLA? PENEPOLARIS, *S. Wood*. Supplement, Tab. IV, fig. 20.

Locality. Cor. Crag, near Orford and Sutton.

A fragment of a shell from the Cor. Crag of Sutton, that I was unable to refer to any existing species, has long been in my possession, but it was too imperfect for determination. The late Dr. S. P. Woodward obtained a somewhat corresponding specimen from the Cor. Crag of the Orford neighbourhood, which has the aperture rather more perfect than my own ; and I am now enabled with the two fragments to give a probable representation for a species. Dr. Woodward's specimen was accompanied with the name of *Mesalia polaris*. I have carefully compared these fragments with that species, and I think that they are quite distinct from it. The sculpture of *polaris* shows four or five broad threads in a spiral direction, with corresponding spaces between them, whereas the Cor. Crag shell is covered with numerous fine threads and narrow depressions. I have placed it provisionally in the above-named genus, which I think it more nearly resembles than that of *Mesalia* ; the shells of this latter genus have a somewhat emarginate base, or rather a reflexion of the lower part of the inner lip, which the Crag shell does not appear to possess.

TURRITELLA PLANISPIRA, *S. Wood.* Crag Moll., vol. i, p. 76. Tab. IX, fig. 11.

M. Mayer in 'Journ. de Conch.,' vol. xiv, p. 173, Pl. III, fig. 2, has described a species of *Turritella*, with the specific name of *Sandbergeri*, and to this he has given as a synonym, *T. planispira*, S. Wood, non Nyst. All the Coralline Crag forms of this genus possess a wide range of variation in the external ornament, but I believe my species *T. planispira* forms a group as well defined and constant as any.

It has been identified with *subangulata* of Brocchi by Mr. Jeffreys ('Quart. Journ. Geol. Soc.,' vol. xxvii, p. 146), but I think it is not that shell, as it agrees with it neither in the form of the whorls (which are invariably flat, and not subangular), nor in the external sculpturing. The figure of it in Tab. IX of 'Crag Moll.,' shows only six or seven spiral threads, but it has always eight, and more frequently nine, all of equal size. In some individuals the central thread thickens, and so produces a faint subangulation, which is scarcely perceptible without a magnifier. This shell, and that called *T. incrassata* var. *bicincta* (fig. 7d of Tab. IX of 'Crag. Moll.'), appear to me to constitute the most marked forms of this genus that lived in the Cor. Crag sea ; but inasmuch as a series may be selected, showing a gradation of all the forms into each other, I doubt whether the whole of the Coralline Crag varieties and species, *triplicata*, *vermicularis*, *bicincta*, and *planispira*, are not merely individual and inconstant variations of the one species *incrassata ;* and the same remark might be applied to Brocchi's species, *duplicata*, *bicarinata*, *subangulata*, *marginata*, and *replicata*.

Before dismissing the genus Turritella, I may mention that Mr. Busk in his beautiful 'Monograph of the Crag Polyzoa,' at p. 59, gives *Cellepora edax*, "habitat Cor. Crag, S. Wood, on a specimen of Natica and Turritella."

The living analogue found by the Rev. Mr. Hinks on the coast of Devonshire, was attached to a *Turritella*, but the late Mr. Alder told me he had a specimen of *C. edax* upon

Trochus exiguus. I believe the shell which the Crag Polyzoon (*C. edax*) has selected for its support is a species of *Turritella;* at least in all the numerous specimens I have seen. The form it has assumed is unlike that of a turriculated shell, but I think the burden imposed upon the animal in its growth, by the *Cellepora* occupying the base, has compelled the Mollusc to expand while the shell was increasing, so as to be deflected from its proper angle of volution, and depressed into a turbinated form. Mr. Busk has very justly pointed out in the case of *Alysidota catena*, as in that of other adherent Polyzoa, that the animal has the power of eroding the surface of the shell upon which it lived, though by what means this is effected is not said.

I am however inclined to think that the destroyer of the *Turritella* was not the Polyzoon, but that the shell has been absorbed or removed by the Mollusc itself in order to lighten its heavy and inconvenient incumbrance, for whenever a portion of the shell is visible, it has retained its exterior ornament without any apparent abrasion, and in all the instances that I have seen, the shell has been a *Turritella.*

I have figured a specimen partially uncovered, obligingly lent to me for that purpose by the directors of the Museum at Norwich, which shows a deflection from the normal angle of volution (*Turritella incrassata*, Supplement, Tab. V, fig. 25 *b*), and another of my own, fig. 25 *a*, wholly enveloped, but distorted. Fig. 3, Pl. XXII, of Mr. Busk's work is, I believe, a *Turritella.*

ALVANIA SUPRANITIDA, *S. Wood.* Crag Moll., vol. i, p. 99, Tab. XII, fig. 11 *a b.* (as *A. ascaris*).

ALVANIA SUPRANITIDA,	*S. Wood.*	Ann. and Mag. Nat. Hist., vol. ix, p. 534, tab. 5.
ACLIS	—	*Loven.* Ind. Moll. Scan., p. 17, 1846.
—	—	*Forbes and Hanl.* Brit. Mol., vol. iii, p. 320, tab. 90, figs. 2 & 3.
—	—	*Jeffreys.* Brit. Conch., vol. 4, p. 103.

Localities. Cor. Crag, Sutton.

This species was described by me as *supranitida* in my catalogue in the ' Ann. and Mag. of Nat. Hist.,' of 1842, but in the ' Crag Mollusca,' I (as there stated), in deference to the opinion of the late Mr. Alder, referred it to *A. ascaris*, Turt. The shell *supranitida* has, however, been recognised by Loven, Forbes and Hanley, and Jeffreys as a living species distinct from *ascaris*, and I have accordingly restored the name.

Mr. Jeffreys in ' Brit. Conc.,' vol. iv, p. 103, mentions *ascaris* from the Cor. Crag as being in the collection I gave to the British Museum, mixed with *supranitida*, but I do not recognise it there, nor have I, though it is inserted in the list of Gasteropoda of Mr. Prestwich's Cor. Crag paper, seen it anywhere from the Crag.

ALVANIA ALBELLA, *Leach*, M.S.   Crag Moll., vol. i, p. 99.   Tab. XII, fig. 11*c*.

ALVANIA ALBELLA, *S. Wood*.   Annals and Mag. of Nat. Hist., vol. ix, p. 534, 1842.
ACLIS WALLERI, *Jeffreys*.   Brit. Conch., vol. iv, p. 105, pl. lxxii, fig. 4, 1867.

Locality.   Cor. Crag, Sutton.

This species was given by me in my catalogue in the 'Annals and Mag. of Nat. Hist.,' for 1842, under the name *albella*, a manuscript name of Dr. Leach for the recent shell since called *Walleri*, which, as stated at p. 99 of the 'Crag Moll.,' I found on a tablet in the Brit. Museum, with the name in Dr. Leach's handwriting attached. This tablet, I regret to say, cannot now be found, and some confusion has arisen in consequence, as Mr. Jeffreys appears to think ('Brit. Conc.,' vol. iv, p. 104) that *albella* of Leach was *ascaris* of Turt.   Mr. Jeffreys recognises *Walleri* in a single specimen in my collection in the British Museum, but there are twenty-six specimens there on a separate tablet which appear to me identical in all respects with some specimens of his recent *Walleri* which he kindly gave me.   I fell into the mistake in the 'Crag Moll.' of supposing this to be an eroded form of the shell *there* described as *ascaris* (*supranitida*), but I have now reverted to my more correct views of 1842.   Since my collection went to the British Museum, both this shell and *supranitida* have become extremely rare in the Crag.

Some Conchologists have adopted the name of *Aclis* for this genus, and it may require a remark from me in explanation for not following the same plan.   In the first place I used the name of *Alvania* in 1842, when the Crag shell was fully described, and this was previous to the name of *Aclis*, proposed by Dr. Lovén in 1846.   In the second place I merely employed a name that had been given by Leach, and attached by him to a specimen in the Brit. Mus.; and in the third place the shells included under the name of *Alvania*, by Weinkauff and others, have no characters by which they could be distinguished generically from those called *Rissoa*.

MENESTHO, *Möller*, 1842.

Diagn. Gen. "*Animal pede elongato, angusto, ore simplici, membrana linguali destituto ; tentaculis brevioribus, crassiusculis, oculos perparvos ad basin internum ferentibus. Operculo pauco-spirato.   Testa conica turrita.*"—(*Möll.*)

The above is a description of this proposed genus, and the author has referred to *Turbo albulus* of the Fauna Greenlandica, as his intended type, the shell of which presents a different form of aperture from either *Rissoa* or *Turritella*, and approaches nearest to *Pasithea*, Lea.; but we do not know the species intended as the type of this latter genus in which several very different forms have been included.

MENESTHO LÆVIGATA, *S. Wood*.   Supplement, Tab. IV, fig. 19.

Spec. Char. Testa elongata turrita, lævigata, apice obtusiusculo; anfractibus (8—9) *planulatis, apertura ovata posterius angulata quinquepartem teste æquante; columella incurva, labro simplice, acuto.*

Length, ½ an inch.

Locality.   Coralline Crag, Sutton.

A few imperfect specimens, and one perfect, were found by myself in the Cor. Crag of Sutton.   The perfect one figured was destroyed while in the hands of the engraver (1866).   Some years after that Mr. Bell sent to me a very perfect individual of what possibly may be the same species, but with the name of *M. Britannica*.   Mr. Bell's shell was found at Sutton.   The shell figured had about eight volutions, the upper three or four more conical than the lower, which were nearly cylindrical; the apex was obtuse and glossy, and the rest of the shell free from striæ or sculpture of any kind; the volutions slightly convex, the upper part being a little contracted; it had a distinct and rather depressed suture; the aperture ovate, acuminated at the junction of the whorl, and it was an elegantly formed shell.   It much resembles *Pyramis striatula*, Couthouy ('Boston Journ. Nat. Hist.,' vol. ii, p. 101, Pl. I, fig. 6, described also by Gould, 'Inv. Mass.,' p. 269, fig. 174, but that shell is said to be covered "with revolving lines," and is probably the same as *Menestho albula;* my shell is smooth.

PYRAMIDELLA LÆVIUSCULA, *S. Wood*.   Crag Moll., vol. i, p. 77, Tab. IX, fig. 2.

Localities.   Cor. Crag, Sutton, and near Orford.   Red Crag, Walton Naze.

Pyramidella læviuscula of the Crag has, according to Mr. Jeffreys, been obtained recent in the Mediterranean.   It is also a fossil in the Belgian Crag, and in the Vienna beds, figured by Hörnes, vol. i, p. 492, Tab. XLIV, fig. 20, and there referred to *P. plicosa*, Bronn.   M. Nyst figured it as *P. terebellata*.   It is probably *terebellata*, Broc., but not of Lamarck.   Whether this be the *unisulcata*, Dujardin, I do not know.   There are two or three species in the Bordeaux beds of nearly the same size; *Pyramidella mitrula*, 'Bast. Bord. Foss.,' Pl. I, fig. 5, is probably another species.   I must refer to M. Deshayes, 'Par. Foss.,' vol. ii, p. 583, who has given full particulars of these fossil *Pyramidellæ*. I have obtained a specimen from the Cor. Crag, near Orford, of this species, which is more elongated than any of my Sutton specimens, and I obtained the shell from the Red Crag of Walton very soon after the publication of the 'Crag Mollusca.'   This shell has been found abundantly in the Coralline Crag at Sutton.

SCALARIA TURTONI, *Turton*.   Supplement, Tab. IV, fig. 7.

> TURBO TURTONIS, *Turt*.   Conch. Dict., p. 208, fig. 97, 1819.
> SCALARIA TURTONIS, *Forb*. and *Hanl*.   Brit. Moll., vol. iii, p. 204, pl. lxx, figs. 1, 2.

Localities.   Chillesford Bed, Sudbourn Church Walks, and Beccles Waterworks.

The above figure represents a specimen obtained by Mr. A. Bell from Sudbourn Church Walks, and I have very recently found another specimen at the same locality.   M. Weinkauff ('Conch. des Mittel,' vol. ii, p. 234) gives two varieties of this species.   One he calls *gracillissima*.   Our fossil is the less elongated one, and corresponds with the variety from the Irish seas; it has a deep suture, and the whorls are nearly circular. *Scalaria trinacria*, Phil., vol. ii, p. 145, looks like, and is probably a short variety of this species.   Mr. Crowfoot sent me a fragment obtained from the Waterworks Well at Beccles, which pierced probably the Chillesford bed; and this fragment I think may be referred to the above-named species.   Fig. 5, Tab. IV, represents the portion of a shell also obtained by Mr. Bell from the bed at Sudbourn Church Walks, which I believe is a variety of the above species.   It has a faint spiral ridge at the base of the volution, similar to that upon *Sc. pseudo-scalaris*, Risso.   I have therefore called it var. *pseudo-Turtoni*.   In the 'Ann. and Nat. Mag. Hist.,' September, 1870, Mr. Bell gives the name of *Sc. pseudo-scalaris* from the same deposit.   I have not seen that species as a British fossil; probably the specimen I have figured as var. *pseudo-Turtoni* may be that on which he has made this reference.   *Scalaria fimbriosa* is given in the list of Mr. Prestwich's paper as from the Red Crag of Woodbridge.   This is probably through some mistake.   *Scalaria communis* is also there given from Waldringfield, but I have seen nothing answering to this species or I would have had it figured.

SCALARIA TREVELYANA, *Leach*.   Crag Moll., vol. i, p. 94, Tab. VIII, fig. 20, and Supplement Tab. IV, fig. 6.

Localities.   Red Crag, Sutton.   Fluvio-marine Crag, Bramerton.   Chillesford Bed, Aldeby.   Middle Glacial, Hopton and Billockby.

I have here given a large representation of a small specimen from Aldeby, found by Messrs. Crowfoot and Dowson.   This young specimen has the three upper volutions free from ribs of any kind.   The specimen I previously figured has lost the upper half in which this character is shown.   I have not met with this species from any other locality of the Chillesford bed, and it is rare in the Fluvio-marine Crag.   A fragment consisting of a single whorl has occurred in the Middle Glacial of Hopton, and another rather more complete at Billockby.

SCALARIA CANCELLATA, *Broc.* Crag Moll., vol. i, p. 95, Tab. VIII, fig. 2, and Supplement, Tab. IV, fig. 2.

Locality. Cor. Crag, near Orford.

A fine specimen of this species having been obtained from the Cor. Crag at Orford, and obligingly given to me by Mr. H. B. Woodward, I have been induced to have it represented in Tab. IV, fig. 2, and I think from this specimen the shell may safely be referred to Brocchi's species.

I regret not having been able to find a better specimen of what I called *Scalaria obtusicostata*, 'Crag Moll,' vol. i, p. 95, Tab. VIII, fig. 21.

SCALARIA GROENLANDICA, *Chemn.* Crag Moll., vol. i, p. 90, Tab. VIII, fig. 11.

Localities. Red Crag, Sutton. Fluvio-marine Crag, Bramerton. Chillesford Bed, *passim.* Lower Glacial, Belaugh. Middle Glacial, Hopton. Upper Glacial, Bridlington.

This shell is not uncommon in the Fluvio-marine Crag, and in some of the localities of the Chillesford bed, and in the Lower Glacial Sand at Belaugh. Two fragments have occurred in the Middle Glacial of Hopton.

In our long list of Crag *Scalariæ* there are several that I have not been able to refer to existing analogues. *Scalaria soluta*, Tiberi, 'Journ. de Conch.,' vol. xi, p. 159, Pl. VI, fig. 3, is probably the young state, and figs. 3, 4, Pl. V of the same Journal, vol. xvi, possibly the full-grown condition of *Scalaria frondosa*, J. Sow., from the Coralline Crag. *Scalaria exima*, Pecchioli, may perhaps be also referred to *frondosa*. The Crag shell is variable, some specimens being more elongated than others, and the spinous fronds more or less produced. This Crag species I have found at Gedgrave.

Scalaria frondicula somewhat resembles *Sc. Algeriana*, Weinkauff, but that Mediterranean shell is said to be covered with spiral striæ; but my Crag shell is quite smooth between the reflected costæ.

CHEMNITZIA CLATHRATA? *Jeffreys.* Supplement, Tab. VII, fig. 18.

ODOSTOMIA CLATHRATA, *Jef.* Ann. and Mag. Nat. Hist., 2nd ser., vol. ii, p. 345, 1848.
CHEMNITZIA — *Forb.* and *Hanl.* Brit. Moll., vol. iii, p. 258, pl. xciv, fig. 4.

Locality. Coralline Crag, Sutton.

A single specimen has lately rewarded my researches, which seems to correspond with a recent British shell described by Mr. Jeffreys as a distinct species. Messrs. Forbes and Hanley gave it under the above name, but with a?, considering it probably a var. of *indistincta* (*curvicostata* of 'Crag Moll.'); it appears to me to differ from that species in being somewhat less cylindrical, and in the ornamentation; the ribs being less curved, and the spiral striæ on the lower part of the volution more prominent or distinct. The outer lip of my specimen is unfortunately not quite perfect.

CHEMNITZIA FILOSA.   Crag Moll., vol. i, p. 82, Tab. X, fig. 7.

This is probably the same as *Parthenia varicosa*, Forbes, from the Ægean, but that specific name having been employed by Basterot for a different shell, apparently of this genus, I have retained the name *filosa* originally given in my catalogue.

CHEMNITZIA ELEGANTIOR, *S. Wood.*   Crag Moll., vol. i, p. 81, Tab. X, fig. 5 (as *Ch. elegantissima*).

Locality.   Cor. Crag, Sutton.   Red Crag, Walton.

This shell was referred to *Chemnitzia elegantissima*, Mont., in the 'Crag Mollusca,' but the ribs are straight, though inclined, but not curved or flexuous, such as those upon *elegantissima*.   I think therefore that it must be treated as distinct, and I propose to call it *Ch. elegantior*.   Mr. Bell has shown me a portion of a specimen from the Red Crag of Walton (see *Odostomia lactea* in Bell's list 'Ann. and Mag.,' May, 1871), which appears to be specifically the same as my own from the Coralline Crag.   I am unable to say whether the *elegantissima* of Woodward's Norwich Crag list be this shell or Montague's *elegantissima*.

CHEMNITZIA INTERNODULA, *S. Wood.*   Crag Moll., vol. i, p. 81, Tab. X, fig. 6.

Localities.   Cor. Crag, Sutton.   Red Crag, Sutton, and Walton, and Butley (Bell). Fluvio-marine Crag, Bramerton.   Middle Glacial, Hopton and Billockby.

This shell has occurred, but very rarely, in the Fluvio-marine Crag at Bramerton, and several imperfect specimens of it have occurred in the Middle Glacial of Billockby and Hopton.   I have not met with it from any of the localities of the Chillesford bed or Lower Glacial Sands.   I understand that it has been found living in the Mediterranean, but have not seen the shell.

CHEMNITZIA RUFA, *Philippi.*   Crag Moll., vol. i, p. 79, Tab. X, fig. 2.

The author of 'Brit. Conch.' (vol. iv, p. 163) seems to doubt the correctness of my assignment, and I have in consequence again examined my specimens.   I still believe my Coralline Crag shell to be identical with the recent British species of that name.   I have also a specimen from the same formation (at Sutton) of *Ch. fulvocincta,* Forb. and Hanl., which Mr. Jeffreys considers only as a variety of *rufa ;* the difference between these two consisting in one having rather more convex volutions than the other. The figure in ' Crag Moll.' has the costæ rather too numerous, and they are not sufficiently erect.

CHEMNITZIA RUGULOSA, *S. Wood.* Supplement, Tab. IV, fig. 15.

Locality. Red Crag, Walton Naze. Fluvio-marine Crag, Yarn Hill?

Two somewhat imperfect specimens, presumedly belonging to this genus, were found by myself many years since, one of which is represented in the above figure. They were not described in the ' Crag Mollusca,' in the hope that something more perfect or capable of better determination might be discovered, but in this I have been disappointed. These are unfortunately much rubbed and their true markings partially obliterated. The volutions are convex on the lower part and flattened above, where there are some obsolete riblets, and there is an indistinct fold upon an upright columella—the characteristic distinction of this genus. The shell it most resembles is *Ch. speciosa,* dredged by Mr. McAndrew in Vigo Bay, but I am not able to say they are the same. Mr. Crowfoot has recently sent me two still more imperfect specimens from the Fluvio-marine deposit at Yarn Hill, which, I think, may be referred to the same species. I have given the name provisionally.

CHEMNITZIA PLICATULA? *Brocchi.* Supplement, Tab. VII, fig. 3.

TURBO PLICATULUS, *Broc.* Conch. Foss. Sub-ap., vol. ii, p. 376, t. vii, fig. 5, 1814.
TURBONILLA PLICATULA, *Hörnes.* Conch. Foss. Wien., vol. i, p. 503, tab. xliii, fig. 33, 1856.

Localities. Red Crag, Butley, Walton Naze (*Bell*). Chillesford Bed, Beccles Waterworks.

A single, but very imperfect specimen (the one figured from Butley) has been put into my hands by Mr. A. Bell, to which the name above was attached. It is too imperfect for diagnosis, and I have referred it as above, though with doubt. The fragment shows about three fifths of what probably was its original size. The plications and ribs are straight and numerous, and the volutions are nearly flat. It differs from the one I have previously called *rugulosa,* in which the volutions are convex on the lower part.

Mr. Crowfoot has recently sent to me two small fragments, which appear to belong to the same species. These, he tells me, were obtained in sinking the Beccles Waterworks well, which, it is to be presumed, pierced the same bed as that not far away at Aldeby, though it has a Fluvio-marine aspect. All these fragments have had their surface more or less decorticated or altered in some degree, and the above reference is not satisfactory.

Mr. Bell ('Ann. and Mag. Nat. Hist.,' May, 1871) gives this also from Walton, but I have not seen that specimen.

ODOSTOMIA INSCULPTA, *Montagu.* Supplement, Tab. IV, fig. 18.

> TURBO INSCULPTUS, *Mont.* Test. Brit., Supplement, p. 129, 1808.
> ODOSTOMIA INSCULPTA, *Forb.* and *Hanl.* Brit. Moll., vol. iii, p. 289, pl. xcvi, fig. 6.
> — — *Jeffreys.* Brit. Conch., vol. iv, p. 139, pl. lxxiv, fig. 4.

Locality. Coralline Crag, Sutton.

An imperfect specimen of this species has recently been obtained by myself from Sutton. It has lost its obtuse apex and a part of the spire, but may, I think, be distinguished by its peculiar ornament. It is the only specimen I have seen.

ODOSTOMIA? GULSONÆ, *Clark.* Supplement, Tab. IV, fig. 26.

> CHEMNITZIA GULSONÆ, *Clark.* Ann. and Mag. Nat. Hist., 3rd ser., vol. vi, p. 459.
> ODOSTOMIA? — *Forb.* and *Hanl.* Brit. Moll., vol. iv, p. 281, pl. cxxxii, fig. 6.
> JEFFREYSIA? — *Jeffreys.* Ann. and Mag. Nat. Hist., Jan., 1859, p. 17.
> ACTIS — *Id.* Brit. Conch., vol. iv, p. 106, pl. lxxii, fig. 5.

Locality. Coralline Crag, Sutton.

A single specimen was found by myself a few years since which I then considered as a new species, and had given to it the name of *O. mitis.* Upon showing the shell to Mr. Jeffreys he considered it as an identity with a British species which he had figured and described under the name of *Odostomia minima* ('Ann. and Mag. Nat Hist.,' 1858, p. 9, Pl. XI, fig. 3); and my shell passed under that name in a report by him upon the dredgings of the Shetland seas ('Brit. Assoc.,' 1863). He has not, however, in 'Brit. Conch.' identified *Od. minima* with any Crag shell, but speaks (vol. iv, p. 107) of *Aclis Gulsonæ* having been found by me in the Coralline Crag at Clacton (corrected to Sutton in vol. v, p. 210). I presume that the shell thus referred to is the single specimen of which I have been speaking, and I have accordingly adopted the specific name for it. The figure above referred to is a copy from a figure of a recent shell, my specimen (which, I believe, is the only one that has been found) having been destroyed while in the hands of the engraver.

ODOSTOMIA CONOIDEA, *Brocchi.* Crag Moll., vol. i, p. 85, Tab. IX, fig. 3 *a* (as
O. plicata).

ODOSTOMIA CONOIDEA, *Broc.* Conch. Foss. Sub-ap., p. 660, c. 16, fig. 2.

Localities. Cor. Crag, Sutton, Ramsholt, and near Orford. Red Crag, Walton?
Fluvio-marine Crag, Bramerton? Middle Glacial, Billockby?

This shell was in the 'Crag Moll.' referred by me to *O. plicata* of Mont. with *O. conoidea*, Broc., as a synonym. The two, however, appear to be, as stated p. 317 of Appendix, distinct. *O. conoidea* is given by Mr. Bell from Walton in 'Ann. and Mag. Nat. Hist.,' September, 1870, but I have not seen the specimen. This is the only Red Crag occurrence of it of which I am aware. An imperfect specimen obtained by Mr. Reeve from the Fluvio-marine Crag of Bramerton, and another by my son from the Middle Glacial sand of Billockby, appear to belong to this species, but they are not sufficiently perfect for certain identification.

ODOSTOMIA PLICATA, *Mont.* Supplement, Tab. IV, fig. 22.

TURBO PLICATUS, *Mont.* Test. Brit., p. 325, tab. xxi, fig. 2.

Localities. Cor. Crag, Sutton. Red Crag.

A single specimen of this shell, which has been obtained by me from the Coralline Crag at Sutton, and is represented in the above figure, enables me to retain the name of Montagu's species as a Cor. Crag shell, as mentioned p. 317 (*Od. plicata*) of Appendix. I had it from the Red Crag, but the name of the locality has not been preserved.

ODOSTOMIA CONSPICUA ? *Alder.* Crag. Moll., vol. i, p. 85, Tab. IX, fig. 3 (as
O. plicata, var.).

ODOSTOMIA CONSPICUA, *Alder.* Trans. Tynes Nat. Field Club, i, 359.

Locality. Cor. Crag, Sutton.

The shell figured in 'Crag Moll.' as variety *b* of *O. plicata*, Mont. seems to me now more to resemble the above shell *conspicua;* that species and *conoidea* being the only two recent British species which, like our Crag shells, are denticulated on the inner side of the outer lip. I have assigned it to *conspicua* with doubt, as it may not improbably be some foreign species unknown to me.

ODOSTOMIA UNIDENTATA, *Montagu*.   Appendix, p. 317, Tab. XXXI, fig. 11.

Localities.   Red Crag, Walton.   Middle Glacial, Billockby?

In addition to the solitary specimen from Walton, mentioned p. 317 of Appendix, a specimen has occurred in the Middle Glacial sand at Billockby that seems referable to this species, but it is not sufficiently perfect to be free from doubt.

ODOSTOMIA OBLIQUA? *Alder*.   Supplement, Tab. IV, fig. 24.

ODOSTOMIA OBLIQUA, *Alder*.   Ann. and Mag. Nat. Hist., vol. xiii, p. 327, pl. viii, fig. 12.

Locality.   Cor. Crag, Sutton.

A single specimen, obtained by me from the Cor. Crag of Sutton, is represented in the above figure, and referred with doubt to the above species; it is, I believe, a young individual, and the form not very well shown.   Mr. A. Bell has since my engraving was made found a full-grown specimen, which I would rather have figured.

ODOSTOMIA? ORNATA, *S. Wood*.   Crag. Moll., vol. i, p. 87, Tab. IX, fig. 6 (as *O. similima*).

This shell is, I now believe, quite distinct from *O. simillima* of Montague, to which I assigned it with doubt in the 'Crag Moll.'   I am unable to refer the Crag shell to any species known to myself, either British or European.   In my 'Catalogue of Crag Mollusca,' 1842, it was called *Rissoa? costellata*, but as the name *costellata* is already appropriated to another shell, I propose to call it *ornata*.   It is so aberrant a form of *Odostomia* that I had intended to erect a new genus for it, but I have thought it better to leave that task to some future author, when some allied forms that can be grouped with it may have been discovered.

ODOSTOMIA ALBELLA, *Lovén*.

Since the figures for this Supplement were engraved Mr. Robert Bell has sent me from the Coralline Crag of Sutton a specimen with the name *albella* attached, but I have not yet had the opportunity of comparing it with Lovén's shell.

Genus.—EULIMENE.

In the 'Crag Mollusca,' vol. i, p. 109, two Red Crag shells are referred to the genus *Paludestrina*, viz. *P. pendula* and *P. terrebellata*.   The first of these was described in my Catalogue of 1842 as *Eulima pendula*.   In external shape both these shells resemble some of the forms of the genus *Niso*, but are distinguishable from them by the absence of the large open umbilicus. They also approach some forms of the genus *Eulima*, but, unlike that genus, the labium (or left lip) is extended over the umbilicus.

Feeling the impropriety of keeping these shells in the genus *Paludestrina*, and pressed by the difficulty of finding a genus suitable to their reception, I propose the above name *Eulimene* (one of the Nereids) for their reception.

EULIMENE PENDULA, *S. Wood.* Crag Moll., vol. i, p. 109, Tab. XII, fig. 6 (as *Paludestrina pendula*).

EULIMA PENDULA, *S. Wood.* Catalogue, 1842.

Locality. Walton Naze.

EULIMENE TEREBELLATA, *Nyst.* Crag Moll., vol. i, p. 109,Tab. XII, fig. 7 (as *Paludestrina terebellata*).

MELANIA TERREBELLATA, *Nyst.* Coq. Foss. de Belg., p. 413, pl. xxxviii, fig. 12, 1844.

Localities. Red Crag, Sutton, and Walton Naze.

Fig 21, *a, b*, Supplement Tab. IV, represents a specimen lately found by myself in the Cor. Crag of Sutton, which, from its imperfect state, I do not feel justified in referring to any particular name. The presence, however, of an open umbilicus shown in the figure, coupled with its tapering form, suggests some affinity with the genus *Niso*.

EULIMA STENOSTOMA, *Jeffreys.* Supplement, Tab. IV, fig. 25.

EULIMA STENOSTOMA, *Jeffreys.* Ann. and Mag. Nat. Hist., 3rd ser., ii, p. 128, pl. v, fig. 7.

Locality. Cor. Crag, Sutton.

A single specimen shown in the above figure has been found by me at Sutton, which, I think, may be referred to this species. The artist has not been very successful in its representation.

EULIMA SIMILIS? *D'Orbigny*, 1847. Supplement, Tab. VII, fig. 16.

MELANIA DISTORTA, *Phil.* En. Mol. Sic., vol. i, p. 158, t. ix, fig. 10, 1836.
— — *Grateloup.* Pl. i, No. 4, fig. 14, 1847.
EULIMA — *Forb.* and *Hanl.* Brit. Moll., vol. iii, p. 232, pl. xcii, figs. 4—6.
— PHILIPPII, *Weinkauff.* Conch. des Mittelm., vol. ii, p. 228, 1868.

q

Length, ¼ of an inch.

Locality.   Red Crag, Walton-on-the-Naze (*A. Bell*).

This specimen was put into my hands as *Eulima distorta* by Mr. Bell.   The aperture appears to me to be too short for the recent species so called, and resembles a young specimen of *polita*.   I therefore assign it to the above species with a note of interrogation.

The Paris basin shell first called *E. distorta* is, I think, with D'Orbigny and Weinkauff, distinct from the existing shell of that name, and I have therefore adopted D'Orbigny's name of *similis*, which has priority to that of *Philippii*, proposed for it by Weinkauff.   The older tertiary fossil differs in size, as also in the proportions of the aperture, to the length of the shell.   The flexure in the spire is present in the young of *E. polita*.

EULIMA SUBULATA, *Donovan*.   Crag Moll., vol. i, p. 97, Tab. XIX, fig. 3.

Localities.   Cor. Crag, Sutton, Ramsholt, and near Orford.

In 'Brit. Conch.,' vol. iv, p. 209, Mr. Jeffreys observes, in reference to the recent *subulata*, "This is not the *Eulima subulata* of Searles Wood nor that of Nyst;" and in the list to the Cor. Crag paper of Mr. Prestwich he repeats this, but at the same time gives the shell as a living West European abysmal form.   I am, however, still unable to perceive any difference between the Crag shell and the living British shell *subulata*, and I have therefore retained the Crag shell under the name I originally assigned to it.

EULIMA BILINEATA ? *Alder*.

EULIMA BILINEATA, *Ald*.   Forbes and Han., vol. iii, p. 238, t. xcii, fig. 9—10.

Locality.   Cor. Crag, Sutton.   Recent, Britain.

A specimen of *Eulima* in my collection, ⅙th of an inch in length, retains the colouring matter on it, but is not otherwise distinguishable from *subulata*.   This colouring matter forms a broad, fulvous spiral band, occupying the centre of the two lower whorls with traces of another narrow band at the upper and lower part of the same whorls.   Mr. Alder in his description of this species says (Forbes and Han., vol. iii, p. 138) that "it has two bands placed close together in the centre of the body whorl, with occasionally a faint indication of another on the upper or lower margin."   If by the obscuration due to the fossilization of my shell the two bands placed close together in the centre of the body whorl have coalesced into apparently one broad band, then the colouring of the recent and fossil shell agrees, but if not, the fossil may be a new species, for which I would propose the name of *E. dubia*. As the sole apparent difference between this and *subulata* consists in this coloration, I have not thought it necessary to figure the specimen.

EULIMA INTERMEDIA, *Cantraine?* Crag Moll., vol. i, p. 96, Tab. XIX, fig. 1 *a* (as var. *vulgaris* of *E. polita*).

Localities. Cor. Crag, Sutton. Red Crag, Walton Naze.

The shell figured 1 *a*, in Tab. XIX, of 'Crag Mollusca,' was there referred to *Eulima polita*, but Mr. Jeffreys, (Brit. Conch., vol. iv, p. 204,) considers the Crag shell to be *Eulima intermedia*, Cantraine, a distinct species. I have therefore upon this authority inserted the species here. Fig. 1 *b* of the same table represents the species to which it was there assigned, viz. *E. polita*, which occurs in the Cor. Crag, at Sutton, and in the Red Crag, at Walton.

EULIMA GLABELLA, *S. Wood.* Crag Moll., vol. i, p. 98, Tab. XIX, fig. 2, and Supplement, Tab. VII, fig. 4, *a, b.*

Locality. Cor. Crag, Sutton.

The shell figured and described by myself under this name is, I believe, quite distinct. The previous figure of my shell does not sufficiently represent the sinuosity in the outer lip, the lower portion having considerable projection; the contraction above causes a slight indentation in the whorl all up the volutions, such as is not possessed by any other British species of this genus. My shell much resembles *Eul. auriculata*, Von Koenen, 'Mittel Oligoc.,' p. 104, Tab. VII, fig. 3, *a—c*, judging from the figure, in having a deeply sinuated outer lip. This Crag shell has a rounded or obtuse apex, unlike the generality of species in this genus, some of which are occasionally decollated, but my shell does not appear to have been mutilated in that way. I have therefore given a fresh view of the mouth and lip of my shell.

EULIMELLA ACICULA, *Phil.* Crag Moll., vol. i, p. 84, Tab. X, fig. 11, *b, c* (as var. *lævigata* of *Chemnitzia similis*).

MELANIA ACICULA, *Phil.* En. Moll. Sic., vol. ii, p. 134, t. ix, fig. 6.

Localities. Cor. Crag, Sutton. Red Crag, Walton (*Bell*). Fluvio-marine Crag, Yarn Hill.

This shell I now consider is no variety of *Chemnitzia similis*, as supposed by me in the 'Crag Mollusca,' but the above *Eulimella (Melania) acicula* of Philippi.

It has been sent to me by Messrs. Crowfoot and Dowson from the Fluvio-marine deposit of Yarn Hill, near Potter's Bridge, Southwold, and Mr. Bell gives it from the Red Crag of Walton ('Ann. and Mag.,' September, 1870).

PALUDINA ? GLACIALIS, *S. Wood.* Supplement, Tab. IV, fig. 14, *a*, *b*., Tab. VII, fig. 25.

Localities. Chillesford Bed, Coltishall. Lower Glacial, Belaugh and Rackheath. Middle Glacial, Hopton.

Several specimens of this shell have occurred in the pebbly sands of Belaugh and Rackheath, and recently another was put into my hands by H. Norton, Esq., of Norwich, from the shell bed beneath the Chillesford Clay, at Coltishall.

The volutions in this shell, which I have referred to the genus *Paludina*, are flat, or rather inclined to be concave externally; the mouth is subcircular, and the inner lip slightly extends over a small umbilicus; the shell is by no means thin, and the apex is very much flattened. The lower glacial sands in which this shell has occurred, as well as the Chillesford Sand at Coltishall, are of Fluvio-marine origin; and in them, in actual association with this *Paludina glacialis*, specimens of *P. vivipara* occur, none of them presenting any sign of departure from their normal form; and I have seen nothing which by connecting this shell with *P. vivipara* would justify the idea that it was a variety of that shell. Moreover, it is difficult to conceive that any species of *Paludina* could thus assume so very distinct a form (which is shown to be common to several individuals) while living in association with the unaltered normal form of that shell.

The Middle Glacial specimen in Tab. VII differs in being more flattened, and it is from a formation which has not only afforded no indication of Fluvio-marine conditions, but whose physical relations indicate it to have been accumulated under several hundred feet of sea depth. The presence however, in abundance of *Littorinæ* in the Middle Glacial sands, renders it probable that some at least of the shells of those sands lived near a shore, and were transported by currents to the place where we now find them, so that the shell may in this way have been a denizen of an estuary or of a river, which was carried into a purely marine and deep-water area. Assuming this, and that it is really the young of *glacialis*, the shell must have undergone a change from its original form in the interval between the commencement of the Lower Glacial formation, where we get the shell at Belaugh and Rackheath, and the accumulation of the Middle Glacial deposit. It is quite possible, however, that the shell may be no *Paludina* at all, and I have assigned its name provisionally, and with doubt.

PALUDINA MEDIA, *Woodward.* Crag Moll., vol. i, p. 110, Tab. XII, fig. 1.

> PALUDINA MEDIA, *S. Woodward.* Geol. of Norfolk, pl. iii, figs. 5 and 6.
> — — *S. P. Woodward.* In Gunn's article on Geology of Norfolk in White's Hist. and Directory of Norfolk Sheffield, 1864.

Locality. Fluvio-marine Crag, Bramerton and Thorpe.

The *Paludina* from Bramerton and Thorpe described in 'Crag Mollusca,' vol. i, p. 110, Tab. XII, fig. 16, was referred to *P. lenta* and *P. unicolor*, but which in vol. ii I changed to *P. parilis.* This Fluvio-marine shell of Bramerton I now think ought to have been called *P. media*, the name the late Dr. S. P. Woodward restored to it, and from his so doing I presume he also considered it a distinct species. This name was rejected by myself because there were two shells figured by S. Woodward, sen., in the 'Geology of Norfolk,' called *media* and *rotundata*, and I did not know which was to be considered the type.

PALUDINA CLACTONENSIS, *S. Wood.* Supplement, Tab. I, fig. 4, *a. b.*

Locality. Post-glacial, Clacton.

The specimens figured were obtained at Clacton, by the Rev. O. Fisher. They were found by him in the marly portion of the deposit at that locality, in association with *Corbicula fluminalis*, and where I have also found some estuarine species *Mytilus edulis*, *Scrobicularia plana*, and *Tellina Balthica.*[1] Under these circumstances the peculiar thickened and somewhat angular form which it possesses, like *glacialis*, suggests the idea whether both it and *glacialis* are not distortions of some known species of *Paludina*, due to their living either in salt or brackish water. I have, however, given it provisionally the specific name of *Clactonensis*.

A shell from the diluvium at Templehof, near Berlin, called *P. diluviana*, Kunth ('Zeitschrift f. Geolog. Geschellschaft,' Berlin, 1865, Tab. VII, fig. 8), very much resembles *Clactonensis*, and may be identical with it.

PALUDINA CONTECTA, *Millet.* Supplement, Tab. I, fig. 6.

> CYCLOSTOMA CONTECTUM, *Millet.* Mollusca of Maine-et-Loire, 1813, p. 5.
> PALUDINA LISTERI, *Forbes* and *Hanley.* Brit. Moll., vol. iii, p. 8, pl. lxxi, fig. 16.

Locality. Pre-glacial Forest Formation, Woman Hythe, Runton.

The above figure represents a shell from the purely freshwater deposit at Woman

---

[1] These three marine species may very possibly, however, not belong to the Post-glacial deposit at all, but be modern shells washed into it; the deposit where these shells occurred being on the present beach.

Hythe. This I have referred to *Paludina contecta* of Millet. The volutions are very convex, with a deep sutural line, and the apex is very acute.

PALUDINA VIVIPARA, *Linné*. Supplement, Tab. I, fig. 5.

HELIX VIVIPARA, *Linné*. Syst. Nat., 12th ed., p. 1247.
PALUDINA VIVIPARA, *Forbes* and *Hanley*. Vol. iii, p. 11, t. lxxi, f. 14, 15.

Localities. Chillesford bed, Easton Bavent Cliff. Lower Glacial, Rackheath.

The specimen figured is that of a small individual from Rackheath, but since it was engraved a full-grown specimen has been obtained by Mr. Cavell from Easton Cliff.

All these fossil forms differ so materially from each other, that I have placed them under separate names. They have in all probability descended from a common ancestor, and most likely *P. lenta* stood in that position, but altered circumstances materially altered their forms, so as to make them permanent varieties, which it is difficult not to call species.

I therefore refer my figured specimens in the following manner :—

Paludina media. Crag Moll., vol. i, Tab. XII, fig. 1.
— *Clactonensis* (*P. diluviana*, Kunth ?). Supplement, Tab. I, fig. 4, *a, b*.
— *glacialis*. Supplement, Tab. IV, fig. 14, *a, b*.
— *vivipara*. Supplement, Tab. I, fig. 5.
— *contecta,* id. Tab. I, fig. 6, *a, b*.

The species of the genus *Paludina* are of difficult determination, and naturalists are far from being in accord respecting their specific limitations. M. Deshayes says ('Hist. des An. sans Vert.,' vol. ii, p. 483) : "Nous devons affirmer n'avoir jamais vu une espèce vivante quelconque, absolument identique avec l'espèce fossil d'Angleterre ou de France;" speaking of the Older Tertiaries. When I applied to Mr. Jeffreys for his opinion respecting the Clacton shell, he replied in letter, March, 1865, (with permission to quote his opinion), "I could find no difference between those (the *Clactonensis*) and crag specimens. I consider the *P. lenta* from the so-called upper Eocene Beds and your *P. parilis* or *P. lenta* of the Crag as the same species, and that this species (including each fossil form) is distinct from *P. unicolor* of the Nile." In the discussion on one of Mr. Prestwich's papers upon the Crag ('Quart. Journ. Geol. Soc.,' vol. xxvi, p. 282), Mr. Jeffreys, however, calls the Crag species *Paludina unicolor;* from which it would seem that this opinion of 1865 has been modified. It must be admitted that the specific determination of the various forms of this genus is in an unsatisfactory state.

HYDROBIA ULVÆ, *Pennant.* Supplement, Tab. IV, fig. 23. Crag Moll., vol. i, p. 109 (as *Paludestrina ulvæ*).

TURBO ULVÆ, *Penn.* Brit. Zool., 4th ed., vol. iv, p. 132, pl. lxxxvi, fig. 120.

Localities. Red Crag, Walton (Bell)? Fluvio-marine Crag, Yarn Hill, near Southwold. Chillesford bed, Aldeby. Post-glacial, Gedgrave.

This species I have not seen as fossil from either of those purely marine formations, the Coralline and Red Crags. The specimen figured is from a formation at Gedgrave, mentioned in vol. i, p. 109, which is, I believe, a post-glacial bed, where land and freshwater shells of an old post-glacial period are intermixed with a re-deposit of Coralline Crag derivatives. It has been found by Mr. Reeve in the Fluvio-marine Crag at Bramerton, and by Messrs. Dowson and Crowfoot at Yarn Hill and Aldeby. This species is given by Mr. Bell from Walton ('Ann. and Mag. Nat. Hist.,' Sept., 1870), but I have not seen the shell.

HYDROBIA SUBUMBILICATA, *Mont.* Crag Moll., vol. i, p. 108, Tab. XI, fig. 2 (as *Paludestrina subumbilicata*).

Localities. Fluvio-marine Crag, Bramerton. Chillesford bed, Bramerton.

This species, so very common in the Fluvio-marine Crag at Bramerton, is, I am informed, very rare in the Marine Chillesford bed at that place, and I have not met with it in the Chillesford bed elsewhere, nor in the Lower or Middle Glacial Sands. I agree in the distinction drawn by Montagu between this shell and its congeners *ulvæ* and *ventrosa*; and distinct from all these is, in my opinion, *Hydrobia thermalis* (*Helix octona*?, Linn.), Crag Moll., vol. ii, p. 319, Tab. XXXI, fig. 12, to which I have referred the shells found in the freshwater (older post-glacial) deposits of Clacton and Grays.

RISSOA PROXIMA, *Alder.* Supplement, Tab. IV, fig. 17.

RISSOA PROXIMA, *Alder MS.* Thompson, An. Nat. Hist., vol. xx, p. 174.

Locality. Coralline Crag, Sutton.

In deference to the British conchologists, I have separated two specimens which I had considered as varieties of *R. vitrea.* They were pointed out to me by Mr. Jeffreys as specifically distinct. The striæ with which they are covered (which are not very visible in my fossils) appear to be the only character by which they can be distinguished.

RISSOA EXIMIA? *Jeffreys.*   Supplement, Tab. VII, fig. 5.

> RISSOA EXIMIA, *Jeff.*   Ann. and Mag. Nat. Hist., n. s., vol. iv, p. 299, 1849.
> CHEMNITZIA EXIMIA, *Forb.* and *Hanl.*   Brit. Moll., vol. iv, p. 278, pl. xc, fig. 1, as
> *Rissoa eximia.*
> ODOSTOMIA EXIMIA, *Jeff.*   Brit. Conch., vol. iv, p. 155, pl. lxxv, fig. 4.

Localities.   Coralline Crag, Sutton.   Living, Shetland Seas.

The single specimen of this shell, which is represented in the figure as above, is all that I have yet met with, and was found by myself in the Coralline Crag of Sutton.   It is strongly ribbed, and has three somewhat broad spiral striæ on the lower part of the whorl, in which respect there is a slight inaccuracy in the figure which shows but two such striæ. These striæ crossing the ribs give rise to two rows of depressions or cavities between them.   The specimen is slightly worn.

RISSOA SEMICOSTATA, *Woodward* (non *Mont.*).   Crag Moll., vol. i, p. 102, Tab. XI, fig. 10, and also fig. 9 (as *R. pulchella*).

Localities.   Red Crag, Butley, Sutton, and Kesgrave.   Middle Glacial, Billockby.

This shell was described by Woodward in his ' Geology of Norfolk ' (1833) under the name of *Turbo semicostatus*, auctorum, but the authors of the 'British Mollusca' and the author of the 'British Conchology' agree that Montagu's *Turbo semicostatus* is identical with *R. striata* of Adams.   As, however, I consider the Crag fossil to be distinct from any known recent species, I have retained it under the name of *semicostata*, Woodward, that name being unoccupied now that Adams' older name of *striata* is applied to the other (Montagu's) shell; the synonym *Turbo semicostatus*, Mont., given at page 102 of Crag Moll.,' being an error. His son, Dr. Woodward, in the list of Norwich Crag shells in White's Directory of Norfolk, refers this species to *inconspicua*, Alder; but the Crag forms invariably have the outer lip thickened and dentated within, which the recent *inconspicua* has not; and if this be a specific character, the two shells must be distinct.

My identification of the shell shown in fig. 9, of Tab. XI, of Crag Moll., with *pulchella*, Phil., has been dissented from; and in such dissent I agree, as I now believe it to be the same as *semicostata*, Woodward; I therefore unite these species under the name of *semicostata*.   Since the publication of the 'Crag Moll.' I have obtained three specimens of the original *semicostata*, Woodward (fig. 10 of Tab. XI), from the Red Crag of Butley, which may be synchronous with the Fluvio-marine Crag of Bramerton, from which the shell shown in fig. 10 came.   Those figured as *pulchella* (fig. 9, *a, b*) were, from what I consider to be, an older portion of the Red Crag, viz. that at Kesgrave and Sutton. Possibly in the interval the shell may have undergone a slight change. Both forms of the shell, with the denticulations well shown, are not uncommon in the Middle Glacial of Billockby, but I have not seen it from either the Chillesford bed or the Lower Glacial sand.

Rissoa senecta, *S. Wood.* Supplement, Tab. V, fig. 15.

Locality. Cor. Crag, Sutton.
Length, $\frac{1}{10}$th of an inch.

A single specimen of this genus, shown in the figure above referred to, has lately been found by myself, which I cannot refer to a known species, and I have therefore given to it provisionally the above name. The volutions (about five) are nearly flat; the costæ few (ten or eleven), large, coarse, and wrinkled; suture distinct, but not deep; spiral striæ large and distant; body whorl two thirds the length of the shell; aperture large and ovate.

Rissoa reticulata? *Mont.* Crag Moll., vol. i, p. 103, Tab. XI, fig. 5.

In the 'Crag Mollusca' I referred this shell to *R. reticulata*, Mont., with a doubt.

Dr. Hörnes makes this Crag shell identical with *R. Montagui*, Payr., and in the list in Mr. Prestwich's Coralline Crag paper Mr. Jeffreys refers the Crag shell to " *R. calathus*, F. and H., not *R. reticulata*, Mont.," while in his 'Brit. Conch.,' vol. iv, p. 12, he says *calathus* is but a very doubtful species, and in his view only a variety of *R. reticulata;* and then adds (p. 13) that S. Wood's Crag shell called *reticulata* more resembles *calathus* than *reticulata*, and may be an intermediate variety. In this chaos of opinion I have thought it wisest to retain my shell under the designation given to it in the 'Crag Mollusca,' and with the same doubt.

Rissoa Stefanisi, *Jeffreys.* Crag Moll., vol. i, p. 106, Tab. XI, fig. 12 (as *R. costulata*, S. Wood).

Rissoa Stefanisi, *Jeff.* Brit. Conch., vol. iv, p. 36.

This shell was described by me in the 'Crag Mollusca' as a new species in ignorance that the name was preoccupied for other shells, as has been pointed out by Mr. Jeffreys in vol. iv, p. 36, of 'Brit. Conch.,' who there proposed for it the name *Stefanisi*. This name I have therefore adopted.

10

NATICA PROXIMA, *S. Wood.*   Crag Moll., vol. i, p. 143, Tab. XVI, fig. 4 ; Supplement, Tab. IV, fig. 12.

Localities.   Cor. Crag, Ramsholt.   Red Crag, Butley.

This shell has been described as living in British and Mediterranean seas under the name *N. sordida* of Philippi, by Forbes and Hanley ('Brit. Mol.,' vol. iii, p. 334), and by Jeffreys ('Brit. Conch.,' vol. iv, p. 219).   As, however, Philippi's name was two years subsequent to that (*proxima*) given by me in my 'Catalogue' of 1842, I have retained it under my own name.   I have now obtained a specimen from the Red Crag, Butley, which is represented in Supplement, Tab. IV.   *N. proxima*, S. Wood, is referred by Weinkauff to *N. fusca*, De Blain, to which species the author of 'Brit. Conch..' in vol. v, p. 215, also refers *sordida.*

NATICA HELICINA ? *Brocchi.*   Supplement, Tab. IV, fig. 8, *a, b.*

> NERITA HELICINA, *Broc.*   Foss. subap., p, 279, t. i, fig. 10, 1814.
> NATICA   —   *Hörnes.*   Foss. Moll. des Wien. Beck., vol. i, p. 525, t. xlvii, figs. 6, 7.

Localities.   Cor. Crag, Sutton.   Red Crag, Walton Naze, Essex.   Bentley, Suffolk.
A few specimens of what I imagine to be the species above referred to have recently been found by myself at both the above localities.

NATICA ALDERI, *Forbes.*   Supplement, Tab. VII, fig. 27.

> NATICA ALDERI, *Forbes.*   Malac. Monen., p. 31, pl. ii, figs. 6, 7.
> —   NITIDA, *Forb.* and *Hanl.*   Brit. Moll., vol. iii, p. 330, pl. c, figs. 2—4.
> —   —   *Rose.*   Geol. Mag., vol. ii, p. 11.

Localities.   Cor. Crag, Gedgrave.   Fluvio-marine Crag, Yarn Hill.   Middle Glacial, Hopton and Billockby.   Post-glacial, March and Kelsea Hill gravels and Nar. Brick-earth, at West Bilney, and East Winch (*Rose*).

Five specimens belonging undoubtedly to this species have been found by Mr. Rose in the Nar. Brick-earth, one of which is figured as above.   These shells have the exterior quite perfect, smooth, and glossy, but I am not able to see the remains of any coloured markings, and they most probably belonged to the white variety.   Some specimens of this shell (but none full grown) have occurred in the Middle Glacial sand of Hopton and Billockby.   I am inclined to think that the shell described in 'Crag Moll.' (vol. i, p, 142, Tab. XVI, fig. 1) from the Red Crag, as *N. Guillemini* belongs to the present species, or it may be the young of *N. catenoides.*

N. Alderi has been found at Yarn Hill by Mr. Crowfoot. I have seen a fine specimen found by Mr. Harmer at March, and it is given (under the name *nitida*) by Mr. Jeffreys from Kelsea Hill. Since my plate has been engraved I have found a specimen of this species in the Coralline Crag of Gedgrave.

NATICA CLAUSA. Crag Moll., vol. i, p. 147, Tab. XVI, fig. 2.

Localities. Red Crag, Sutton and Butley. Fluvio-marine Crag, Bramerton. Chillesford bed, Aldeby, and Horsted. Lower Glacial, Rackheath. Middle Glacial, Hopton, and Billockby. Upper Glacial, Bridlington. Post Glacial, Kelsea Hill.

Specimens of this species have been now obtained from all the above formations. It is not uncommon in the Middle Glacial, but the specimens are all young shells. It seems the reverse of common in all the formations prior to the Middle Glacial. It is given from Kelsea Hill by Mr. Jeffreys.

NATICA GRŒNLANDICA, *Beck.* Crag Moll., vol. i, p. 146, Tab. XII, fig. 5.

Localities. Red Crag, Butley. Fluvio-marine Crag, Bramerton, and Thorpe in Suffolk. Chillesford bed, Chillesford. Upper Glacial, Bridlington. Post Glacial, Kelsea Hill.

Since the publication of the ' Crag Moll.,' I have myself found this shell at Butley and Chillesford. It has been found by Mr. Reeve in the Fluvio-marine Crag, Bramerton, and is given by Mr. A. Bell from that of Thorpe in Suffolk.

Mr. Jeffreys gives it from Kelsea Hill. It appears to me that the *N. borealis* of Mr. A. Bell's list in the ' Ann. and Mag. Nat. Hist.' for May, 1871, is *N. Grœnlandica*, Beck.

NATICA HEMICLAUSA, *Sow.* Crag Moll., vol. i, p. 144, Tab. XIV, fig. 5.

Localities. Red Crag, Walton, Sutton, and Butley. Chillesford bed, Easton ?

In my Monograph I gave this shell from Walton and Sutton, and I have since found it at Butley. Mr. A. Bell informed me that he had observed it at Easton Cliff, but he has not so given it in either of his lists in the ' Ann. and Mag. of Nat. Hist.' Dr. Woodward gives it in his list (in ' White's Directory ') as common at Norwich, but I have not met with an instance of its occurrence in that neighbourhood.

NATICA CIRRIFORMIS, *Sow.*　Crag Moll., vol. i, p. 145, Tab. XVI, fig. 7.

I only know this shell as a Cor. Crag species. It is given, however, by Mr. Bell ('Ann. and Mag. Nat. Hist.,' September, 1870) from the Red Crag of Sutton, and of Waldringfield, but I have not seen the specimens.

NATICA MULTIPUNCTATA, *S. Wood.*　Crag Moll., vol. i, p. 148, Tab. XVI, fig. 9.

This shell Mr. Bell informed me he had obtained from the Chillesford bed of Easton Cliff, but I have not seen the specimen, nor does he give it in either of his lists in the 'Ann. and Mag. of Nat. Hist. ;' personally I only know the species from the Red Crag of Walton, Sutton, and Butley, and from the Cor. Crag.

NATICA OCCLUSA, *S. Wood.*　Crag Moll., vol. i, p. 146, Tab. XII, fig. 4, Supplement, Tab. IV, fig. 11.

Localities. Red Crag, Butley. Chillesford bed, Easton Bavent? Upper Glacial, Bridlington.

This species was proposed by me for the Bridlington shell, figured in 'Crag Moll.' Since then I have obtained the specimen represented in fig. 11 of 'Supplement,' Tab. IV, from the Red Crag at Butley, which appears to resemble *N. occlusa* from Bridlington very closely, though the spire is not quite so much elevated, but more so than that in *N. clausa*. I have therefore referred it to the Bridlington species; Mr. Bell gives it from Easton Bavent, but I have not seen the specimen.

NATICA CATENA, *Da Costa.*　Crag Moll., vol. i, p. 142, Tab. XVI, fig. 8.

Localities. Red Crag, Sutton and Butley. Fluvio-marine Crag, Bramerton. Chillesford bed, Horstead and Coltishall. Lower Glacial Sand, Belaugh. Middle Glacial, Billockby and Hopton. Upper Glacial, Bridlington.

In none of the fossil specimens of this shell that I have seen are there any remains of the marks resembling those upon the recent shell. In 'Brit. Conch.,' vol. iv, p. 222, is the following observation, "The coloured markings of this species (*catena*) are not exhibited in the Crag shells so named by Mr. Wood, although they are retained in his *N. millepunctata*"—a remark that seems to imply a doubt as to the correctness of my reference of *catena*. The cause of this difference I imagine is that the *red* spots of the one are more durable than the *brown* chain-like markings of the other, as in the dead or beach laid specimens they are generally invisible. *Red* spots appear to be permanent upon some of the Older Tertiary *Pleurotomæ*, and I believe that colour is preserved upon some of the older Secondary Fossils. So far as the uncertainty which always attaches from the resemblance to this species which other species, when fossil, may from decortication put on, will allow me to say, *N. catena* seems common in the Fluvio-marine Crag of Bramerton, and in the Chillesford bed at Horstead and Coltishall, but rare in the Lower Glacial sands, and is common in the young state in the Middle Glacial of Hopton Billockby. I have not seen it from the Post-Glacial beds of East Anglia.

NATICA PUSILLA? *Say.* Supplement, Tab. IV, fig. 9.

> NATICA PUSILLA, *Say.* Journ. Acad. Nat. Sci. Phil., ii, p. 257, 1822.
> — — *Binney.* Gould. Inv. Massach., 2nd ed., p. 344, fig. 613, 1870.

Locality. Coralline Crag, near Orford.

The above represents a shell I have lately obtained, but, like the generality of specimens of the Crag *Naticæ*, the glossiness of the exterior is gone. I have given it the above name with doubt, as my shell differs in some respects from the existing one. On comparing my specimen, it seems to have rather more convex volutions than the recent *pusilla*, and the pad over the umbilicus larger and more extended in the Crag shell. It resembles somewhat *Nat. occulta*, Desh., 'Des. An. sans Vert.,' Pl. LXVIII, figs. 11—13, and it appears intermediate between the two, as if the Crag one descended from the Paris basin shell with alteration, and the recent species from the Crag one with still further alteration. I feel much disposed to consider it specifically distinct, and to call it *N. occultata*, but having only two or three specimens, and those with the exterior not perfect, I have preferred giving it the above name.

NATICA CATENOIDES, *S. Wood.* Crag Moll., vol. i, p. 141, Tab. XVI, fig. 16, Supplement, Tab. IV, fig. 13 *a, b.*

Localities. Red Crag, Waldringfield, Sutton, and Walton. Chillesford bed, Easton Bavent. ?

The shell represented in fig. 13 of Supplement, Tab. IV, was obtained at Waldringfield by Mr. Canham, which being much larger than the one represented in 'Crag Moll.,' I thought it desirable that it should be figured; especially as considerable uncertainty has existed, and, indeed, still exists, respecting the correct appropriation of this Red Crag shell. The late Edward Forbes considered it identical with *N. glaucina* (*catena*), 'Mem. of the Geol. Survey,' 1846, p. 430; while in the 'Brit. Moll.,' vol. iii, p. 306, 1853, the authors refer this Crag shell to *Nat. sordida*. Again, M. Thuden has still more recently placed it as a doubtful synonym with *Nat. nitida* ('Om. de J. Bohus. Postplioc. eller. glac. format.,' p. 56, 1866). I cannot say that I agree with any of these references; the shell more resembling *Nat. heros,* Say, to which Mr. Jeffreys refers it. I believe however that it is distinct, as that shell has a deeper suture. I have found some imperfect specimens in the Cor. Crag of Sutton which may possibly belong to *catenoides* but more probably to *N. helicina,* Broc. *N. catenoides* is given also by Mr. Bell from Easton Cliff, (' Ann. and Mag. Nat. Hist.,' September, 1870) but I have not seen the specimen.

NATICA MONTACUTI, *Forbes.* Supplement, Tab. IV, fig. 10.

NATICA MONTACUTI, *Forb.* Malac. *Monen.*, p. 32, pl. ii, figs. 3, 4.

Locality. Upper Glacial, Bridlington.

My figure represents a specimen in the British Museum, among the Bridlington Fossils, to which the above name is attached, and this I think may be fairly referred as above. This name is introduced by the late Dr. S. P. Woodward in his list of Bridlington Shells, ' Geol. Mag.,' vol. i, p. 53.

NATICA HELICOIDES, *Johnston.* Crag Moll., vol. i, p. 145, Tab. XVI, fig. 3.

Localities. Red Crag, Sutton and Butley. Fluvio-marine Crag, Bramerton. Chillesford bed, Horstead, Coltishall, and Aldeby. Lower Glacial, Belaugh. Middle Glacial, Hopton. Upper Glacial, Bridlington. Post Glacial, March and Kelsea Hill.

This shell appears to be rare in the Red and Fluvio-marine Crags, but common in the Chillesford bed at certain localities. It is rare in the Lower Glacial, and a single young specimen only has occurred in the Middle. It is very abundant and of large size in the March gravel. Mr. Jeffreys gives it as rare at Kelsea Hill.

Three small specimens were found by myself in the Coralline Crag of Sutton, which I once thought belonged to the genus *Natica,* and I called them *Natica depressula* (' Crag Moll.,' vol. i, p. 149), but which I afterwards described in vol. ii, p. 319, as *Jeffreysia patula.* I am sorry to say no other specimen resembling them has since come into my possession. Mr. Jeffreys says they are the fry of *Velutina virgata.* They are probably the fry of some species, but I think not of *virgata,* as my Crag specimens of that species have a larger and more obtuse apex.

AMAURA CANDIDA, *Möller.* Supplement, Tab. I, fig. 3 *a, b.*

AMAURA CANDIDA, *Möll.* Ind. Moll. Groenl., p. 7, 1842.
— — *H.* and *A. Adams.* Genera, vol. i, p. 214, pl. xxii, fig. 9, 1858.
— — *A. Bell.* Ann. and Mag. Nat. Hist., September, 1870.

Locality. Red Crag, Butley.

This was found by Mr. Bell, to whom I am indebted for the use of the specimen for the above figure; it is the only one I have seen from the Crag.

LITTORINA RUDIS, *Maton*.   Supplement, Tab. V, figs. 9 and 10 *a*, *b*.

In my Monograph, vol. i, p. 118, Tab. X, fig. 14, *a—k*, I have given figures and descriptions of a variety of forms of shells belonging to the genus *Littorina*, from the Fluvio-marine Crag of Norfolk and Suffolk, which I had considered as all belonging to one species, *littorea*, and I have here had figured two or three more to show the extraordinary range in variation to which they had been subject. The British Conchologists, although they have kept separate several forms of this genus, have no accordance respecting specific division.

I have obtained a very large number of specimens of *Littorinæ* from the Fluvio-marine Chillesford bed, at Horstead and Coltishall, and from the Lower Glacial sand at Belaugh, and these correspond principally with the form usually seen in our markets. With these, as might be supposed, are small or young specimens strongly marked with spiral striæ and with a well-defined suture, while the large or full-grown specimens are nearly smooth. Two or three of the distorted figures of *Littorina*, 'Crag Mol.,' vol. i, Tab. X, fig. 14, may probably be referred to what is called *rudis*. Mr. Jeffreys says (vol. iii, p. 367) that *rudis* is viviparous, and *littorea* oviparous. If this be so, that distinction might entitle them to be considered as more than specifically distinct; but as far as their testaceous covering goes there seems so much intermingling of character between *rudis* and *littorea*, not to speak of the numerous other forms treated as species or varieties, that I confess to the greatest uncertainty in assigning shells to these respective species separately; what seems to be *L. rudis* from Bramerton is shown in Tab. V, fig. 9; and a distortion of the same species from the same place, put into my hands by Mr. Horace Woodward, is shown in fig. 10. *L. littorea* occurs also in the Middle Glacial at Hopton and Billockby, in the Upper Glacial at Bridlington, and in the Post-Glacial Gravels of March, Hunstanton, and Kelsea Hill, and in the Nar Brickearth.

LACUNA RETICULATA, *S. Wood*.   Crag Moll., vol. i, p. 122, Tab. XII, fig. 10.

MACROMPHALUS RETICULATUS, *S. Wood*.   (Catalogue of shells from the Crag.)   Ann. Nat. Hist., 1842, p. 537.

This shell is excessively rare to my researches, and I am unable to give to it any additional particulars; it does not strictly conform to the characters of the genus as generally described, but it has a broad and flattened pillar lip or elongated umbilicus; it nearly resembles *Lacuna elegans*, Deshayes, 'An. sans Vert. du Bas. de Par.,' vol. ii, p. 371, Pl. XVII, figs. 4—6, but it is probably distinct. Another species, somewhat resembling it, is placed in the same genus by Dr. von Koenen, 'Mittel Olig. Norddeutschlands,'

Tab. VII, fig. 10, called *L. striatula*, which, however, I think is also different from the Crag shell. *Lacuna cliona*, Raincourt and Munier, 'Journ. de Conch.,' vol. xi, p. 201, Pl. VII, fig. 1, is a similar Eocene fossil with a reticulated surface ; this group might, I think, be separated from *Lacuna*, the only resemblance to which is the elongated umbilicus. Fig. 23 of Supplement Tab. V was engraved under an impression I had that it might be *Lacuna quadrifasciata*. It, however, differs so little from some of the least elongated forms of *Littorina suboperta*, that it may be an immature form of that shell.

LACUNA VINCTA, *Mont.* Crag Moll., Appendix, p. 316, Tab. XXXI, fig. 13.

Localities. Fluvio-marine Crag, Bramerton. Post Glacial, March, and Kelsea Hill.

This shell seems to be confined to the Fluvio-marine Crag (where, according to Mr. Reeve, it is rare) and not to have occurred in any other Crag, or East Anglian Glacial bed. It is abundant and in good preservation in the March Gravel, and is given by Mr. Jeffreys as abundant at Kelsea Hill.

LACUNA CRASSIOR, *Mont.*?

This shell is given as occurring at Kelsea Hill, by Mr. Jeffreys, in 'Quart. Jour. Geol. Soc.,' vol. xvii, p. 450, but he is silent as to it in his 'Brit. Conch.' I insert it, therefore, with doubt as an East Anglian fossil.

TROCHUS TURGIDULUS ? *Brocchi*. Supplement, Tab. V, fig. 8.

TROCHUS TURGIDULUS, *Broc.* Conch. foss. Subap., vol. ii, p. 353, t. v, fig. 16, 1814.

Alt. ¼ of an inch.

Locality. Coralline Crag, Sutton.

Two small specimens found by myself are here referred with doubt to the above-named species ; they differ from *T. Montacuti* in having a sharper angle to the lower part of the volution, and they have been somewhat abraded, by which a great part of the outer coating has been removed, obliterating some of its character. They are smaller than the representation by Brocchi, but they may not be full grown ; they resemble the figure by Dubois, 'Coq. Foss. Volh. Pod.,' Pl. II, figs. 29, 30, but are less strongly striated. It is, however, an unsatisfactory identification.

TROCHUS TUMIDUS, *Mont.* Crag Moll., vol i, p. 130, Tab. XIV, fig. 2.

Localities. Red Crag, Sutton. Fluvio-marine Crag, Bramerton. Chillesford bed, Bramerton, and Aldeby.

This shell has been found by Mr. Reeve in the Chillesford bed at Bramerton, as well as in the Fluvio-marine Crag below, and by Messrs. Crowfoot and Dowson at Aldeby.

TROCHUS ZIZYPHINUS, *Linné.* Crag Moll., vol. i, p. 124, Tab. XIII, fig. 9.

Localities. Cor. Crag, Sutton, Ramsholt, and near Orford. Red Crag, Sutton. Fluvio-marine Crag, Bramerton. Middle Glacial, Hopton?

This species is given in Dr. Woodward's Nor. Crag list in 'White's Directory,' but I have not myself seen it from that Crag. Fragments of a large, finely striated Trochus are common in the Middle Glacial sand of Hopton, which, there can be little doubt, are of this species.

TROCHUS CINERARIUS, *Linné.* Crag Moll., vol. i, p. 131, Tab. XIV, fig. 7.

Localities. Red Crag, Walton and Sutton. Middle Glacial, Hopton. Post Glacial, March.

I have not met with any certain trace of this shell in the Fluvio-marine Crag or at any of the localities of the Chillesford bed, at which I am somewhat surprised. One tolerably perfect specimen and several imperfect have occurred in the Middle Glacial sand of Hopton, and I have it from the March Gravel. Some very injured apices of a Trochus sent me from Bramerton may not improbably belong to this shell, but they are not sufficient to justify my inserting it from that locality.

TROCHUS NODULIFERENS, *S. Wood.* Crag. Moll., vol. i, p. 126, Tab. XIII, fig. 6 (as
T. papillosus); and Supplement, Tab. V, fig. 14.

TROCHUS GRANOSUS, *S. Wood.* Catalogue, 1842.
— PAPILLOSUS, *Da Costa.* Crag. Mollusca, vol. i, p. 126.

Localities. As in 'Crag Moll.' and Fluvio-marine Crag, Bramerton and Thorpe? Middle Glacial, Hopton?

In speaking of this shell in the 'Crag Mollusca' I pointed out the differences which existed between it and the recent *papillosus*, Da Costa (*granulatus*, Born), but I did not consider that they justified my referring it to a distinct species. The author of the 'Brit.

11

Conchology,' however, insists that this Crag shell is distinct from the living *papillosus* (*granulatus*).

In 'Supplement,' Tab. V, fig. 14, I have shown a specimen from Walton Naze, which represents another of the various forms of this shell, and which in its granulations seems undistinguishable from the recent *papillosus*, but is much flatter, and without any convexity in the whorls. Under these circumstances I have thought it best provisionally to place the shell shown in the above figure, and in figures 6, *a*, *b*, *c*, of Tab. XIII of 'Crag Mollusca,' as a new species under the name *noduliferens*.

Whether *T. granulatus*, from Walton Naze, in Mr. Bell's list ('Annals and Mag. Nat. Hist.,' September, 1870), presents any nearer approach to the living shell I know not, as I have not seen the specimen.

Dr. Woodward gives *Trochus granulatus*, Born (*similis* Sby.), in his list of Norwich Crag shells in 'White's Directory,' as from Bramerton and Thorpe, on more than one authority. But whether the recent *granulatus*, or the above shell now called *noduliferens* is meant I am unable to say.

Fragments of a granulated Trochus occur in the Middle Glacial of Hopton that may belong either to this shell or to *papillosus* (*granulatus*).

TROCHUS BULLATUS, *Phil. ?* Crag Moll., vol. i, p. 124, Tab. XIII, fig. 4 (as *T. zizyphinus*, var. *monstrosa*) ; Suppl., Tab. VII, fig. 20.

TROCHUS BULLATUS, *Philippi.* Moll. Sic., p. 226, vol. ii, t. xxviii, fig. 8.

Localities. Cor. Crag, Sutton and near Orford. Red Crag, Walton Naze.

The specimen figured in 'Suppl.,' Tab. VII, was obtained by me some years ago from the Reg Crag of Walton. It differs altogether from *zizyphinus* in the tumidity of the whorls and in the presence on them of faint granulations (which, however, rather resemble diagonal sculpturing than granulations), and in these particulars the shell seems identical with the figure · and description of Philippi's species *bullatus*. I have also a half-grown specimen from the same locality. The shell figured in 'Crag Mollusca,' Tab. XIII, fig. 4, as *zizyphinus* var. *monstrosa*, from the Coralline Crag, Sutton, appears to be the same species, and Mr. Bell observes 'Ann. and Mag. Nat. Hist.,' May, 1871, that " he has obtained two specimens, one decorticated (similar to the shell figured in the 'Mon. Crag Moll.,' Tab. XIII, fig. 4), from the Cor. Crag, Gedgrave, and that Prof. Sequenza had sent him a series of Philippi's *T. bullatus* in all stages of growth and preservation, and a close comparison of their sculpture and form enabled him to correlate the Italian and Crag shells." Mr. Bell showed me the series thus obtained, and, as it seemed to bear out the view thus expressed, I have inserted the species from both the Coralline and Red Crag.

MARGARITA GROENLANDICA, *Chemnitz.* Supplement, Tab. V, figs. 11, *a, b.*

> TROCHUS GROENLANDICUS, *Chemn.* Conch. Cab., vol. v, p. 108, t. clxxi, fig. 1671.
> MARGARITA UNDULATA, *Gould.* Invert. Massach., p. 254, fig. 172.*
> — — *S. P. Woodward.* Norwich Crag Shells, p. 5, 1864.

Diameter, ¼ of an inch.

Locality. Fluvio-marine Crag ?, near Norwich. Middle Glacial, Hopton.

A specimen of this species has been obligingly forwarded to me by Mr. Bayfield, of Norwich, and he thinks it was found either at Thorpe or Postwick, but he is not able to say whether it is from the upper or lower bed. The present specimen is not a full-grown individual, as it possesses little more than three volutions. G. B. Sowerby describes this shell (*M. undulata*) as having four, Messrs. Forbes and Hanley describe it as having five, and Mr. Jeffreys as having six, volutions.

Three varieties of this species are given by Mr. Jeffreys, and our present shell seems to correspond with the one he has named *lævior,* which is said to be smooth. The Crag shell has visible lines of growth with one or two nearly obsolete spiral ridges, but there are no nndulations upon the upper part of the volution.

The pullus of this species, as well as those of *M. maculata* and *M. trochoidea,* are free from striæ or ornament of any kind.

Several small specimens, more or less imperfect, of a *Margarita* have occurred in the Middle Glacial sand of Hopton. Their rubbed condition will not allow one to say positively that they belong to this species, but there can be little doubt of their belonging to one of the varieties of it.

MARGARITA MACULATA, *S. Wood.* Crag. Moll., vol. i, p. 135, Tab. XV, fig. 3.

Localities. As in 'Crag. Moll.'

This elegant species would be a shell of some importance if the one found fossil in America and also recent upon the Coast of California should be identical with it. I have not been able to see the fossil from Williamsburg, spoken of by Sir Charles Lyell "*Solarium,* nearly allied to *Solariella maculata,*" 'Proc. Geol. Soc.,' 1845, p. 555, but I have compared a recent shell from the Cataline Islands *Solariella peramabilis,* Carpenter, "Rep. Moll., West Coast of N. America," 'Brit. Assoc.,' p. 653, 1864, with the Crag species. The recent shell has a more elevated spire, and in consequence a more contracted umbilicus, while the striæ in the Crag shell are rather more distinct.

The shell from the Cataline Islands is of a rufous brown, and possibly the spots remaining upon the Crag fossil may be some of its original colour.

The little recent shell from the British seas, *Trochus pusilla*, is given by Dr. P. Carpenter as a synonym to *Margarita Vahlii*; the young state of *M. trochoidea* from the Crag could scarcely be separated from it.

MARGARITA ARGENTATA ? *Gould.*   Supplement, Tab. V, fig. 12.

MARGARITA ARGENTATA, *Gould.*   Invertebrata Mass., p 256, fig. 174.

Locality.   Chillesford bed, Aldeby.

In a packet of shells found at Aldeby, and obligingly sent to me by Messrs. Crowfoot and Dowson, was the small and not quite perfect specimen figured as above. It is a young individual, and very distinctly striated all over, and I have (but with doubt) referred it to *Marg. argentata*, Gould. It has a small apex, differing in that respect from *M. trochoidea* of 'Crag Moll.'

Length, 0·1.

Mr. Jeffreys has, in vol. v, 'Brit. Conch.,' p. 202, given *argentata* as a synonym of *Margarita glauca*, Möller, 'Ind. Moll. Groenl.,' p. 8.   The specimen now figured is more coasely striated than the representation of *Trochus glaucus* in 'Brit. Conch.,' vol. v, Pl. CI, fig. 6.   Gould speaks of his shell as " composed of four convex whorls, the last of which is slightly angular."   His figure does not show this angularity, but it is perceptible in the Crag shell.

ADEORBIS PULCHRALIS, *S. Wood.*   Crag Moll., vol. i, p. 139, Tab. XV, fig. 4.

In the 'Quart. Journ. Geol. Soc.,' vol. xxvii, p. 495, it is said by Mr. Jeffreys, " *Adeorbis pulchralis.*   Swedish expedition, 320—600 fathoms. *Margarita trochoidea*, S. Wood, is the same species."   On a re-examination of my specimens I still believe they are distinct.

ADEORBIS STRIATUS, *Phil.*   Crag Moll., vol. i, p. 137, Tab. XV, fig. 7 (as *Adeorbis striatus*, S. Wood.)

Locality.   As in Crag Moll. only.

In 'Brit. Conch.,' vol. iii, p. 315, *Adeorbis striatus*, *A. supranitidus*, and *A. tricari-natus* are considered as varieties of a British shell called there *Trochus Duminyi*, Requien. A specimen from the Oligocene of Cassel sent me by Dr. von Koenen, with the name *carinatus* attached, approaches as near, or even nearer, to the three Crag species as (judging from Mr. Jeffreys' figure and description) does the recent *Duminyi*.   While partaking, however, of a portion of the characters of the three Crag species, *carinatus* yet differs from all of them. *A. striatus* is finely striated, without carinæ; *A. supranitidus* is smooth, with one to three sharp carinæ, and coarsely striated within the umbilicus; and *A. tricarinatus* has three very elevated carinæ, with striations between them. These are the characteristics of the three Crag species, but *A. carinatus* from the Oligocene has three carinæ, of which two are more conspicuous than the third, and all are less elevated than in *tricarinatus*, while the recent *A. Duminyi* is described as having eight to ten sharp and narrow striæ on the upper part of the whorl, the lowest of which, being more prominent, forms one faint carina. *A. tricarinatus* is certainly nearer to the Oligocene species than to any other, but yet is, I think, distinct, the carinæ being so much more elevated. We have here five different forms, and I see no reason for not retaining my three Crag species under the names I gave them. Dr. Speyer unites Philippi's *dubius* with *carinatus*.

SOLARIUM VAGUM, *S. Wood.* Supplement, Tab. VII, fig. 29, *a*, *b*.

Locality. Red Crag, Waldringfield.
Diameter, ½ an inch.

The figure above referred to represents another shell sent to me by Mr. Canham from the nodule workings at Waldringfield.

This I have placed in the genus *Solarium*, as it seems best entitled to that position, though somewhat of an aberrant character. I am unable to refer it specifically to anything known to me. The specimen is rubbed and worn, but it was probably, when perfect, nearly, or perhaps quite, smooth on the upper surface, with an obsolete ridge around or above the periphery; the under side has a moderate-sized umbilicus, edged with a sharp crenated, or rather nodulous, margin, like that of many species in this genus; its nearest resemblance is *Solarium simplex*, Bronn, but that species has only one ridge on the under side around the umbilicus, while upon our shell there are three or four spiral striæ. The present shell is the only representative of the genus *Solarium* that I know of from the Upper Tertiaries of England, and I suspect that it is derived from some older formation, for which reason I propose the name *vagum*. The shell spoken of as *Solarium pseudo-perspectivum*, from the mud deposit at Selsea,[1] 'Geol. Mag.,' vol. vi, p. 41, on which the name was founded, is a specimen of *Bifrontia Laudinensis*, washed into this mud bed from the Eocene formation beneath, where specimens of that species are abundant.

The genus *Phorus* is given by Philippi as fossil at Palermo, which deposit seems to belong to the Upper Tertiaries, but I have not seen this genus as fossil from beds newer than the Eocene in England. Mr. Whincopp showed me the cast of a species belonging

---

[1] This specimen is in the collection of Dr. Reed, of York.

to this genus which was obtained from the nodule workings in the Red Crag near Woodbridge, the matrix of which resembled that of an older Tertiary bed. It was evidently a derived specimen.

CYCLOSTREMA LÆVIS, *Philippi*.   Supplement, Tab. V, fig. 13, *a, b*.

> DELPHINULA LÆVIS, *Phil.*   En. Moll. Sic., vol. ii, p. 146, pl. xxv, fig. 2, 1844.

Diameter, one line, nearly.

Locality.   Coralline Crag, Sutton.

I have two specimens which correspond so closely with the figure and description of the Mediterranean shell above referred to that I have adopted for them the name given by Philippi.   I have compared my fossils with recent specimens of *serpuloides*, and I think the coarse and prominent ridges surrounding the umbilicus, of which there are no traces in *serpuloides*, is sufficient for specific distinction.   The fine striations which cover the under side of *serpuloides* do not appear in my shell, but they may possibly have become obliterated.

CYCLOSTREMA? SPHÆROIDEA, *S. Wood*.   Crag Moll., vol. i, p. 122, Tab, XV, fig. 9 (as *Turbo*).

Locality.   As in 'Crag Mollusca.'

This shell has, I find, beeen obtained by Mr. Jeffreys in the recent dredgings in the Bay of Tangiers, and in his 'Report,' 1870, p. 161, is referred by him to the genus *Cyclostrema* of Marryat, of which *Helix serpuloides* is supposed to be the type.   I have here adopted that generic name, though not without misgivings, as the peretreme of my shell is not continuous.   It does not seem far removed from *Adeorbis subcarinatus*.

HOMALOGRYA ATOMUS, *Phil*.   Supplement, Tab. VII, fig. 28.

> SKENEA NITIDISSIMA, *Forb.* and *Han.*   Vol. iii, p. 158, pl. lxxiii, f. 7, 8.
> HOMALOGYRA ATOMUS, *Jeff.*   Brit. Conch., vol. iv, p. 99.

Locality.   Cor. Crag, Sutton.

This represents a very minute shell which I have lately found in the Cor. Crag of Sutton, and I am anxious to preserve its likeness on account of the dangers attending such a minim.   It resembles the young state of *Valvata cristata*.   My specimen is probably not full grown, as it has only two volutions.   It is like the spiral portion of *Cæcum*, but it differs from that shell in having an upper and under side, whereas in *Cæcum* the whorls are perfectly horizontal.   I have referred my present specimen to a recent shell, which Mr. Jeffreys considers as *Truncatella ulomus*, 'Phil. En. Moll. Sic.,' vol. ii, p. 134, Tab. XXIV, fig. 5. There is a figure of the animal and shell by Mr. Jeffreys in the 'Ann. and Mag. Nat. Hist.,' January, 1859, p. 18, Pl. III, fig. 16, where it is described under the name of *Euomphalus nitidissimus*.

CÆCUM TRACHEA, *Mont.* Crag Moll., vol. i, p. 115, Tab. XX, fig. 5.
— MAMMILLATUM, *S. Wood.* Crag Moll., vol. i, p. 116, Tab. XX, fig. 4.
— GLABRUM, *Mont.* Crag Moll., vol. i, p. 117, Tab. XX, fig. 6.
— INCURVATUM, *Walker.* Crag Moll., vol. i, p. 117, Tab. XX, fig. 7, *a, b.*

The above four species were given by me in the 'Crag Mollusca,' but *incurvatum* only provisionally, as being possibly the young of one of the other species.

Dr. Philip Carpenter in his "Monograph on the *Cœcidæ*" 'Proc. Zoo. Soc.', 1858, dissents from my determination of *C. trachea*, and regards the shell figured by me under that name as a new form, to which he assigns the name of *tumidum*, rejecting the living species *trachea* from the Crag category.[1] My species *C. mammillatum* he recognises, as well as *C. glabrum*, Mont., while *C. incurvatum* he also thinks may be the young of one of the other species. He also recognises among my specimens placed in the British Museum a new form, which he names *liratum ;* and he makes new genera for the reception of all.

As Dr. Carpenter has made the family of *Cœcidæ* a special object of study, I think it desirable (though I do not fully agree with him) to give the Crag *Cœcidæ* according to his views, which are thus :

CARPENTER'S NAMES.	NAMES IN CRAG MOLLUSCA.
Elephantulum liratum	Not given.
Anellum tumidum	*Cæcum trachea*, vol. i, p. 115, Tab. XX, fig. 5.
Fartulum mammillatum	*C. mammillatum*, vol. i, p. 116, Tab. XX, fig. 4.
Brochina glabra	*C. glabrum*, vol. i, p. 117, Tab. XX, fig. 6.
Young of *mammillatum*	*C. incurvatum*, vol. i, p. 117, Tab. XX, fig. 7.

Dr. P. Carpenter does not recognise any of these forms as living, with the exception of *glabra*. Mr. Bell gives ('Ann. and Mag.,' May, 1871) *C. mammillatum* from the Red Crag of Walton Naze.

[1] In my Catalogue in 'Mag. of Nat. Hist.,' 1842, I place a note of interrogation against *C. trachea*, and pointed out that the recent shell is regularly annulated and smooth, and that my Crag shell differed from it in having the annuli more irregular and rugose.

CAPULUS UNGARICUS, *Linn.* Crag Moll., vol. i, p. 155, Tab. XVII, figs. 2 *a—g*.

Localities. Cor. Crag, Sutton, Ramsholt, and near Orford. Red Crag, *passim.* Fluvio-marine Crag, Bramerton (Woodward). Middle Glacial, Hopton.

All the above localities for this shell are within my own knowledge except the Fluvio-marine of Bramerton, where, according to Woodward's list in White's 'Directory,' is said to occur small and rare. The specimens from the Middle Glacial are very young ones.

CAPULUS RECURVATUS, *S. Wood.* Crag Moll., vol. i, p. 156, Tab. XVII, fig. 3 *f* (as *C. militaris*, Mont.).

Localities. Cor. Crag, Sutton, and near Orford. Red Crag, Walton, Sutton, Newbourn, and Waldringfield.

The name of *militaris* given by Montagu to our species being posterior to that given by Linné to a different shell inhabiting the West Indies must be abandoned. I, therefore, fall back upon the name *recurvatus* given in my Catalogue of 1842, for the specimen shown in fig. 3 *f* of Tab. XVII, of 'Crag Moll.' Figs. 3 *b, c, d*, may be the young of *C. ungaricus*.

Mr. Bell has described in the 'Annals and Mag. Nat. Hist.' for September, 1870, a shell from the Red Crag of Waldringfield as a new species under the name of *C. incertus*, the specimen of which he kindly submitted to me. He has also ('Ann. and Mag.,' 1871) given the name *Brocchia sinuosa* to the shell shown in 'Supplement,' Tab. VII, fig. 26 *a, b*, which may be the same as *Patella sinuosa*, Brocchi. In the monograph of the 'Crag Moll.' I showed one of these sinuous forms of the *Capulidæ*, under the name var. *partim sinuosus* of *C. militaris*, regarding it as an accidental variation due possibly to the adherence of the shell to a *Pecten*. Looking at the various forms figured by Prof. Salvatore Bionde ('Estr. dagl Atti dell Acad. Gioenia de Sc. Nat.,' Vol. XIX, Sec. Series, 1864) in his monograph of the so-called Genus *Brocchia*, and at the specimen figured in 'Supplement,' Tab. VII, I must admit that the idea of an adherence to a *Pecten* will not explain these features. Bionde's figures of some twelve forms under the generic name *Brocchia* show one or more sinuosities in each, but they are not all in the same *part* of the shell nor in the same direction. Neither do they appear in the young shell, but only upon that part of the shell which must have been formed after the animal was half grown; and, however caused, suggest the idea that these peculiar features are due to some accidental circumstances besetting the growth of certain individuals of the genus *Capulus*. Under these circumstances I do not see my way to the adoption of the genus *Brocchia* until further investigations, especially on living ·forms if such be discovered, have demonstrated that this testaceous covering pertained to an animal generically differing from those producing the shell *Capulus*. For a similar reason I have not adopted as a separate species Mr. A. Bell's *C. incerta*, thinking that it is probably only a distortion of this nature of one or other of the species of Capulus described and figured in the 'Crag Mollusca.'

CALYPTRÆA CHINENSIS, *Linn.*   Crag Moll., Vol. I, p. 159, Tab. XVIII, fig. 1.

Localities. Cor. Crag, Sutton, Ramsholt, and near Orford. Red Crag, *passim.* Fluvio-marine Crag, Bramerton. Chillesford Bed, Aldeby. Middle Glacial, Hopton.

This species has occurred within my knowledge at all the above localities. The apices of the shell are not uncommon in the Middle Glacial. The large squamose and imbricated form of this shell appears to be confined to the Coralline Crag, and to the Walton Red Crag, the specimens from all the other localities being the small living British form.

EMARGINULA FISSURA, *Linné.*   Crag Moll., vol. i, p. 164, Tab. XVIII, fig. 3 *a.*

Localities. Cor. Crag, Sutton, and near Orford. Red Crag, Walton, Sutton, and Butley. Middle Glacial, Hopton?

This is one of the most abundant shells of the Coralline Crag at Sutton. Among these specimens may be seen very great variation both in the radiatory lines and the cancellation, also in the comparative height and in the position of the vertex. This point or apex is in some subcentral, in others it nearly overhangs the base of the shell; in some this vertex is much elevated, in others depressed with every intermediate gradation. A fragment of a shell of this genus, the sculpture on which seems to agree with *fissura*, has occurred in the Middle Glacial of Hopton.

EMARGINULA ROSEA (?), *Bell.*

EMARGINULA ROSEA, *Bell.* Zool. Journ., vol. i, p. 52, pl. iv, fig. 1, 1824.

Locality. Cor. Crag, Sutton.

Mr. Jeffreys gives this species as present in my collection in the British Museum ('Brit. Con.,' Vol. III, p. 261), and I have, therefore, inserted the species as a Cor. Crag shell with a note of interrogation, for I am not able myself to detect a sufficient difference among the many variable forms to justify their separation; *rosea, cancellata, elongata,* and *decussata,* may, perhaps, all be found among my specimens, but they so graduate into each other, and having all lived together, that I cannot venture on my own authority to call them specifically distinct from *E. fissura.* The most distinct form and

12

the one which if we had the evidence of fossil specimens only for a guide I should be disposed to regard as most entitled to specific distinction, is that figured by me in Tab. XVIII, fig. 3 *b*, under the name of var. *punctura;* and to show this better I have given in Tab. VII, fig. 24 of this 'Supplement' an enlargement of the sculpture on this var. *punctura.* As to the proportional length of the fissure and shape of the shell, I find it to vary so irregularly as to be no guide for specific distinction. *E. capuliformis*, ' Phil. Mol. Sic.,' Vol. I, p. 116, Tab. VII, fig. 12, is merely a distortion of one or other of these forms, and I have the same among my specimens.

EMARGINULA CRASSA, *J. Sow.* Crag Moll., vol. i, p. 165, Tab. XVIII, fig. 2.

Localities as in ' Crag. Moll.'

The shell with this name from the Coralline Crag differs materially from that found in the Red Crag, being much smaller; it also has rounded and more prominent rays, is more conical or elevated, and has comparatively a deeper sinus at the margin, while it differs as much from the Red Crag shell of this name as any of the five forms, *rosea, fissura, cancellata, elongata,* and *decussata* do from each other, and if they are to remain specifically distinct I would call my Coralline Crag shell a separate species also under the name *E. crassalta.* Specimens of *E. crassa* from the Red Crag measure $2\frac{1}{4}$ inches in length.

FISSURELLA COSTARIA, *Basterot.* Supplement, Tab. VII, fig. 19.

> TISSURELLA COSTARIA, *Bast.* Mem. Geol. Bord., p. 71, 1825.
> — NEGLECTA, *Desh.* Exp. Sci. de Morie., p. 134, 1844-8.
> — MEDITERRANEA, *Gray.* Apud. Sow. Conch. Ill., fig. 30.
> — ITALICA, *Hörnes.* Vien. Foss. p. 641, t. 50, fig. 80.

Length, $1\frac{3}{4}$ inch, breadth $1\frac{1}{8}$ inch.

Locality. Coralline Crag, Sutton. Red Crag, Waldringfield (A. Bell).

A fine specimen has been obtained from Sutton by Mr. Bell and put into my hands for description with the name of *F. neglecta.* This name was first given by Deshayes, ' Coq. Foss. des Env. de Par.,' Tome II, p. 20, Pl. II, figs. 10—12, but rejected in his second work as not belonging to the Paris Basin, and Basterot adopted the above name.

This shell much resembles *F. græca,* and its principal difference appears to be that the decussating lines, or lines of growth, are stronger and closer in this species, and the rays or radiating ridges are more uniform, and not alternate, as in *F. græca;* but the young of *græca* is very variable in its markings, and the keyhole opening is broader in the very young shell, near the recurved vertex, than it is on the anterior side.

Tab. 483, figs. 1—3, ' Min. Conch.,' belongs, I believe, to *F. græca,* Linn., and not to the present species.

In the list of Crag shells by Mr. Alfred Bell, 'Ann. and Mag. Nat. Hist.,' Sept., 1870, as also in the list by Mr. Jeffreys, that accompanies Mr. Prestwich's Cor. Crag paper, is the name of *Puncturella* (*Cemoria*) *Noachina*, from the Cor. Crag of Sutton. On applying for a sight of his specimen Mr. Bell tells me it was a very small one, and has, unfortunately, been lost, and my application to Mr. Jeffreys for a sight of the specimen on the authority of which he has inserted this species in his list, has also been unsuccessful.

As *Cemoria Noachina* (*Patella Noachina*, Linn.) at the present day is a very northern form, I was anxious for clear evidence of its existence in the Cor. Crag. I thought possibly it might be the young state of *Fissurella græca*, which has a recurved apex, such as is represented at fig. 4 *c*, Tab. XVIII, of 'Crag Moll.,' and in this state of things I do not venture to give it as a Crag shell.

TECTURA ? PARVULA, *Woodward.* Crag Moll., vol. i, p. 162, Tab. XVIII, fig. 8.

Locality. Fluvio-marine Crag, Bramerton.

In the 'Crag Mollusca' I observed that this shell might possibly be the young state of *Patella vulgata* which occurs, though rarely, in the Fluvio-marine Crag. Mr. Reeve tells me there are only five specimens in the Norwich Museum, and these have been in my hands for examination. They are all small, and very thin, which I imagine must be from a loss of part of the shell. They are elongated in form, as if they lived upon the leaf or stem of a Fucus. The vertex is very excentric, like that of *Tectura*, but they are more distinctly rayed or costated. I confess not to be able to determine their true position. They much resemble *Lottia alveus*, Gould, 'Inv. Mass.,' p. 154, fig. 13, in form; but that shell is said to be ornamented with very fine radiating striæ, while the Crag shell has raised radiating costulæ. Altogether, *Tectura parvula* must be regarded as a very doubtful species.

TECTURA FULVA, *Müller.* Crag Moll., vol. i, p. 161, Tab. XVIII, fig. 7.

Localities. Cor. Crag, Sutton. Red Crag, Walton? Middle Glacial, Hopton.

In his 'Brit. Conch.,' vol. iii, p. 251, Mr. Jeffreys observes that my specimens from the Cor. Crag, which I described as this species, appear to belong to *Lepeta cæca*; but as in his list accompanying Mr. Prestwich's Cor. Crag paper in 'Quart. Jour. Geol. Soc.,' vol. xxvii, p. 145, Mr. Jeffreys inserts *Tectura fulva*, and omits *Lepeta cæca*, I infer that he has abandoned that opinion. In his list accompanying the Red Crag paper, however, he inserts *Lepeta cæca* from the Red Crag of Walton as does Mr. Bell also ('Ann. and Mag.,' Sept., 1870). I have not, however, yet been able to see anything to justify the insertion of *Lepeta cæca* as a Crag shell.

Two small specimens of *T. fulva*, imperfect, but comprising the worn apex and larger part of the shell, have occurred in the Middle Glacial of Hopton.   In the ‘ Crag Mollusca ’ I stated that this shell was by no means rare in the Cor. Crag, but that is incorrect ; it seems rare there, at least, it has now become so.

DENTALIUM RECTUM, *Linn*.   Supplement, Tab. V, fig. 19, *a, b*.

> DENTALIUM RECTUM, *Gmelin*   Syst. Nat. ed. 13, pp. 37, 38.
> —   —   *Poli.   Test. utriusque* Sic., vol. iii, t. lvi, fig. 28.
> —   ELEPHANTINUM, *Desh*.   Monog. du Gen. Dent., p. 27, pl. iii, fig. 7.

Locality.   Red Crag, Sutton, Waldringfield (Bell).

The specimen figured was given to me by the late Mr. Acton, of Grundisburg, who said it was obtained from the Coprolite diggers at Sutton.

There are three species described by Linné so closely resembling each other that it is difficult to say what character will satisfactorily separate them, viz. *D. elephantinum, arcuatum*, and *rectum ;* the straight form of the present shell is the greatest, perhaps the only distinction.   In my fossil the sculpture appears to be generally about twelve large costæ, with a smaller intermediate one.   Mr. Bell gives this from Waldringfield.

Dentalium costatum, vol. i, Tab. XX, fig. 1 *d*, from the Red Crag, may probably belong to this species.

DENTALIUM ENTALIS, *Linn*.   Supplement, Tab. VI, fig. 20.

> DENTALIUM ENTALIS, *Linn*.   Syst. Nat., p. 1263.

Localities.   Cor. Crag, Sutton, and near Orford.   Post Glacial, Kelsea Hill.

The figure above referred to is the representation of two or three fragmentary specimens lately found by myself in the Coralline Crag at Orford and Sutton, and these are all that I have seen from that formation.   They show a perfectly smooth and glossy surface, but the terminal portion is not perfect.   Rubbed and worn specimens of a ribless *Dentalium* are occasionally found in the Red Crag, but they are not perfect enough for determination.

I have referred my shell as above, conceiving it most probably to be the same as the Mediterranean species.   *D. tarentinum, D. abyssorum*, and the present species so much resemble each other that it is difficult to point out a sufficient difference for specific separation in the recent and perfect shells ; while in the fossil it is even more difficult.

Dentalium entale is given as a fossil of Kelsea Hill in the list of shells by Mr. Jeffreys, in Mr. Prestwich’s paper, ‘ Geol. Journ.,’ vol. xvii, p. 449.

DENTALIUM ABYSSORUM, *Sars.* Crag. Moll., vol. i, p. 189, Tab. 20, fig. 2 (as *D. entale*).

Locality. Upper Glacial, Bridlington.

The shell figured and described by me from Bridlington appears to belong to the arctic and deep water species or variety, *abyssorum.*

DENTALIUM DENTALIS, *Linn.* Crag Moll., vol. i, p. 188, Tab. XX, fig. 1 (as *D. costatum*).

DENTALIUM COSTATUM, *J. Sow.* Min. Con., t. 70, fig. 8, 1814.

Localities. Cor. Crag, Sutton, and near Orford. Red Crag, Sutton. Middle Glacial, Billockby.

In the 'Crag Moll.' (vol. i, p. 189) I mentioned that I thought this Crag shell the same as the living Mediterranean shell, *D. dentalis*, Linn., and as this seems now the general opinion, I have here placed it under Linné's name. Two imperfect specimens, showing the strong costæ, but flattened by wear, have occurred in the Middle Glacial of Billockby, and I have no doubt of their belonging to this species.

ACTÆON NOÆ, *J. Sow.* Crag Moll., vol. i, p. 169, Tab. XIX, fig. 6.

Localities. Red Crag, Walton, Brightwell, Newbourn, Butley. Fluvio-marine Crag near Norwich?

This shell is given by S. Woodward, in his 'Geology of Norfolk' (1833), as occurring rarely near Norwich, but his son Dr. Woodward, in his list in 'White's Directory,' treats this as a mistake for *Actæon* (*Tornatella*) *tornatilis*, which also occurs in the Fluvio-marine Crag of Bramerton. I however find that *Actæon Noæ* was among a set of shells from Norwich submitted to me and the late G. B. Sowerby by Sir Chas. Lyell, and published at p. 328 of vol. III of the 'Mag. of Nat. Hist.,' New Series, which were by us carefully compared and considered. Therefore, although I have not since met with the shell from the Fluvio-marine Crag, I give it from that Crag with a doubt. Mr. Jeffreys has stated, 'Quart. Journ. Geol. Sci.,' vol. xxvi, p. 283, that this shell has been found fossil in Iceland by Prof. Steenstrup. Dr. O. Mörch also gives this species in his list of Crag Shells ('Geol. Mag.,' vol. viii, p. 395), from Hallbjarnastadir, in Iceland.

ACTÆON TORNATILIS, *Linn.* Crag Moll., vol. i, p. 170, Tab. XIX, fig. 5.

Localities. Cor. Crag, Sutton. Red Crag, Sutton and Butley. Fluvio-marine Crag, Bramerton. Chillesford bed, Burgh and Aldeby.

ACTÆON SUBULATUS, *S. Wood.*   Supplement, Tab. V, fig. 16 ; Crag Moll., vol. i, p. 170
Tab. XIX, fig. 7.

Localities.   Cor. Crag, near Orford.   Red Crag, Sutton, and Butley.

The figure in Supplement, Tab. V, represents a specimen from the Coralline Crag at Orford.   It much resembles, and is probably, the first incoming of the shell *A. subulatus,* figured in Tab. XIX of 'Crag Moll.,' from the Red Crag, although the present shell is a little more elongate than the last-mentioned figure.   I have, however, recently obtained a specimen from the Red Crag of Butley, which is similar to the Cor. Crag one, but larger, being perhaps an older shell.   The apex of the Cor. Crag fossil is obtuse, not an uncommon character with many of the Coralline Crag species ; in other respects it much resembles *Tornatella sulcata,* Lam., but the pullus portion of that shell has a sinistral volution which I do not see in my own, and there is a difference in the size and position of the fold upon the columella.   The striations on *sulcata* are also carried equally over the whole whorl, which is not the case in *subulata.*   The resemblance of this shell to the Eocene *sulcata* is closer than is that of several Crag shells to those living species with which they have been considered by some as identical.   I believe my shell figured in Crag Moll., Tab. XIX, fig. 7, to be specifically distinct from *tornatilis.*

ACTÆON LEVIDENSIS, *S. Wood.*   Crag Moll., vol. i, p. 171, Tab. XIX, fig. 4.

This shell has become exceedingly rare to my researches, and, so far as I know at present, is restricted to the single locality of Sutton in the Cor. Crag.   Whether this be the same as the Belgian fossil *T. elongata,* Nyst., I have not been able to ascertain, but judging from the figure by that author, I should scarcely think it was.   It is, however, quite distinct from the Older Tertiary *elongatus* of J. Sowerby.

ACTÆON ? ETHERIDGEI, *A. Bell.*   Supplement, Tab. V, fig. 17.

ACTÆON ? ETHERIDGII, *A. Bell.*   (N. S.) Ann. and Mag. Nat. Hist., Sep. 1870.
—   EXILIS ? *Jeffreys.*   Quart. Journ. Geol. Soc., vol. xxvii, p. 486, Nov., 1871.

Locality.   Red Crag, Walton Naze (*A. Bell*).

The specimen figured was found by Mr. A. Bell.   It was forwarded to me with the above name in commemoration of the able palæontologist of the ' Geological Survey,' and I have great pleasure in adopting that specific name for our shell.

This is a strange form, and I know not in what genus it ought to be placed.   Species in *Actæon* are all more or less covered with spiral striæ, but this shell seems to be quite smooth (which, however, may be accidental), and it has a projecting shoulder to the volution; perhaps it would be more correctly placed either in *Tornatina* or *Actæonina*.

I presume that this is the shell called *exilis* by Mr. Jeffreys in his list to Mr. Prestwich's Red Crag paper, p. 486.

CHITON DISCREPANS? *Brown.* Supplement, Tab. IV, fig. 27.

> CHITON DISCREPANS, *Brown.* Conch., Gt. Brit., pl. xxxv, fig. 20, 1827.
> — — *Forb.* and *Hanl.* Brit. Moll., vol. ii, p. 396, pl. 58, fig. 4.
> — CRINITUS, *G. B. Sow.* Desc. Cat. Brit. *Chitones*, p. 2, fig. 88—93 ;
> Ency. Method., pl. 163, fig. 11—17.

Locality. Coralline Crag, Sutton.

The figure above referred to represents the valve of a *Chiton*, which I have lately found, and which I have assigned to the above-named British species, though with some slight hesitation ; my specimen appears to be the third or fourth valve with the sustentacula quite perfect, and these have a peculiar form. The length of the valve is about equal to the breadth of one of its sides, and the ornamentation appears to correspond to that of the shell to which it is here assigned. Unfortunately, in the works upon British Conchology we have neither represented, nor described, the form or magnitude of these processes to the valves, which are, in my opinion, good auxiliary characters. This central valve now figured seems to differ from those of *fascicularis* in being more rounded at the lateral posterior termination of the valve, and the sustentacula are rather larger. We are not likely for some time to obtain a fossil with all the valves in position, so that we must do the best we can with the materials we possess, and I have given my figure as one step towards a correct determination.

CHITON RISSOI, *Payr.* Crag Moll., vol. i, p. 186, Tab. XX, fig. 11.

My Crag shell, which I referred as above, is regarded as *Ch. cinereus*, Linn., by Mr. Jeffreys ('Quart. Journ.,' vol. 27, p. 143). My Crag valves form a perfectly semicircular arch, without any angularity or pointed keystone like those of *cinereus*, and I have therefore retained my original name.

CHITON STRIGILLATUS, *S. Wood.* Crag Moll., vol. i, p. 186, Tab. XX, fig. 10.

This may probably be *Chiton Hanleyi* of the 'British Conchologists.'

In the 'Geol. of Norfolk' by S. Woodward, published in 1833, is the name of *Chiton octovalvis* (p. 44), and this name is repeated by his son Dr. S. P. Woodward, in his list of Norwich Crag shells, as from Thorpe, "a single valve," and is marked unique. This specimen cannot now be found, and as no confirmation of it appears in any of the collections known to me, it is impossible to say what species was intended.

SCAPHANDER LIBRARIUS? *Lovén.* Supplement, Tab. V, fig. 18 *a, b.*

SCAPHANDER LIBRARIUS, *Lovén.* Ind. Moll. Scandin., p. 10, 1846.
— — *Jeffreys.* Brit. Moll., vol. v, pl. 102, fig. 9, 1869.

Locality. Coralline Crag, Sutton.

My cabinet contains two or three specimens of the above genus, which appear to differ from *lignarius,* Linn., in being much less expanded at the lower part, and I have referred them, but with doubt, to the new species of Dr. Lovén. My specimens are not quite perfect.

BULLÆA VENTROSA, *S. Wood.* Crag Moll., vol. i, p. 182, Tab. XXI, fig. 11.

This is still a rare shell, and found only in the Cor. Crag of Sutton, so far as I know. In 'Brit. Conch., vol. iv, p. 425, Mr. Jeffreys has described a new species of *Bullæa,* under the name of *Urtriculus (Bullæa) ventrosus.* This does not appear to be the same as the Crag *ventrosa,* and his use of the name *ventrosus* may give rise to confusion. I propose that his shell be called *ventriculosus.*

Amphisphyra globosa, Lovén, is closely allied to *Bullæa ventrosa,* but it is described in 'Brit. Conch.' (*Urtriculus globosus,* vol. v, p. 223, Pl. CII, fig. 8) as having "slight, indistinct, and irregular spiral lines, which are only discernible with the aid of a magnifying power, and in certain lights." If this be an essential character, *globosa* must also differ from the Crag *ventrosa,* as the latter has very distinct and regular spiral striæ, otherwise I should have been disposed to regard the two species as identical.

In the address of the President of the Geological Society for 1871, at page liv, is a "List of Mollusca known hitherto as fossil only, and now discovered to be living in the depths of the Atlantic." In this list is the name of *Cylichna ovata* as a Coralline Crag fossil. I do not know this shell, nor can I ascertain to what the name refers.

RINGICULA BUCCINEA, *Broc.* Crag Moll., vol. i, p. 22, Tab. IV, fig. 2.

Localities. Cor. Crag, Sutton, and near Orford. Red Crag, Sutton.

In a paper by the Rev. O. Fisher in the twenty-second volume of the 'Quarterly Journal of the Geological Society,' p. 26, *R. buccinea* is given among a list of shells obtained by him from the Fluvio-marine deposit of Yarn Hill, near Potter's Bridge, Southwold. The specimen fortunately was preserved, and on re-examination it turns out to be one of *R. ventricosa,* so that *buccinea* is not known from any newer deposit than the Red Crag, in which, moreover, it is extremely rare, and may, even in that Crag, be only derivative from the Coralline.

RINGICULA VENTRICOSA, *J. Sow.* Crag Moll., vol. i, p. 22, Tab. IV, fig. 1.

Localities. Coralline Crag, Sutton. Red Crag, Sutton and Butley. Fluvio-marine Crag, Bramerton (Woodward), Yarn Hill (Fisher). Chillesford bed, Aldeby (Crowfoot and Dowson).

In the 'Crag Mollusca' *Ringicula* was placed in the section *Solenostomata* of Fleming (*Canalifera,* Lam.), depending upon the peculiar construction of the shell. Recent observations have removed it near to *Actæon,* in which position I have here placed it in deference to the Malacologists. It is, however, of a very aberrant character, possessing as it does a deep siphonal canal, very unlike its present associates. *R. ventricosa* still remains very rare in the Coralline Crag.

Since my communication in 1870 to the 'Ann. and Mag. of Nat. Hist.,' respecting the peculiarity of the Crag shells of this genus, I have found more than a hundred other specimens of *R. buccinea,* all in the presumed full-grown condition, that is, with a thickened outer lip. Of course this outer lip, while the animal is growing, must necessarily have a plain or simple margin, but the peculiarity is that it has so rarely[1] died in that condition I had imagined, and do so still, that these animals, as also those of *Trivia,* completed their shell in anticipation of their decease, and that many of the small specimens we find are young individuals that have thus assumed the adult form.

RISSOA ABYSSICOLA? *Forbes.* Supplement, Tab. VII, fig. 2.

RISSOA ABYSSICOLA, *Forbes.* Brit. Moll., vol. iii, p. 86, pl. lxxviii, figs. 1, 2.

Locality. Cor. Crag, Sutton. Living Britain, Scandinavia, and Mediterranean.

A single specimen which I have quite recently found in the Cor. Crag of Sutton is shown in the above figure. It may, I think, be referred to *abyssicola* of Forbes.

CANCELLARIA VIRIDULA, *Fab.* Crag Moll., vol. i, p. 66, Tab. VII, fig. 21, as *C. costellifer.*

CANCELLARIA VIRIDULA, var. COUTHOUYI. Supplement, Tab. VI, fig. 12.

This species is now considered to be the *Tritonium viridulum* of Fabricius' Fauna

---

[1] I, indeed, doubt whether a perfect specimen without the thickened lip has occurred in the Crag; the few such that I possess appear all of them to have had the thickened lip broken off.

13

Grœnlandica, 1780. A specimen of the more ovate variety, *Couthouyi*, has been obtained from the Red Crag of Butley by Mr. A. Bell, and is represented in the figure in Tab. VI of this 'Supplement.'

SCALARIA VARICOSA, *Lamarck*. Crag Moll., p. 90, Tab. VIII, fig. 14.

Localities. Cor. Crag, Sutton. Red Crag, Walton, Waldringfield, and Sutton (Bell).

This shell is evidently the same as *S. interrupta*, J. Sow., 'Min. Conch.,' tab. 577, fig. 3. The specimen there figured is said to have come from the Eocene of Barton Cliff· Mr. F. E. Edwards, whose Eocene collection is unrivalled, tells me that he does not know the shell as an Eocene species, *S. interrupta* of Dixon's 'Geol. of Sussex' being obviously a different shell. It is therefore most probable that the specimen figured in 'Min. Conch.' was from the Coralline Crag, the colouring of the figure being that of the Cor. Crag specimens. I should have been disposed to refer the Crag shell to Brocchi's *pumicea*, but it does not quite agree with either his or Hörnes' figure of that species, and I have had no opportunity of comparing the shells themselves. Should it prove distinct then, inasmuch as Lamarck's *varicosa* is generally regarded as the same as *pumicea*, I would suggest for our Crag shell the name of *Scalaria funiculus*.

Mr. Bell ('Ann. and Mag.,' Sept., 1870) gives the shell from the Red Crag of Walton, Waldringfield, and Sutton.

Mr. Jeffreys, in his List to Mr. Prestwich's Coralline Crag Paper, says that *Scalaria subulata* is a variety of *S. foliacea*, and this again a variety of *frondosa*, but in the Red Crag Paper (p. 496) he corrects this and says that *subulata* is a distinct species, and that Mr. McAndrew had dredged it off Teneriffe. It would seem from this that if a distinct Crag form is found living it is a species, but if not, then it is only a variety.

SCALARIA SEMICOSTATA, *J. Sowerby*.

A specimen of this, since the foregoing 'Supplement' went to press, has been sent to me from the Red Crag by Mr. Charlesworth. Although in the greatest perfection, this specimen can only, I think, be a derivative from the Eocene. I shall figure it in the concluding part of the 'Supplement.'

TROPHON ELEGANS, *Charlesworth*.

Very recently the fragment of a shell from the Coralline Crag near Orford has been sent to me by Mr. Cavell, and this, I think, may be referred to *Trophon elegans* of

Charlesworth. If this be so the specimens hitherto spoken of under this name, and which were found on the beach at Felixstow were probably derived from the Cor. Crag.

I will figure this fragment (unless a better specimen in the meantime should be found) in the concluding part of this 'Supplement.'

PTEROPODA.

Cleodora infundibulum, *S. Wood.* Crag Moll., vol. i, p. 191, Tab. XXI, fig. 14.

I have obtained only one specimen of this little Pteropod since it was described as above, and this, like my others, is imperfect. It was from the Cor. Crag of Sutton.

At p. 120 of vol. v 'Brit. Conch.,' it is said that my Crag shell is probably *Clio caudata* of Linné, *Cleodora compressa* of Souleyet. This, however, wants confirmation. Of the recent shell it is said that "the apex is globular." I have not heard of *C. infundibulum* having been found anywhere but in the Cor. Crag of Sutton.

It may possibly be the shell so called figured in Barbut's 'Worms,' Tab. VII, fig. 7, but I have not been able to satisfy myself on the point.

BIVALVIA.[1]

ANOMIA EPHIPPIUM, *Linné.* Crag Moll., vol ii, p. 8, Tab. I, fig. 3.

Localities. Cor. Crag passim. Red Crag, Sutton. Fluvio-marine Crag, Bramerton. Chillesford bed, Aldeby and Bramerton. Middle Glacial, Hopton and Billockby. Upper Glacial, Bridlington (*Woodward*).

In the Cor. Crag young specimens are very abundant, as they are also in the Middle Glacial sands of Hopton and Billockby, and, unlike the general condition of the fossils of those sands, they are usually uninjured. This species occurs in the Fluvio-marine Crag of Bramerton, though rarely, but it is common in the Chillesford bed, where that bed is not in the Fluvio-marine condition, as at Aldeby and Bramerton; but I have not met with it where it is in the Fluvio-marine condition, as at Horstead, Burgh, or Coltishall, nor in the Lower Glacial sands. It is rare in the Red Crag.

ANOMIA EPHIPPIUM, var. ACULEATA, *Müller.* Crag Moll., vol. ii, p. 9, Tab. I, fig 2.

Localities. Cor. Crag, Sutton. Chillesford bed, Aldeby and Bramerton. Middle Glacial, Hopton.

Specimens of this shell have been sent to me by Messrs. Crowfoot and Dowson from Aldeby, and two young specimens have occurred at Hopton.

ANOMIA STRIATA, *Brocchi.* Crag Moll., vol. ii, p. 11, Tab. II, fig. 3.

Localities. Cor. Crag passim. Red Crag, Sutton and Butley. Chillesford bed, Aldeby. Middle Glacial, Hopton.

A specimen of this shell was sent to me from Aldeby by Messrs. Crowfoot and Dowson, and a solitary specimen was obtained by me from the Red Crag of Sutton,

---

[1] A system now prevails of restoring to a species the earliest name given to it from the date of Linné's 12th edition of his 'Syst. Nat.,' however long such name may have been in disuse. This is only fair and just, and I have endeavoured to conform to such a rule; but why this custom should be adopted with respect to species, yet disregarded with respect to the more comprehensive divisions of the Animal Kingdom, I am at a loss to understand, and I still see no reason why Linné's term of BIVALVIA should be superseded.

one from the Lower Glacial sand of Belaugh, and a fragment from the Middle Glacial sand of Hopton, and these are all the instances that have occurred to my knowledge. It is by no means uncommon in the Cor. Crag, where specimens have a diameter of two inches.

ANOMIA PATELLIFORMIS, *Linné*. Crag Moll., vol. ii, p. 10, Tab. I, fig. 4.

Localities. Cor. Crag passim. Red Crag passim. Fluvio-marine Crag, Bramerton. Chillesford bed, Sudbourn Church Walks, Aldeby and Bramerton.

This species has been sent to me by Mr. Reeve from both beds at Bramerton, and by Messrs. Crowfoot and Dowson from Aldeby. Mr. Bell ('Ann. and Mag. Nat. Hist.,' 1870) gives it from Sudbourn. In the other localities I have obtained it myself.

At p. 323, Appendix to 'Crag. Moll.,' Tab. XXXI, fig. 24, a Cor. Crag fossil was figured erroneously under the name of *Aplysia?* This I have since found to be the more solid portion of the lower or adherent valve of some species of *Anomia*, probably *A. ephippium*.

OSTREA EDULIS, *Linné*. Crag. Moll., vol. ii, p. 13, Tab. II, fig. 1.

Localities. Cor. Crag passim. Red Crag passim. Fluvio-marine Crag, Bramerton and Thorpe. Post Glacial, March, Kelsea Hill, Hunstanton, and Nar Valley.

In 'Quart. Journ. Geol. Soc.,' vol. xxvi, p. 94, a list of shells from the Middle Glacial sands is given on my authority, and in it this species is given from Stevenage, in Herts. Since then I have doubted whether the specimens upon the strength of which the name was inserted may not be those of *Gryphæa dilatata*, and I have therefore omitted the species as a Middle Glacial shell. So far as the testaceous remains are a guide, I cannot in many instances distinguish between the forms of the secondary *Gryphæa* and those of the recent *Ostrea*. The species *O. edulis* is given in Dr. Woodward's list (in 'White's Directory') as occurring at Bramerton and Thorpe, but I have not seen it from thence, nor from any of the localities of the Chillesford bed. It is very profuse in the Post Glacial gravels of March, Kelsea Hill, and Hunstanton, as well as (according to Mr. Rose) in all the Nar Brickearth localities.

OSTREA COCHLEAR, *Poli*. Crag Moll. vol. ii, p. 14, Tab. II, fig. 1*c*, as *O. edulis*, var. *spectrum*.

Mr. Jeffreys, in his list accompanying Mr. Prestwich's Cor. Crag paper, and also Mr.

Bell ('Ann. and Mag. Nat. Hist.,' 1870) give *O. cochlear* as a Cor. Crag fossil, the former referring my var. *spectrum* to that species.

This probably may be so. I have therefore inserted it here as a species.

OSTREA PLICATULA, *Gmel.* Supplement, Tab. VIII, fig 10.

Locality. Cor. Crag, Sutton.

I have represented a specimen of the genus *Ostrea* obtained by myself which varies considerably from the normal condition of *O. edulis*, but it is difficult to determine the limitation of a species in this variable genus; however, if *O. cochlear* be entitled to that distinction, I think my present shell may have the same honour. It may possibly be *O. ungulata*, Nyst., but I have given to it the prior name by Gmelin. This is very variable, adhering sometimes by a very small part of the shell near the umbo, but I have specimens with similar rays in which the attachment of the lower valve had been effected by much the larger part of the surface. It resembles, and is probably the same as *O. cristata*, Born (tab. 7, fig. 3), which is plicated only near the margin.

OSTREA FLABELLULA, *Lamarck.*

Locality. Red Crag, Sutton?

Many years ago I noticed in the late Mr. Edward Acton's collection of shells from the nodule workings in the Red Crag near Woodbridge a specimen of this species; but I have been unable since his death to find where it has gone to, or I would have had it figured. It was mentioned in my paper on the extraneous fossils of the Red Crag ('Quart. Journ. Geol. Soc.,' vol. xv, p. 32), as probably introduced into the Red Crag from an older formation. I perceive that it is given as a species from Biot, which is classed among the Upper Tertiaries, so it is not clear but that it may have lived in the Cor. Crag sea. The lithological condition of the Red Crag specimen was, however, like that of *Rostellaria lucida* and others, regarded as derivative from older Tertiaries.

HINNITES CORTESYI, *De France.* Crag Moll., vol. ii, p. 19, Tab. III.

Localities. Cor. Crag, Ramsholt. Red Crag, Sutton, Trimley (*Bell*). Fluvio-marine Crag, Thorpe?

In the 'Ann. and Mag. Nat. Hist.,' September, 1870, this shell is given by Mr. Bell from the Red Crag of Trimley as *Hinnites giganteus*, and in the list to Mr. Prestwich's Cor. Crag paper ('Quart. Journ. Geol. Soc.,' vol. xxvii, p. 139), this name is also inserted for the Crag shell by Mr. Jeffreys, and it is given in the list by Dr. P. Carpenter in his ' Brit. Assoc. Reports,' 1863. I have compared my Crag shell with recent specimens of *H. giganteus* in the British Museum, and I cannot agree with those opinions. M. Fischer has, in ' Journ. de Conch.,' vol. x, p. 205, referred the fossil to *Ostrea crispa*, Broc. (' Conch. foss. sub-Apen.,' vol. ii, p. 567), which may possibly be correct ; but there is no figure given of that species by Brocchi, and I have therefore left the Crag shell as originally described.

I have myself also found a single valve in the Red Crag at Ramsholt, subsequent to the publication of the ' Crag Mollusca.' In the ' Geology of Norfolk ' this species is given by S. Woodward as occurring rarely and in fragments at Thorpe, and the same thing is repeated in the Norwich Crag list of his son Dr. Woodward, but I have not seen it myself from the Fluvio-marine Crag.

PECTEN PRINCEPS, var. PSEUDO-PRINCEPS, *S. Wood.* Supplement, Tab. VIII, fig. 9.

Localities. Fluvio-marine Crag, Bramerton and Yarn Hill.

When describing *Pecten princeps* from the Coralline Crag (' Crag Moll.,' vol ii, p. 32, Tab. VI, fig. 1), I had occasion to refer to ' The Geology of Norfolk,' where at p. 44 that name is inserted, and against which is the letter *a* signifying abundant. I could not then obtain the sight of a specimen, or ascertain from any of my collecting friends at Norwich that they possessed this shell, and I thought possibly, from the general character of the Fauna of the Norwich Crag, that fragments of *P. Islandicus* might have been mistaken for it. Since then the late Dr. S. P. Woodward informed me that two specimens of *P. princeps* had been found near Norwich, and one of these (fig. 9 *b*), by the kindness of the Committee of the Norwich Museum, has been transmitted to me for examination. I have also had sent to me by Mr. Valentine Colchester (a son of my old friend Wm. Colchester, on whose land at Sutton I have obtained so many Crag species), a specimen which he found in association with *Voluta Lamberti* in the Fluvio-marine deposit at Yarn Hill, and this also I have had represented, as it is the opposite valve to the one found at Bramerton, while another specimen from the same place has been obtained by Mr. E. Cavell. These different specimens present very considerable variation from the typical Coralline Crag shell in the exterior ornament, so much so that I thought at first sight they must be distinct; my specimens of *princeps*, however, from the Cor. Crag differ in the number and size of the rays from the one figured in ' Min. Conch.,' tab. 545, which I presume to be correct, and these both differ essentially from the specimens now figured, more especially so from the one from Yarn Hill, which possesses nearly two hundred imbricated rays, while those of mine from the Cor. Crag have not half that number, even assuming an intermediate ray to be elevated into a primary one.

In the list in the Cor. Crag paper of Mr. Prestwich ('Quart. Journ. Geol. Soc.,' vol. xxvii, p. 140) Mr. Jeffreys gives this species as identical with *P. Islandicus*, a Clyde fossil ('Crag Moll.,' vol. ii, p, 40, Tab. V, fig. 1); but from that I dissent, as the difference between them is not inconsiderable, both in the sculpture and the form of the shell, as also in that of the ears.

Pecten princeps is not included by M. Nyst in his 'Descript. foss. de Belgique,' 1843; but in his later 'Listes des Fossiles des divers Étages' the species is given from both the Sables gris and from the Sables jaunâtres.

If the living *Islandicus* be, as is not improbable, the modified descendant, through the variety *pseudo-princeps*, of the older Pliocene form *princeps* of the Coralline and Belgian Crags, this again must have descended from some yet older form, and in this way the identification of species must go on indefinitely, unless such a line as seems necessary to me in this case be drawn.

PECTEN GERARDII, *Nyst.* Crag Moll., vol. ii, p. 24, Tab. V, fig. 5.

I know this shell only from the Cor. Crag. Mr. Jeffreys, in 'Brit. Conch.,' vol. ii, p. 68, speaking of *Pecten Testæ*, says that it resembles *P. Gerardii*, but in his list of Crag shells in the 'Quart. Journ.,' vol. xxvii, p. 140, *P. Gerardii* is referred to *P. Grœnlandicus*, Chemn. My own opinion is that the Crag shell is distinct from either of those existing species. As before stated ('Crag Moll.,' vol. ii, p. 25,) it more resembles an American fossil, but I believe it to be distinct from all, although the concluding remark which I have made in the case of *P. princeps* applies, *mutatis mutandis*, to the case of *P. Gerardii* and *P. Grœnlandicus*.

PECTEN VARIUS, *Linn.* Crag Moll., vol. ii, p. 41; Supplement, Tab. VIII, fig. 7.

Localities. Cor. Crag, Sutton? Middle Glacial, Hopton. Post Glacial, Nar Brick-earth (*Rose*).

The figure represents a small specimen found in the Cor. Crag, which I have provisionally referred as above, but it is a doubtful identification. Fragments of *varius* are not uncommon in the Middle Glacial sands of Hopton, but I do not know it from the Red or Fluvio-marine Crag, the Chillesford bed, or the Lower Glacial sands.

PECTEN NIVEUS? *Macgillivray.* Supplement, Tab. VIII, fig. 8.

PECTEN NIVEUS, *Macgill.* Edin. Phil. Journ., vol. xiii (1825), p. 166, pl. iii, fig. 1.

Locality. Cor. Crag, Sutton.

A single specimen, as above represented, is all that I have obtained or seen from the Upper Tertiaries of East Anglia which could be referred to this species. The artist has not made the figure sufficiently elliptical, and it possesses thirty-eight fine, delicate, closely set ribs. Since the engraving was made, however, I have obtained some intermediate forms, which induce me to doubt whether the present specimen, as well as that figured from the Coralline Crag as *varius*, may not be merely extreme modifications of *P. opercularis*.

PECTEN PUSIO, *Pennant*. Crag Moll., vol. ii, p. 33, Tab. VI, fig. 4.

Localities. Cor. Crag passim. Red Crag, Walton and Sutton. Middle Glacial, Hopton. Upper Glacial, Bridlington?

Some fragments of this species have occurred in the Middle Glacial sand, but they are rare. Its occurrence in the Fluvio-marine Crag seems doubtful, and I have not met with it from the Red Crag of Butley. There is a specimen in the British Museum presented by Dr. Murray with the locality of Bridlington attached to it, but it seems so unlike in colour and condition to the Bridlington fossils that its occurrence at that locality seems doubtful. Although very abundant in the Cor. Crag, I have never seen a specimen that showed marks of attachment by the exterior of the valve.

PECTEN OPERCULARIS, *Linn.*[1] Crag Moll., vol. ii, p. 35, Tab. VI, fig. 2; and Supplement, Tab. VIII, fig. 6.

Localities. Cor. Crag passim. Red Crag passim. Fluvio-marine Crag, Bramerton, Whitlingham, Thorpe, and Bulchamp. Chillesford bed, Aldeby and Chillesford. Middle Glacial, Billockby and Hopton.

The figure in Supplement represents a small shell from the Cor. Crag, which I think belongs to this species; it is about half an inch in diameter. It appears to be a distortion, and resembles the figure of *Ostrea arcuata*, Broc. ('Conch. foss. Subapen.,' tab. xiv, fig. 2), which is probably a distortion. This species is common in some of the localities of the Fluvio-marine Crag, and rare at others. It occurs, though not commonly, in some of the localities of the Chillesford bed, but I have not met with it from either Horstead, Burgh, or Coltishall, or with any trace of it in the Lower Glacial sands. In a fragmentary state it is very abundant in the Middle Glacial sands at Billockby, Clippesby, and Hopton.

[1] In the 'Ency. Method.,' pl. 314, fig. 1 *b*, is the figure of a shell with its animal inhabitant. This figure which seems very fanciful, was probably intended for *P. opercularis*, but the animal is represented as having a large protruding foot on one side with two projecting siphons on the other.

PECTEN SEPTEMRADIATÙS, *Chemn.*   Crag Moll., vol. ii, p. 30, Tab. 4, fig. 2 (as *Pecten Danicus*).

Locality.   Red Crag, Sutton and Foxhall.

This species was figured in the 'Crag Mollusca' from a Clyde bed specimen, and I mentioned (p. 31, note) that a worn specimen of my own from the Red Crag might belong to this species.   Mr. A. Bell gives it from the Red Crag nodule pits at Foxhall ('Ann. and Mag. Nat. Hist.,' May, 1871).   I therefore retain it as a fossil of the East Anglian Upper Tertiaries, but with doubt.   I have adopted the above name in obedience to the rule of priority of nomenclature.

PECTEN WESTENDORPIANUS, *Nyst.*   Crag Moll., vol. ii, p. 323, Tab. XXXI, fig. 25 (as *P. maximus*, var. *larvatus*), Supplement, Tab. VIII, fig. 1.

Localities.   Red Crag, Sutton and Waldringfield.

In the 'Crag Moll.' at the above reference this was thought by me to be possibly a variety of the common recent British shell, *P. maximus*, I having then only the flat valve, while M. Nyst's work, in which *Westendorpianus* is figured, showed only its convex valve ('Foss. Belg.,' pl. xviii, fig. 10).   This did not justify me in referring my specimen to Nyst's species, and I was unwilling, as mentioned at the time, to call it a new species.   Since this, however, M. Nyst, in the 'Bull. de l'Acad. Roy. des Sc. de Belgique,' has referred to my figure in the 'Crag. Moll.,' and observed that it represents the upper valve of his species *Westendorpianus*.   I am, however, now able to introduce a figure of the lower or convex valve from a specimen which Mr. Canham has lately obtained from the nodule pits in the Red Crag at Waldringfield, which agrees with Nyst's figure.   The rays are broader than those either of *maximus* or *Jacobæus*, and the depressions are deep and narrow, corresponding to the elevations of the flat valve (Tab. XXXI, fig. 25), which interlock when they are closed.   Mr. Canham's specimen of the lower valve is the only one known to me, and the one I figured as the flat valve is, I believe, also unique.

Both of the specimens are probably derivative from some older Pliocene bed, though as yet no trace of the species has occurred in the Coralline Crag.

LIMA SUBAURICULATA, *Mont.*   Crag Moll., vol. ii, p. 47, Tab. VII, figs. 3 *a* and *b*.

  LIMA SUBAURICULATA, *Forbes* and *Hanley*.   Brit. Moll., vol. ii, p. 263, tab. 53, **figs.**
           4—5.
  — ELLIPTICA, *Jeff.*   Brit. Con., vol. ii, p. 81 ; and vol. v, pl. 25, fig. 2.

Localities.   Cor. Crag, Sutton.   Recent, Mediterranean, British, and Scandinavian Seas.

There is much confusion in reference to this species and the following (*elongata*), owing to the uncertainty as to which form Montague's figures were intended to represent and his description to attach, the two figures of Montague (one of the interior and the other of the exterior) being inconsistent with each other.   Both were united by me in the 'Crag Mollusca' under the name *subauriculata*, the following species (*elongata*) being shown only as var. *elongata* (fig. 3 *c*).   Mr. Jeffreys has, in 'Brit. Conch.,' given a new specific name (*elliptica*) to one of the two forms, which, he says, is that one from which Forbes and Hanley appear to have taken their description, and he has assigned Montague's name of *subauriculata* to the following species (*elongata*) ; and he adds that the species to which he thus assigns the name *elliptica* has not, in his belief, been found south of the Hebrides.   Mr. Hanley, however, assures me that the specimen from which the figure and description in the 'Brit. Moll.' were taken came from the British Channel.   As the other form has long been described under other names, it seems to be adding to complexity to introduce the name *elliptica* for either ; and I have therefore retained Montague's name for the least elongated of the two forms which occur in the Crag, and in doing this I adhere to Forbes and Hanley's view of the matter.

LIMA ELONGATA, *Forbes.*   Crag Moll., vol. ii, p. 47, Tab. VII, fig. 3 *c* (as *L. sub-auriculata*, var. *elongata*).

  LIMA ELONGATA, *Forbes.*   Rep. Ægean Invertebrata, Brit. Ass., 1843.
  — SULCULUS, *Leach.*   M.S. adopt. Loven. Ind. Moll. Scan., p. 32, 1846.
  — SUBAURICULATA, *Forbes* and *Hanley*.   Brit. Moll., vol. ii, p. 263 (**not in**
           figure).
  — — *Jeffreys.*   Brit. Conch., vol. ii, p. 82 ; vol. v, pl. 25, fig. 3.

Locality.   Cor. Crag, Sutton.   Recent, Arctic and British Seas, Mediterranean, Ægean, and Canaries.

This elongated form, figured by me as a variety only in 'Crag Moll.,' must, I presume, now be regarded as a distinct species, the only question being that of name.   As it is clear

15

what was meant by Forbes in his ' Ægean Report' and by myself in the ' Crag Mollusca' (p. 48), there seems no reason why the name of Forbes, first published for the elongated form, should not be adopted as a specific one for it.

LIMA OVATA, *S. Wood*. Crag Moll., vol. ii, p. 48, Tab. VII, fig. 5.

This shell somewhat resembles *L. crassa*, of Forbes, of which two specimens from the Mediterranean were given by Forbes to me, but it is much more elongated and has less imbricated costæ than *crassa*. It appears to me also different from *L. Sarsii*. *Ostrea nivea*, Ren. (figured by Broc., tab. 14, fig. 14), was given by me as a synonym for *L. subauriculata*. Mr. Jeffreys, in his Cor. Crag list to Mr. Prestwich's paper, identifies it with *L. ovata*, but *O. nivea* is more than double the linear dimensions of *ovata*, and is, moreover, described by Philippi as " *lateribus compressa*," which is not the case with *ovata*. Philippi, moreover, adds to the description of *L. nivea* " *medio longitudinaliter sulcata*," which description, as well as the size, assimilates it to *subauriculata*. The Belgian Crag shell called *nivea* by Nyst is also described as " *lateribus compressis*."

LIMA EXILIS, *S. Wood*. Crag Moll., vol. ii, p. 43, Tab. VII, fig. 6.

Localities. Cor. Crag, Ramsholt, Sudbourn, and Sutton. Red Crag, Walton and Butley.

This shell is by Mr. Jeffreys in ' Quart. Jour.,' vol. xxvii, p. 139, referred to *L. inflata*, Lam., as also by Mr. Bell in ' Ann. and Mag. Nat. Hist.,' May, 1871. The authors of the ' Brit. Moll.' (vol. ii, p. 268) place the Crag shell *exilis* as a synonym to *L. hians*. I have re-examined my specimens, and find a difference between the Cor. Crag shell and the recent *inflata*, the former having a narrower hinge-line, and being less inflated than the latter. At the same time, however, it must be admitted that the specimens of *exilis* from the Red Crag approach rather nearer to the living *inflata*. In ' Crag Moll.,' vol. ii, p. 43, it was said of the shell that it somewhat resembled *L. inflata*, but is flatter and undeserving that name. Specimens of *inflata* from the Italian Tertiaries are much larger and more tumid.

LIMA HIANS, *Gmelin*. Crag Moll., vol. ii, p. 44, Tab. VII, fig. 2.

Localities. Cor. Crag, Ramsholt. Middle Glacial, Hopton?

Some hinges and other fragments of a species of *Lima* have occurred in the Middle Glacial sands. The coarseness of the striation on one of the fragments suggests their belonging to this species.

LIMA SQUAMOSA, *Lamarck.* Supplement Tab. X, fig. 1 *a, b.*

> OSTREA LIMA, *Linn.* Syst. Nat., edit. 12, p. 1147 (pars).
> — — *Poli.* Test. Utrius. 9, Sec., pl. xxviii, figs. 22—24.
> LIMA SQUAMOSA, *Lam.* Hist. des An. sans Vert., t. vi, p. 156.
> — — *G. Sowerby.* Genera *Lima*, fig. 2 ; Ency. Method., pl. 296, fig. 4.
> — — *A. Bell.* Ann. and Mag. Nat. Hist., 1871.

Locality. Cor. Crag, near Orford.

The two specimens figured were obtained from a dealer in Orford, one of which has been kindly lent to me for figuring by Dr. Reed, of York, and the other by Mr. Cavell, of Saxmundham. These differ slightly from the recent species in having only eighteen to twenty rays, but in other respects there is sufficient resemblance to justify their reference to it.

In 'Crag Moll.,' vol. ii, p. 46, Tab. VII, fig. 4, and in the ' Ann. and Mag. Nat. Hist.' for 1839, p. 335, a small shell from the Cor. Crag of Sutton was figured and described by me under the name *L. plicatula,* and this (of which I have some other examples) may possibly be the young of *squamosa.* My little specimens, however, appear to differ from *squamosa* in being much less elongated and more ornamented between the rays, so that if they be young of *squamosa* the shell must alter materially in its growth. All these specimens are very rare, and until some of intermediate growth can be obtained the union of the two names under one cannot be adopted.

AVICULA PHALÆNOIDES, *S. Wood.* Crag Moll., vol. ii, p. 51 ; as *A. Tarentina* ? Supplement, Tab. VIII, fig. 12. Addendum Plate, fig. 23.

Locality. Coralline Crag, Gedgrave.

In the ' Crag Mollusca' I introduced the name of *Avicula Tarentina,* from some fragments which I have had figured as above. A comparison of these with the recent *Tarentina* shows the Crag shell to have been possessed of a much thicker and broader and more solid hinge, resembling that of the Bordeaux fossil, *A. phalænacea,* Bast. As, however, my fragments, though apparently belonging to a larger shell than *Tarentina,* are those of shells possessing scarcely half the thickness or hinge dimensions of *phalænacea,* I have thought it best to give the Crag species provisionally the name of *phalænoides.*

PINNA PECTINATA? *Linn.*   Crag Moll., vol. ii, p. 50, Tab. VIII, fig. 11.

Localities.   Cor. Crag, Ramsholt and Sutton.   Lower Glacial, Cromer Cliff?

The specimen figured as above referred to was found by myself in the Cor. Crag at Ramsholt, and, so far as its imperfect condition will permit of comparison, I still think it may retain the name previously given to it, but with the same doubt.   Abundant fragments of a fibrous shell, that I think can only belong to some species of *Pinna*, were sent to me by Mr. Gunn from the Contorted Drift of the Cromer Cliff.

PINNA RUDIS? *Linn.*   Supplement, Tab. IX, fig. 11.

Locality.   Cor. Crag, Aldbro.

This specimen was obtained from the Polyzoan or Coralline Bank at Aldbro by the late Rev. T. Image, and placed in the Museum at Bury St. Edmunds.   I am indebted to the kindness of Mr. H. Prigg, jun., of that town, and to the Museum Committee, for the use of it for the present figure.   The specimen is embedded in the hardened matrix, and a portion of the shell at the smaller end has been removed.

Philippi says " Genus Pinnarum valde intricatum."   That remark, perhaps, cannot be restricted to the genus *Pinna ;* still, with the very imperfect materials afforded by the Cor. Crag, I think it desirable to give these specific names provisionally.

Fragments of *Pinna* are not uncommon both in the Cor. and Red Crags, but they are wholly indeterminable.

MYTILUS EDULIS, *Linn.*, var. GIGANTEUS?   Supplement, Tab. VIII, fig. 4.

Locality.   Red Crag, Sutton.

The above represents a specimen of *Mytilus* in a very mutilated condition, but it indicates a magnitude for the shell when perfect of its having been six inches in length, and as this far exceeds the size of *M. edulis* of our own shores, I have thought it deserving of being figured.   It came from the nodule pits in the Red Crag at Sutton, and may possibly be another species derived from some older formation.   *M. edulis*, either whole or in fragments, is common to all the East Anglian beds except the Coralline Crag, in which only *M. hesperianus* occurs.

The variety *ungulatus* I have not seen from the Coralline Crag, but it is given from it by Mr. Bell in 'Proceedings of the Geologists' Association,' April, 1872.

Modiola discors, *Linn.* Crag Moll., vol. ii, p. 63, Tab. VIII, fig. 5.

Locality. Chillesford Bed, Chillesford.

In vol. ii of 'Crag Moll.' I assigned with doubt to this species a specimen obtained from the Chillesford Bed, but I have not seen it from any other East Anglian formation. In the list to Mr. Prestwich's 'Cor. Crag' paper this species is entered (*Modiolaria discors*) as a shell of that Crag from Sutton, but I have not seen the specimen upon which that entry is based. I have re-examined my own specimens of *Modiola* in the British Museum and do not recognise the species among them.

Modiola Petagnæ, *Scacchi.* Crag Moll., vol. ii, p. 60, Tab. VIII, fig. 6 *b* (as *M. costulata*, var. *Petagnæ*).

Localities. Cor. Crag, Sutton; and Red Crag, Walton Naze.

In the 'Crag Moll.' I figured this shell as a variety only of *costulata*, following in this respect Philippi's earlier rather than his later view, and not desiring to swell the number of species where it could be avoided. The Mediterranean authors, however, keep the two forms as distinct species, and therefore I, not from my own judgment, but in deference to these authors, have here assigned *Petagnæ* as a separate species.

Modiola phaseolina, *Phil.* Crag Moll., vol. ii, p. 59, Tab. VIII, fig. 4.

Localities. Cor. Crag, Sutton, Ramsholt, and near Orford.

This is another shell which I figured at the above reference, and I have since then found several additional specimens; but whether they be specifically distinct from the young of *Modiola modiolus* is very doubtful (see 'Crag Moll.,' p. 59). There is not, I fear, any sufficient difference in their forms, or any other character by which *phaseolina* can be specifically distinguished from *modiolus*. Both forms are present in the Crag, and I have permitted their names to remain separate.

Nucula Cobboldiæ, *J. Sow.* Crag Moll., vol. ii, p. 82, Tab. X, fig. 9; and Supplement, Tab. X, fig. 2.

Localities. Red Crag, Sutton, Bawdsey, Felixstowe, Waldringfield, and Butley. Fluvio-marine Crag, Bramerton, Thorpe, Bulchamp, Thorpe by Aldbro, and Yarn Hill. Chillesford bed, Chillesford, Aldeby, Easton Cliff, and Bramerton. Lower Glacial,

Belaugh, and Weybourn.   Middle Glacial, Billockby, Clippesby, and Hopton.   Upper Glacial, Dimlington, and Bridlington.

This shell has recently been identified with *N. Lyalli* from the north-west coast of America, and by others with *N. insignis* from Japan, while there is a third shell that has equal pretensions to identity, viz. *N. mirabilis* (Adams and Reeve) ; all these having the exterior ornamentation like that upon the upper part of *N. Cobboldiæ*, but none of these living shells, so far as they are known, approach in size any of the fossil specimens, those of *A. Lyalli* that I have seen being not more than half in linear dimension ; and, as pointed out by me originally ('Crag Moll.,' vol. ii, p. 83), two Cretaceous species are similarly ornamented.   All the specimens of these Pacific shells, which I have yet seen, are destitute of that broad exterior belt extending from a fourth to a third of the shell's diameter, which is free from the oblique or zig-zag markings, and forms the margin of the full-grown specimens of *N. Cobboldiæ* from the Red Crag at Butley, and from the successive Glacial beds of East Anglia.   In order to show this belt I have had a specimen figured from Belaugh.

If the recent shells which I have seen and which are all destitute of this plain belt should prove to be only young individuals, which, when full grown, would acquire it, *N. Cobboldiæ* would, nevertheless, not agree with them, because the tumidity and more elevated umbo of all these Pacific shells is such, that if they grew to the size necessary to add on the belt they would assume a very different form than that of *Cobboldiæ ;* and these recent shells are also more angular than our fossil.   I, therefore, still retain this well-known fossil as a distinct species.

No trace of this shell has yet occurred in that part of the Red Crag which exists at Walton Naze and which I regard as the oldest part of that formation, and as possessing Mediterranean affinities, but it gets more common in the newer portions of the Red Crag with northern affinities.   In the Butley Red Crag *Cobboldiæ* is common, as it is also at some localities of the Fluvio-marine Crag and of the Chillesford bed.   In the Lower Glacial sands it is rare where these sands are Fluvio-marine, as at Belaugh, but commoner where they are more marine, as at Weybourne.   It is very abundant, in a fragmentary condition, in the Middle Glacial sands, and it appears to be characteristically the shell of the Pre-glacial and earlier Glacial periods, and to have disappeared from British seas towards the later part of the Glacial period.[1]

The shell called *Nucula (Acila) Lyalli*, by Mr. Bell ('Ann. and Mag. Nat. Hist.,' 1871), was placed in my hands, and I believe it to be only the young state of *Cobboldiæ*.

---

[1] In the foot-note to p. 26 of the Introduction to this Supplement, mention is made of a thin band of sand intercalated in the upper glacial clay at Dimlington, which contained mollusca with valves united. Since that note was published some specimens from this bed were kindly forwarded by Sir Charles Lyell, and among these were several of *Nucula Cobboldiæ*.   Mr. Leonard Lyell's description, which Sir Charles forwarded with the specimen, speaks of this sandband having been literally packed with perfect specimens of *Nucula Cobboldiæ*.

It was a small flat specimen, with neither the angularity of outline nor the elevated umbo of the recent Pacific shells.

NUCULA LÆVIGATA, *J. Sow.* Crag Moll., vol. ii, p. 81, Tab. X, fig. 8, and var. *calva.*
Supplement, Tab. VIII, fig. 5 *a, b.*

Localities. Red Crag passim.
Var. *calva.* Cor. Crag, Sutton and Orford.
The Coralline Crag variety differs from the typical form found in the Red Crag of Walton Naze in being smaller and more decidedly truncated, without the projecting part of the margin at the shorter or siphonal side of the valve. I propose to distinguish it as a well-marked variety under the above name, *calva.*

This variety much resembles an Oligocene shell represented by Dr. Speyer ('Die Ober Oligocänen Tertiar gebilde,' p. 42, Tab. V, figs. 3—5), called *N. peregrina,* Desh., which, judging from the figure, seems to be of a form somewhat intermediate between the Red and Cor. Crag shells, but I have not been able to procure a specimen of the German fossil for comparison.

NUCULA PROXIMA, *Say.* Crag Moll., vol. ii, p. 87, Tab. X, fig. 7 (as *N. trigonula*).

NUCULA PROXIMA, *Say.* Journ. Acad. Nat. Sc., 2, p. 270, 1822.

Localities. Cor. Crag, Sutton and near Orford.
At the above reference I observed that *N. proxima* of the American authors was probably the same as the Crag shell *trigonula,* and as I have seen no reason to alter that opinion it is only right, as the name *proxima* has priority to mine, that the shell should be here placed under that designation.

NUCULA NITIDA, *G. Sow.* Supplement, Tab. X, fig. 12.

NUCULA NITIDA, *G. Sow.* Conch. Illustr., *Nucula,* No. 29, fig. 20.
— — *Hanley.* Thesaurus, p. 46, pl. 229, fig. 120.

Localities. Cor. Crag, Sutton and near Orford.
As the British conchologists seem to consider this shell to be a distinct species I have here followed their example, although at page 87 of vol. ii of the 'Crag Mollusca' I considered it to be only a variety of *trigonula,* and as my specimens correspond with the recent form I have had one figured. It is less tumid than *trigonula* and has a smooth

exterior. Our present shell seems intermediate between *nucleus*, which is more extended posteriorly, and *proxima*.

NUCULA TENUIS, *Mont.* Crag Moll., vol. ii, p. 84, Tab. X, fig. 5.

Localities. Red Crag, Bawdsey. Fluvio-marine Crag, Bramerton (*Reeve*). Chillesford bed, Chillesford, Bramerton (*Reeve*), and Aldeby (*Crowfoot & Dowson*). Middle Glacial, Hopton? Upper Glacial, Bridlington.

I have seen the specimens from all the above localities. Those from Hopton are fragmentary and may belong to *nucleus*, but more probably to the present species.

LEDA OBLONGOIDES, *S. Wood.* Crag Moll., vol. ii, p.90,Tab. X, fig. 17 (as *Leda myalis*, Couth.).

NUCULA OBLONGOIDES, *S. Wood.* Mag. Nat. Hist., New. Ser., vol. iv, p. 297, tab. 14, fig. 4, 1840.

Localities. Red Crag, Sutton and Butley, Fluvio-marine Crag, Bramerton, Thorpe, Bulchamp, Thorpe by Aldbro, and Yarn Hill. Chillesford bed, Chillesford, Aldeby, Easton Cliff, Bramerton, Horstead, Coltishall, and Burgh. Lower Glacial, Belaugh, and Rackheath. Middle Glacial, Hopton and Billockby.

The shell figured No. 17 in Tab. X of vol. ii of 'Crag Moll.' as *L. myalis*, Couthouy, can, I think, hardly be referred to that species. It is the same as that figured No. 4 of tab. 14 of vol. iv of 'Mag. Nat. Hist.,' New Series, to which I gave the name of *oblongoides*. This shell is common in the newer part of the Red Crag (that of Butley) and in the Fluvio-marine Crag at Thorpe by Aldbro, Bulchamp, and Bramerton, and also in the Chillesford bed at all its localities, and in all it maintains well its form and characters. I for some time thought it to be identical with the North East American shell, *limatula*, Say, but it is less attenuated, and has not the preponderance in length of the hinder (or posterior) part of the shell over the anterior which characterises that species, and the ligamental socket is larger. It also approaches, in some respects, all of the following, viz. *hyperborea*, Torel; *amygdalea*, Valenciennes; *sapotilla*, Gould; *myalis*, Couth.; and *Woodwardi*, Hanley, which, perhaps, are all varieties of one species; but I cannot satisfactorily identify it with any of them, though it seems to come nearest to *hyperborea*. I have, therefore, fallen back upon my original name *oblongoides*. The figure in the 'Ann. and Mag. of Nat. Hist.' is a most accurate representation of it. The shell occurs in the Lower Glacial sands of Belaugh and Rackheath, and fragments (sometimes large) of what seems to be the same species are very common in the Middle

Glacial of Hopton Cliff.[1] The *Leda hyperborea* of Mr. A. Bell's list in 'Ann. and Mag.,' Sept., 1870, from Butley, is, I think, the above *oblongoides*.

LEDA SEMISTRIATA, *S. Wood.* Crag Moll., vol. ii, p. 91, Tab. X, fig. 10.

In his list in White's 'Directory,' Dr. Woodward gave, on the authority of a single valve in the Middleton collection, this species from the Norwich Crag. I suspect that this was a spurious specimen. I know this species from no newer bed than the Coralline Crag.

LEDA LANCEOLATA, *J. Sow.* Crag Moll., vol. ii, p. 88, Tab. X, fig. 16.

Localities. Red Crag, Bawdsey. Fluvio-marine Crag, Bramerton. Chillesford bed, Chillesford. Middle Glacial, Hopton.

A considerable fragment of this shell, exhibiting the wavy line ornamentation of the exterior, has occurred in the Middle Glacial sand of Hopton, and a similar fragment in the Fluvio-marine Crag of Bramerton.

LEDA MYALIS? *Couthouy.* Supplement, Tab. IX, fig. 2 *a, b.*

NUCULA MYALIS, *Couth.* Bost. Journ. Nat. Hist., vol. ii, p. 61, pl. 3, fig. 7, 1838.

Locality. Fluvio-marine Crag, Postwick. Lower Glacial, Runton.

The specimen figured is one in the British Museum, and agrees with the recent *myalis* of Couthouy in all respects, but I question whether after all it is anything more than an extreme form of the preceding species *oblongoides*. The specimen figured is, with another valve in the British Museum, marked "Postwick," and a third there is marked "Runton" which can only be from the Lower Glacial sands, which are fossilliferous at that place.

Fig. 13 *a, b,* of 'Crag Moll.,' Tab. X, may retain the name of *L. caudata,* Donovan, and fig. 12 *a, b,* of the same plate I will refer to *minuta,* Mont. This latter is from the Red Crag, and the *caudata* is the Bridlington form.

---

[1] It is the shell referred to under the name *limatula* in the list in 'Quart. Journ. Geol. Soc.,' vol. xxvi, p. 94.

16

ARCA TETRAGONA, *Poli.*   Crag Moll., vol. ii, p. 76, Tab. X, fig. 1.

Localities.   Cor. Crag, Sutton, Ramsholt, and Sudbourn.   Red Crag, Sutton.

The specimen represented in figs. 1 *a* and *b* ('Crag Moll.') is from the Red Crag at Sutton, and it was there considered as belonging to the above-named species; figs. 1 *c* and 1 *d* are from the Coralline Crag.   In the 'Brit. Conch.,' vol. ii, p. 181, *A. tetragona* is stated to be in the Red and Cor. Crag, but in the list accompanying Mr. Prestwich's paper this species (*tetragona*) is given from the Cor. Crag only, while my shell from the Red Crag is referred by this author to *A. imbricata*, Poli.   In consequence of this statement I have again examined and compared my specimens with the Mediterranean shell, but must still retain my previously formed opinion.

PECTUNCULUS GLYCIMERIS, *Linn.*   Crag Moll., vol. ii, p. 66, Tab. IX, fig. 1.

Localities.   Cor. Crag passim.   Red Crag passim.   Fluvio-marine Crag, Bramerton. Chillesford bed, Aldeby.   Middle Glacial, Billockby and Hopton.   Upper Glacial, Bridlington.

The genus *Pectunculus* is one in which the species are of very difficult determination, as may be seen by the number of synonyms given by various authors, and the little accordance there is as to specific separation.   In the Cor. Crag the prevailing form is the thick variety, corresponding in this respect with the Mediterranean shell or rather with the Sicilian fossils, but I have also found in the Cor. Crag the thinner and less tumid shell, like that of the recent British variety.   In the 'Crag Moll.' I gave *P. glycimeris* and *P. pilosus* as varieties of one species, following the authors of the 'Brit. Moll.,' and I see no sufficient reason to alter that opinion.   All I can say is that, if there be two species, as appear to be made out of these two varieties by some authors, they are both present in the Coralline Crag as well as in the Red.   Mr. Canham has obtained from the Red Crag at Waldringfield a specimen which measured $3\frac{3}{4}$ of an inch in one direction and $3\frac{1}{2}$ in the other, and this of great solidity, and with these are specimens thin and oblique, while some have prominent umbones; in others this part is much depressed, with every intermediate shade of difference.   It is difficult to say whether the thick shells in the Red Crag are, like the thinner ones, natives of the Red Crag sea, or whether they be derived from the Cor. Crag, as I think there can be no doubt that a large proportion of the *Pectunculi*, which make up the mass of the thin bed at the base of the Red Crag, from which the phosphatic nodules are extracted, have been thus derived.   *P. stellatus* and *P. insubricus* are, I believe, mere varieties of this species.   *P. glycimeris* (I presume the thin variety) was for many years almost unique in the Fluvio-marine Crag, but it has been found somewhat plentifully in a newly worked pit at Bramerton. A single worn specimen of small size, found by Messrs. Crowfoot and Dowson, is the only instance of its occurrence in the Chillesford Bed. It has not occurred in the Lower Glacial sands, but it abounds in the Middle Glacial, both at Billockby and Hopton, where specimens of all sizes occur, from the smallest fry up to the ordinary size of the British shell. I have not seen even a fragment from that formation of those larger or more solid forms occurring in the Cor. and Red Crags.

Fig. 1 *d* of Tab. IX of ' Crag Moll.' corresponds with what Brocchi has called a species under the name of *A. inflata*, and Mr. A. Bell gives *P. insubricus*, Broc. (*violascens,* Lam.), also as in Cor. Crag. Some specimens which I have seen that somewhat resemble these so-called species are, I believe, only varieties. Mr. Jeffreys, in his list of the Crag shells, gives *glycimeris* only from the Cor. Crag, but introduces *P. pilosus* as a new Red Crag shell. If these be really two distinct species I would rather reverse this determination, and while giving *pilosus* to the Cor. Crag, regard that form as only derivative in the Red.

LIMOPSIS PYGMÆA, *Phil.* Crag Moll., vol. ii, p. 71, Tab. IX, fig. 3.

. Localities. Cor. Crag, Sutton, and near Orford. Red Crag, Walton, Waldringfield, and Felixstow. Middle Glacial, Billockby.

This shell, according to M. Meyer (' Cat. Syst. Ter.,' p. 120), is *Trigonocælia anomala*, Eichwald, 1830, not *pygmæus*, Munster, and, according to Mr. Jeffreys, it has recently been dredged living near Corsica, and in abysmal depths in the Atlantic. Mr. Bell gives the species as occurring at Walton (' Ann. and Mag. Nat. Hist.,' 1870), also from Waldringfield and Felixstow. A single perfect valve has occurred in the Middle Glacial sand of Billockby. It is a most abundant shell in the Cor. Crag of Sutton, where the two valves are frequently united.

LIMOPSIS AURITA, *Brocchi.* Crag Moll., vol. ii, p. 70, Tab. IX, fig. 2.

This shell is very abundant in the Cor. Crag, near Orford, but I have never found it in the Red Crag, with one doubtful exception. Mr. Jeffreys gives it in his list from the Red Crag at Waldringfield (I presume on Mr. Bell's authority). This is given by Mr Jeffreys as a shell living in the Shetland seas, and though no uncommon fossil in the Italian beds, it has not, I believe, been yet found living in the Mediterranean.

UNIO PICTORUM, *Linné.* Supplement, Tab. VIII, fig. 3.

Locality. Post Glacial, Grays.

In the 'Crag. Moll.,' vol. ii, p. 99, Tab. XI, fig. 13, is the representation of a shell from Stutton, which is there called *M. tumidus*, but the specimen was not in very perfect condition, and its correct reference is by no means certain. The specific line of separation between *tumidus* and *pictorum* is not easily determinable even in the recent state, and less so with fossils. The specimen here figured was found by Mr. Pickering at Grays, and to which he has given the above name.

UNIO TUMIDUS, *Philippsson.* Supplement, Tab. VIII, figs. 2 *a, b.*

Locality. Post Glacial, Grays.

This is one of a group which is exceedingly abundant; indeed, it was the only form I could find at that locality after many days' search. It is a peculiar variety, being more inequilateral than any recent specimens of that name which I have seen, the pedal side being peculiarly short. The specimens at Grays lie there in myriads, but I have never seen one approaching the dimensions of the recent shell, which is said to have reached four and a half inches in length. The Grays specimens never exceed three inches, and rarely attain to that size. In deference to those authors who have distinguished *tumidus* from *pictorum* I have kept the two forms as distinct species, but in my opinion the two graduate into each other, and are not distinguishable.

In 1864 I found an imperfect specimen of *Unio* in the Fluvio-marine Crag at Bramerton, but unfortunately this was not perfect enough for representation or determination. It appeared to resemble *M. tumidus*, and I have provisionally referred it to that species. This same inequilateral form *tumidus* also occurs at the base of the bed E of Sect. V (Kessingland) of the map sheet accompanying the introduction to this Supplement.

UNIO LITTORALIS, *Lam.* Crag Moll., vol. ii, p. 98, Tab. XI, fig. 12.

I have hunted at Grays for several years without finding this species, and Mr. Pickering tells me he has not been able to find it there, but Sir Charles Lyell has shown me some specimens which he obtained at that locality many years ago, that undoubtedly belong to this species.

ANODONTA CYGNEA, *Linn.* Crag Moll., vol. ii, p. 102, Tab. II, fig. 10.

This is also an abundant shell at Grays, and by no means scarce at Clacton, but from extreme fragility specimens are difficult to obtain. I have also found in the bed at Runton (C of Sect. III of the map sheet) the variety *anatina* or *paludosa*, 'Turt. Brit. Biv.,' p. 240, Tab. 15, fig. 6, in which the dorsal and ventral margins are nearly parallel. My specimen from Runton measures nearly five inches in length.

Unio margaritifer is given by Mr. Prestwich in the 'Quart Journ. Geol. Soc.,' vol. xxvii, p. 467, as a species from near Runton Gap on the Norfolk Coast, but this appears to be an error, as pointed out by Mr. A. Bell in 'Geol. Mag.,' vol. ix, p. 214. I do not know this shell as a British fossil.

CORBICULA FLUMINALIS, *Müller.* Crag. Moll., vol. ii, p. 104, Tab. XI, fig. 15, as *Cyrena consobrina.*

Localities. Red Crag, Waldringfield (*A. Bell*). Fluvio-marine Crag. Thorpe, near Aldbro (*S. Wood*). Dunwich (*Crowfoot*). Bramerton and Postwick (*Woodward*). Bulchamp (*Dowson*). Lower Glacial, Belaugh (*Harmer*). Post Glacial, Bramwell near Cambridge (*Bell*). Kelsea Hill, Gedgrave near Orford, Stutton-on-Stour, Grays, and Clacton (*Fisher*).

This shell (like several European freshwater bivalves) has a great number of synonyms,[1] and it is an important species as concerns the post glacial sequence of deposits. It is somewhat variable, but not more so than other of our freshwater inhabitants. It has been said that freshwater shells vary more than marine, but I have never seen greater variation among them than is exhibited by the varieties and distortions shown by fossil specimens of *Trophon antiquus* and *Littorina littorea.*

The Red Crag appears to be the oldest deposit in which *Corbicula fluminalis* has been met with in this country. The specimens of this species mentioned by me as having been found on the top of the Cor. Crag at Gedgrave ('Crag Moll.,' vol. ii, p. 105) belong to what is probably one of the older Post Glacial deposits, into which some of the Cor. Crag fossils have been washed. It appears to have lived in Britain before the very severe conditions of the Glacial Period had set in, and we find it again an inhabitant of our waters in deposits more recent than those of that epoch, but all the specimens that I have seen from the beds of Crag age, as well as the solitary specimen I have seen from the Lower Glacial sands at Belaugh, are small; the largest of them scarcely more than half

---

[1] This species, including the fossil in Europe and the recent shell from the Nile and China, has had given to it five generic and sixteen specific names. Mr. Gregory ('Geol. Mag.,' vol. vi, p. 81) gives this as living in the Vaal River, South Africa.

the linear dimensions of the shells that occur in the Post Glacial valley deposit of Stutton.[1]

This *Corbicula* lived here in association with a few other Molluscs which have disappeared from the British Isles, viz. *Unio littoralis*, *Hydrobia marginata*, *Helix fruticum*, and *Helix ruderata*, the two former of which have survived not far to the southward of us, while the two latter have gone north, and are now found in Siberia and North America, extending in a northerly direction up to the Arctic Circle.

FRAGILLIDÆ.

There are several small and tender bivalves in the Coralline Crag, figured and described in the ' Crag Moll.,' which I think might be united in one family under the name of *Fragillidæ*, separating them into genera or sections, the formula for each of which is here given.   The specific name annexed may be considered the type of each.

Lepton, Turton (*squamosum*, Mont.).   Shell thin, roundedly ovate or obtusely triangular, equivalved, equilateral, slightly compressed; surface of valves pitted or ornamented; beaks generally acute; margins plain; hinge with two lateral teeth in each valve, and one small cardinal tooth.   Palleal line simple; connexus cartilaginous.

Lasæa, Leach (*rubra*, Mont.).   Shell small, thin, inequilateral, equivalved, oval or oblong, closed, externally smooth or covered with fine lines of growth; beaks depressed; hinge with two prominent lateral teeth and one small cardinal tooth; palleal line simple; connexus cartilaginous.   The animal is said to have a tubular prolongation in front of the mantle.

Bornia, Philippi (*ovalis*, S. Wood).   Shell smooth, thin, ovate, or subtriangular equivalved, slightly compressed, closed; hinge with two diverging teeth in each valve; palleal line simple; connexus cartilaginous.

Scacchia, Philippi (*elliptica*, Scac.).   Shell small, thin, ovate or elliptical smooth, inequilateral, closed; two cardinal teeth in one valve and one in the other; palleal line simple; connexus cartilaginous.

Scintilla, Deshayes (*ambigua*, Nyst).   Shell ovately transverse, thin, equivalved, equilateral, smooth, closed; two teeth in one valve and one prominent obtuse tooth in the other; palleal line simple; connexus cartilaginous.   *Sportella* resembles this, but it has a ligamentous connector. This latter I have not seen from the Crag, but it is not rare in the Older Tertiaries.

Kellia, Turton (*suborbicularis*, Mont.).   Shell thin, suborbicular, or slightly transverse equilateral, smooth, closed; hinge with two teeth in each valve, one before and the other behind the connector; palleal line simple; connexus cartilaginous.

[1] The Stutton Post glacial deposit is in the low ground skirting the estuary of the Stour, opposite Manningtree, and is not to be confounded with the well-known Crag locality of Sutton in the same county

Montacuta, Turton (*bidentata*, Mont.). Shell small, thin, transverse or oblong smooth; equivalved, inequilateral; two cardinal teeth in each valve, the hinder one formed by the backward pressure of the connector. Palleal line simple. Connexus cartilaginous.

Sphenalia, S. Wood (*substriata*, Mont.). Shell small, thin, transverse, oblong, or wedge-shaped; equivalved; very inequilateral, compressed, smooth; hinge edentulous or with one obsolete tooth in each valve formed by the elevated side of the pit for the connector. Palleal line simple. Connexus cartilaginous.

Crag Species.

Lepton squamosum, Mont.   Crag Moll., vol. ii, p. 114, Tab. XI, fig. 8.
—   *nitidum*, Turt.   Crag Moll., vol. ii, p. 116, Tab. XI, fig. 7.
—   *deltoideum*, S. Wood.   Crag Moll., vol. ii, p. 115, Tab. XI, fig. 9.
—   *depressum*, Nyst.   Crag Moll., vol. ii, p. 116, Tab. XI, fig. 6.
Lasæa rubra, Mont. (*Kellia*).   Crag Moll., vol. ii, p. 125, Tab. XI, fig. 10.
—   *pumila*, S. Wood (*Kellia*).   Crag Moll., vol. ii, p. 124, Tab. XII, fig. 15.
—   *Clarkiæ*, Clark.   Supplement, Tab. IX, fig. 10.
—   *intermedia*, S. Wood.[1]   Supplement, Tab. X, fig. 22.
Bornia ovalis, S. Wood.   Supplement, Tab. IX, fig. 3.
Scacchia elliptica, Phil. (*Kellia*).   Crag Moll., vol. ii, p. 121, Tab. XII, fig. 13.
—   *cycladia*, S. Wood (*Kellia*).   Crag Moll., vol. ii, p. 122, Tab. XI, fig. 4.
—   *orbicularis*, S. Wood (*Kellia*).   Crag Moll., vol. ii, p. 120, Tab. XII, fig. 9.
Scintilla ambigua, Nyst (*Kellia*).   Crag Moll., vol. ii, p. 120, Tab. XII, fig. 11.
—   *compressa*, Phil. (*Kellia coarctata*).   Crag Moll., vol. ii, p. 123, Tab. XII, fig. 10.
Kellia suborbicularis, Mont.   Crag Moll., vol. ii, p. 119, Tab.XII, fig. 8.
Montacuta bidentata, Mont.   Crag Moll., vol. ii, p. 126, Tab. XII, fig. 17.
—   *elliptica*, S. Wood.   Supplement, Tab. X, fig. 21.
—   *truncata*, S. Wood.   Crag Moll., vol. ii, p. 127, Tab. XII, fig. 16.
—   *ferruginosa*, Mont.   Crag Moll., vol. ii, p. 129, Tab. XII, fig. 14.
Sphenalia substriata, Mont. (*Montacuta*).   Crag Moll., vol. ii, p. 128, Tab. XII, fig. 12.
—   *donacina*, S. Wood (*Montacuta*).   Crag Moll., vol. ii, p. 131, Tab. XI, fig. 3.
Cryptodon rotundatum, S. Wood.   Crag Moll., vol. ii, p. 135, Tab. XII, fig. 19 (as *Cryptodon ferruginosum.*
—   *sinuosum*, Donovan.   Crag Moll., vol. ii, p. 134, Tab. XII, fig. 20.

[1] This is from Aldeby.

LEPTON NITIDUM, *Turton.* Supplement, Tab. IX, figs. 7 *a, b, c.*

Localities. Cor. Crag, Sutton. Fluvio-marine Crag, Bramerton. Chillesford bed, Aldeby and Beccles.

The specimen figured in 'Crag Moll.,' vol. ii, Tab. XI, fig. 7, under the above name, is, I now believe, the same as *L. depressum,* Nyst (fig. 6 of the same Tab.), which I think is distinct from *nitidum.* I have recently found in the Cor. Crag at Sutton the specimen represented as above, which I have no doubt is the shell called *nitidum* by Turton, and Mr. Canham showed me a similar one from the same locality. A specimen was sent to me by Mr. Reeve from Bramerton, and Messrs. Crowfoot and Dawson have sent one from Aldeby and several from the Waterworks Well at Beccles, which may be referred to the same species. These latter are from the Chillesford bed.

LEPTON DELTOIDEUM, *S. Wood.* Crag Moll., vol. ii, p. 115, Tab. XI, fig. 9.

Localities. As in 'Crag Moll.'

This shell is given by Dr. Hörnes ('Vienna Foss.,' p. 249, and by M. Weinkauff, 'Conch. des Mitt.,' p. 178) as a synonym to *Bornia corbuloides,* but that shell has a smooth exterior, while the Crag shell is covered with a kind of wavy granular ornament. Mr. Jeffreys, in his list to Mr. Prestwich's paper, 'Geol. Journ.,' vol. xxvii, p. 139, referred the Crag shell (*deltoideum*) to *Erycina Geoffroyi,* Payr. He has since, however, written me that he was in error in so doing, and that he now considers the shell to be identical with *corbuloides;* he also sent me two specimens of that species for comparison. These proved to be entirely destitute of any external ornament, the presence of which is so marked a characteristic of the Crag *deltoideum.* Under these circumstances I have retained my Crag shell as a distinct species. The specimens from the Red Crag are slightly worn, but the ornament may be detected by means of a lens; these have a slight depression in the central part of each valve.

LASÆA CLARKIÆ, *Clark.* Supplement, Tab. IX, fig. 10.

> LEPTON CLARKIÆ, *Clark.* Ann. Nat. Hist., 2nd ser., 1852, p. 293, and vol. ix, p. 191.
> — — *Forbes* and *Hanl.* Brit. Moll., vol. vi, p. 255, pl. cxxxii, fig. 7.
> — — *Jeffreys,* Brit. Conch., vol. ii, p. 202, pl. xxxi, fig. 5.

Locality. Cor. Crag, Sutton.
Length, $\frac{1}{16}$ of an inch; *height,* $\frac{1}{20}$ inch.

This shell I have placed in the genus *Lasæa*, as it corresponds in its dentition with *Lasæa rubra*, and it has not the peculiar ornament of *Lepton* upon the exterior, but is covered with concentric striæ or fine lines of growth. Mr. Jeffreys describes the recent shell as being "marked with longitudinal radiating lines" ('Brit. Conch.,' vol. ii, p. 203); these are not visible in my fossil, but my specimens are rare.

LASÆA INTERMEDIA, *S. Wood.* Supplement, Tab. X, fig. 22.

Localities. Chillesford Bed, Aldeby. Middle Glacial, Hopton.

The specimen figured was sent to me by Messrs. Crowfoot and Dowson, who found it with some others at Aldeby, and a single perfect specimen of the same species has been found by my son in the Hopton Sand.

It resembles *Lasæa pumila* of the Cor. Crag in size and outline, but it is much flatter or more compressed, and it differs essentially in its dentition. Our present shell is transversely ovate, very inequilateral, compressed, and with a smooth exterior; the right valve has a rudimentary cardinal tooth and a very elongated lateral tooth on the longer or pedal side, with a small one on the other nearly at right angles; the left valve has corresponding denticles, but not so prominent.

At first I thought the present shell might be referred to *M. Dawsoni*, 'Brit. Conch.,' vol. ii, p. 178, but Mr. Jeffreys, who examined the Aldeby specimens, said " this little bivalve is, I consider, the younger state of your *Kellia pumila. Montacuta Dawsoni* differs from it in being flatter and having no cardinal tooth, the lateral teeth being very much shorter and stronger." The denticles of my shell are much longer than in either *pumila* or *Dawsoni ;* and I have, therefore, for the present, kept them distinct. My shell is by no means thin, and the anterior muscle-mark is large and deeply impressed, from which I think it is full grown.

BORNIA OVALIS, *S. Wood.* Supplement, Tab. IX, figs. 3 *a, b.*

Locality. Cor. Crag, Sutton.

Length, $\frac{5}{16}$ths of an inch.

Two specimens, both unfortunately of the same value, have lately been found by myself, and I am unable satisfactorily to refer them to a known species. My shell has two short but prominent diverging teeth, one on each side of the depression for the cartilaginous connector. There is an indentation in the umbo which much resembles that present in some species of *Cochlodesma*, through which the cartilage protrudes. This opening may have been made in a similar manner, or it may be accidental.

I have, from these imperfect materials, declined giving a diagnosis.

17

In the 'Crag Moll.,' vol. ii, p. 132, I described a small shell under the provisional name of *Cyamium eximium*. I am sorry to say that I have seen nothing since that will assist in its correct determination; and the specimen itself has been subsequently much injured. It may probably be a small or young individual of the above. Under these circumstances I have thought it best to suppress the name *Cyamium eximium* in the general list which accompanies this Supplement.

SCACCHIA CYCLADIA, *S. Wood*.   Crag Moll., vol. ii, p. 122, Tab. XI, fig. 4 (as *Kellia cycladia*).

Locality.   As in 'Crag Moll.'

This shell is still very rare to my researches. I have not met with a specimen since the one above referred to was engraved, but I have re-examined my specimens given to the British Museum, and think them quite distinct from *Scacchia (Kellia) orbicularis*.

Anatina? pusilla, Phil., 'En. Moll. Sic.,' vol. i, p. 9, Tab. 2, fig. 5, may possibly be the same as my Crag shell, but it will be necessary to compare specimens before such identity can be established.

SCACCHIA ORBICULARIS, *S. Wood*.   Crag Moll., vol. ii, p. 120, Tab. XII, fig. 9 (as *Kellia orbicularis*); and Supplement, Tab. IX, fig. 9.

Localities.   As in 'Crag Moll.'

The shell figured in the present Supplement is a very globose variety of this species, without the obliquity generally observable, from which I at first imagined it was a distinct species. Specimens of *orbicularis* have lately been obtained by Mr. Jeffreys in the living state, and he has referred them to *Scacchia cycladia* (*Kellia cycladia*, 'Crag Moll.'). I, however, believe the species to be distinct.

KELLIA SUBORBICULARIS, *Mont*.   Crag Moll., vol. ii, p. 118, Tab. XII, fig. 8.

Localities.   As in 'Crag. Moll.'

If this be not one of the boring bivalves, it is a shell that is often found in a crypt with a true excavator, as I have found a perfect specimen of it in association with the valves of a *Gastrochæna* in a crypt formed in a fragment of an *Ostrea* from the Cor. Crag of Sutton, and it was of that size that it could not escape through the terminal opening; moreover, it was in the crypt in front of the valves of the *Gastrochæna*. This species (*suborbicularis*) has an extensive range in the living state, and I have a specimen from the Coast of California, given to me by Dr. P. Carpenter, in which I cannot detect the slightest difference from the recent shell of our own seas or from my Crag fossil.   Mr.

Charlesworth has shown me a specimen of *suborbicularis* from the Red Crag in a crypt with a *Pholas*.

The new species of *Kellia* described by Forbes in his 'Ægean Report,' 1843, cannot now be found, and his short descriptions, unaccompanied by figures, are insufficient for specific determination. Perhaps his *Kellia transversa* may be the same as my *Scacchia cycladia*, but I cannot alter the name of my shell upon such uncertainty.

SCINTILLA AMBIGUA, *Nyst.* Crag Moll., vol. ii, p. 120, Tab. XII, fig. 11 (as *Kellia ambigua*).

SCINTILLA AMBIGUA? *Desh.* An. sans. Vert., t. i, p. 700, pl. xlix, figs. 13—15.

Localities. Cor. Crag, Sutton, and near Orford. Red Crag, Walton and Sutton. Chillesford Bed, Chillesford and Aldeby.

In Mr. Jeffreys' list of Cor. Crag shells ('Quart. Journ. Geo. Soc.,' vol. xxvii, p. 139) this species is referred to *Erycina pusilla*, Phil., but in the same Journal subsequently, at p. 493, Mr. Jeffreys says that "*Kellia ambigua* is not *Erycina pusilla*, Phil., but *Scintilla Parisiensis* of Conti." He has since written to me that he now refers it to *Kellia Geoffroyi*, Payr, of which he sent me a specimen for comparison. I find that shell, however, to be covered with fine radiating lines, of which I can detect no trace in the numerous specimens of *ambigua* in good preservation which I have examined. I have therefore retained the name *ambigua*, Nyst, under which I originally described this shell, merely changing the generic appellation.

The name of *Scacchia elliptica* is given in the same list by Mr. Jeffreys, at p. 485, as a species new to the Red Crag. On application to Dr. Reed, of York, in whose cabinet is the specimen on which this introduction of the species as a Red Crag shell is founded, he obligingly sent it to me for examination, and I find it to belong to *Scintilla ambigua*, which I had given as a Red Crag species in 'Crag Moll.,' vol. ii, p. 121. *Scacchia elliptica* is abundant in the Cor. Crag at Sutton, as is also *Scintilla ambigua*, but the latter becomes very rare in the newer Formations (where it appears to have died out), and is, moreover, exceedingly variable in its outline as well as in the degree of its tumidity.

MONTACUTA BIDENTATA, *Mont.* Crag Moll., vol. ii, p. 126, Tab. XII, fig. 17.

Localities. Cor. Crag, Sutton, and near Orford. Red Crag, Walton Naze. Chillesford Bed, Aldeby. Middle Glacial, Hopton. Post Glacial, Nar Brickearth, at Pentney.

This species has been obtained by Messrs. Crowfoot and Dowson from Aldeby, by my son (a single valve) from Hopton, and by Mr. Rose from Pentney. The specimen upon which the species was given as from Bridlington in 'Mem. Geol. Survey,' vol. i, p. 409, 1846, was in the Bowerbank collection now in the British Museum, and had, according to information furnished me by the late Dr. Woodward, a memorandum on the back of the tablet, stating that it came from the Nar Brickearth, where the shell has been found by Mr. Rose. I have, therefore, omitted the locality of Bridlington for it. The figure in 'Crag Mollusca' is not quite accurate, the posterior side being a little too rounded for the general form of the Crag specimens.

MONTACUTA FERRUGINOSA, *Mont.* Crag Moll., vol. ii, p. 129, Tab. XII, fig. 14.

Localities. Cor. Crag, Sutton, and near Orford. Chillesford Bed, Aldeby.

Specimens of this species were among the shells from time to time sent me for examination by Messrs. Crowfoot and Dowson from Aldeby.

MONTACUTA ELLIPTICA, *S. Wood.* Supplement Tab. X, fig. 21.

Locality. Coralline Crag, Sutton.

The above figure represents a specimen of *Montacuta*, which appears to be distinct from the shell called *bidentata*. It has the posterior side more rounded and more extended than *bidentata*, and is less inequilateral. It is also less compressed and more elliptical, and has the denticles comparatively longer.

Genus.—SPHENALIA (see ante, p. 121).

SPHENALIA DONACINA, *S. Wood.* Crag Moll., vol. ii, p. 131, Tab. XI, fig. 3 (as *Montacuta? donacina*).

Localities. As in Crag Moll.

A Coralline Crag shell from Sutton was described in the 'Crag Mollusca' under the above provisional name. It, however, does not in the hinge accord with *Montacuta bidentata*, the type of the genus *Montacuta*, and I have accordingly changed its position. I am sorry to say that the Crag fossil is still very rare in my collection. A recent British specimen has been procured by Mr. Jeffreys, which he has referred to my species, and he is equally at a loss where to place it. He says ('Brit. Conch.,' vol. ii, p. 216), "in shape it is a miniature *Zenatia*, a genus founded by Dr. Gray, but having an external ligament." The shell, however, given by Messrs. H. and A. Adams as the type of *Zenatia* ('Gen. of Recent Moll.,' vol. ii, p. 384, plate cii, fig. 1, *Z. acinaces*) is an aberrant form of *Lutraria*, with a deep sinus in the mantle mark. The nearest approach to this species seems to me to be the succeeding species *substriata*.

SPHENALIA SUBSTRIATA, *Mont.* Crag Moll., vol. ii, p. 128, Tab. XII, fig. 12 (as *Montacuta substriata*); Supplement, Tab. X, fig. 20.

Localities. Cor. Crag, Sutton, and near Orford. Chillesford Bed, Aldeby.

The figure in Supplement, Tab. X, represents a small specimen showing a difference in form from the full-grown shell, being much less inequilateral, the umbo being nearly central, which was found by Messrs. Crowfoot and Dowson at Aldeby. Mr. Jeffreys says ('Brit. Conch.,' vol. v, p. 177), "the fry of *substriata* are nearly globular, like *Kellia suborbicularis*, with the beak in the middle of the dorsal area."

CRYPTODON ROTUNDATUM, *S. Wood.* Crag Moll., vol. ii, p. 135, Tab. XII, fig. 19 (as *C. ferruginosum*).

CRYPTODON ROTUNDATUM, *S. Wood.* Ann. and Mag. Nat. Hist., 1840, p. 247.

Conceiving that I had erroneously referred my Crag shell, I applied to Mr. Jeffreys for the sight of his recent specimens, and he obligingly sent me a suite of *ferruginosum*, as also of *Croulinensis* for comparison, and I now believe my fossil to be distinct from either, although it more closely resembles the latter species. Both *ferruginosum* and *Croulinensis* are united by Forbes and Hanley, but if the two be recognised by conchologists I must consider my shell as distinct from either. The Crag form differs from *Croulinensis* in the hinge region, that shell being destitute of the depression possessed by the Crag one on the anterior side of the umbo, and from *ferruginosum* in its obliquity. It is not improbable that the Crag shell was the common ancestor of both the living species. I have, therefore, restored to my Crag form the name originally given to it in my 'Catalogue of Shells from the Crag of 1840' I know it only from the Coralline Crag at Sutton, and there it is very rare.

LORIPES DIVARICATUS, *Linn.* Crag Moll., vol. ii, p. 137, Tab. XII, fig. 4.

Localities. Red Crag, Sutton. Fluvio-marine Crag, Bramerton and Thorpe. Middle Glacial, Hopton.

This shell has not appeared in the oldest part of the Red Crag, that of Walton, nor is it common in the newer portions. It is not uncommon in one of the pits at Bramerton, but has not yet occurred in the Chillesford bed at any of its localities. One perfect valve and some fragments have been obtained from the Middle Glacial sand.

The fragment represented in Supplemental Tab. X, fig. 18, was figured under the idea that it belonged to *Loripes lacteus*, Linn.; but a re-examination of it has induced me to doubt whether it be anything more than a fragment of some other bivalve in which a small piece has broken out of the hinge.　It is from the Middle Glacial sand of Hopton, and, like most of the specimens in that sand, is worn.

Lucina borealis.　Crag Moll., vol. ii, p. 139, Tab. XII, fig. 1 ; Supplement, Tab. IX, fig. 5.

Localities.　Cor. Crag passim.　Red Crag passim.　Fluvio-marine Crag passim. Chillesford Bed, Chillesford, Bramerton, Aldeby, Easton, and Horstead.　Lower Glacial, Weybourne (*Reeve*).　Middle Glacial, Hopton and Billockby.

This species is exceedingly common in the Cor. and Red Crags, as also in the Middle Glacial Sands at Hopton, but in this latter it is generally in a fragmentary condition, the only perfect valves that have occurred from these sands being those of very young shells.

The shell figured in this 'Supplement' (Tab. IX, fig. 5) is of a specimen sent to me by Messrs. Crowfoot and Dowson, who found it at Aldeby.　In form it strongly resembles *L. spinifera*, but I believe it is merely a distorted specimen of *L. borealis*.

Diplodonta dilatata, *S. Wood*.　Crag Moll., vol. ii, p. 145, Tab. XII, fig. 5.

Localities.　As in 'Crag Mollusca.'

This was so called from a presumption that it was specifically distinct from *rotundata*, which I still believe it to be.　I also, in the 'Crag Mollusca,' placed *D. dilatata*, Sow., from the older tertiaries of Sussex, as a synonym ; and gave as another synonym the shell figured by Nyst under the name of *D. dilatata*, Phil.　The shell figured by Philippi under this name was, however, referred by him in a subsequent volume to *D. rotundata* (' Phil.,' vol. ii, p. 24) ; while the shell figured by Nyst, and referred by him to Philippi's so-called *dilatata*, is now called by Nyst *D. Woodii*, apparently under the impression that Philippi's *dilatata* is a subsisting species, instead of its being the same as *D. rotundata*.　Nyst's figure (' Coq. Foss. de Belg.,' pl. 7, fig. 1) corresponds with the Crag shell, but his description, "son coté postérieur est très elargé et subanguleux," does not.　My shell is also, I now believe, distinct from *D. dilatata*, Sow., of Dixon's 'Geology of Sussex.'　The nearest approach to it that I know is a shell from Grignon (*D. profunda*, Desh.), but that is also, I believe, distinct.　Under these circumstances I have retained the Crag shell under the original name given by me to it in 1840, although it is placed by the author of the 'Brit. Conch.' as a variety only of *rotundata*. *D. dilatata* is given by the late Dr. S. P. Woodward in his list of Norwich Crag shells, but there is no authority attached to this name (unfortunately there are but few authorities for names in that list), and I am not certain whether the species he speaks of be the *rotundata* of Mont., or *dilatata*, S. Wood.

DIPLODONTA ASTARTEA, *Nyst.* Crag Moll., vol. ii, p. 146, Tab. XII, fig. 2.

Localities. Cor. Crag, Sutton. Red Crag, Sutton. Fluvio-marine Crag, Bramerton.

The shell with this name may be the same as *D. trigonula*, Bronn, *D. apicalis*, Phil., although this latter author keeps the two specifically separated, but our Crag fossil seems to have attained to larger proportions.

The figure in the 'Crag Mollusca' was taken from a Red Crag specimen, but I mentioned (p. 146) that the Coralline Crag form differed somewhat from that of the Red, but not sufficiently so to justify a separation of the two, and I am still of the same opinion.

The species is given in Dr. Woodward's Norwich Crag List in White's 'Directory' as from the Fluvio-marine Crag; and this is confirmed by a specimen of it having recently been obtained by Mr. Harmer from Bramerton.

LUCINOPSIS UNDATA, *Pennant.* Supplement, Tab. IX, fig. 4 *a—b.*

> VENUS UNDATA, *Penn.* Brit. Zool., ed. iv, vol. iv, p. 95, pl. lv, fig. 51.
> — INCOMPTA, *Phil.* En. Moll. Sic., vol. i, p. 44, t. iv, fig. 9.
> LUCINA CADUCA, *Scac.* Catal., p. 5, fide *Phil.*
> LUCINOPSIS UNDATA, *Forb. & Hanl.* Brit. Moll., vol. i, p. 435, pl. xxviii, figs. 1, 2.
> — — *Jeffreys.* Brit. Conch., vol. ii, p. 363, 1863.

Spec. char. L. testa tenui, orbiculato-quadrata, æquilaterali, compressiuscula, lævi, tenuissime et concentricè striata, lunula areaque non distinctis.

Diam. $\frac{1}{2}$ an inch.

Locality. Chillesford bed, Aldeby.

Two small specimens have been found by Messrs. Crowfoot and Dowson, one of which is represented in the above figure.

This species resembles *L. Lajonkairii* in outline and dental characters, but differs in being quite smooth or with only lines of growth without any radiating striæ.

LUCINOPSIS LAJONKAIRII, *Payr.* Crag Moll., vol. ii, p. 148, Tab. XI, fig. 14 *a—c.*

This species is very rare in the Red Crag, from which I have never met with more than one specimen, probably derived from the Cor. Crag, where at one time it was somewhat abundant.

This may possibly be referred to *Venus candida*, Gmel., figured by Gualt., 'Test.,' Tab. 75, fig. L, but my shell does not very well accord with those figured by Hörnes and Philippi, which are smaller and much more inequilateral. The representation in the 'Ency. Method.,' pl. 272, fig. 2 *a, b*, corresponds with our Crag shell.

Mr. McAndrew (whose recent death I deeply deplore, and than whom no one has added more to our knowledge of living European and Atlantic Mollusca) sent me for examination two specimens from the Mediterranean of a shell considered to be identical with this species. These scarcely exceed half an inch in diameter, and if full grown would be much more tumid than my Crag specimens. They were also more coarsely striated or radiated. Under these circumstances I am not satisfied that Lajonkairii is a living species.

CHAMA GRYPHOIDES, *Linné.*   Crag Moll., vol. ii, p. 162, Tab. XV, fig. 8.

Localities.   As in 'Crag Moll.'

When describing this shell in the 'Crag Moll.' I stated it to be very rare in the Cor. Crag, and it has ever since continued so to my researches. The specimens found by me in the Cor. Crag have always been very small or young individuals, while those from the Red Crag appear all to be full-grown specimens or nearly so. The solidity of the specimens would well protect them in a removal from one formation into another, and I believe, notwithstanding its present scarcity in the Cor. Crag, that the specimens which have been found in the Red Crag of Sutton are extraneous.

In Mr. Jeffrey's List of Red Crag Shells appended to Mr. Prestwich's paper, 'Geol. Journ.,' vol. xxvii, p. 482, this species is given as from Walton, but upon whose authority is not stated. I have never seen it from that locality, but if it be so I should then be more disposed to regard it as a denizen of the Red Crag Sea, and as additional evidence of the greater antiquity of the Walton bed over the rest of the Red Crag.

VERTICORDIA CARDIIFORMIS, *S. Wood.*   Crag Moll., vol. ii, p. 150, Tab. XII, fig. 18 (as *Hippagus verticordius*).

Locality.   Cor. Crag, Sutton.

Mr. Jeffreys in his list appended to Mr. Prestwich's paper ('Geol. Journ.,' vol. 27, p. 139) has referred this Crag species to what he has called *Pecchiolia acuticostata*, and he has obligingly sent me a single valve for examination. I find this recent shell to differ from the Crag species in being much more tumid and in having the ribs more elevated. In the recent shell these ribs are ornamented with a double row of very fine *spinulæ*, not a trace of which can I discover upon any of my Crag specimens. These may possibly have been rubbed off, but the probabilities are that had they ever existed some trace would remain on one or other of the ribs of the numerous well-preserved Crag specimens that I have examined; but I can detect none. The number of ribs is probably not a reliable character, but while this recent specimen had only fourteen ribs my Crag specimens vary from that number to seventeen. I do not feel justified under these circumstances in adopting the identification of the Crag shell with *acuticostata*. In my 'Eocene Bivalves' (page 138) I have given reasons for recurring to my original generic name of *Verticordia* for the group of Mollusca to which the present shell belongs; and in consequence I have reverted to the specific name of *Cardiiformis*, under which I originally sent it to the 'Min. Con.' in 1844 (Tab. 639).

CARDITA BOREALIS ? *Conrad.* Crag Moll., vol. ii, p. 168, Tab. XV, fig. 6 (as *C. analis ?*).

CARDITA BOREALIS, *Conrad.* Amer. Mar. Conch., 39, pl. viii, fig. 1.

Locality. Chillesford bed, Sudbourn Walks? Upper Glacial, Bridlington.

This shell, which was described by me under the name *analis*, I know from Bridlington only, but Mr. Bell ('Mag. Nat. Hist.' for 1870) gives it from the Chillesford bed of Sudbourn Church Walks, though I have not seen the specimen. In the 'List of Shells from the Norwich Crag,' by the late Dr. Woodward, the Bridlington shell was referred to *Cardita borealis*, Conrad. The specimens from Bridlington in the British Museum (about twenty-five in number) are much smaller, not measuring more than half an inch in diameter, while the American shell reaches at least an inch, and appears to have a more excentric inclination of the umbo; but, according to Gould, in the young shell the beaks are more central, so that this difference may be due to the Bridlington specimens being all young ones. A shell from Labrador, that seems to be fossil, or semifossil, shown to me by Mr. A. Bell, resembles the Bridlington shell; while a recent specimen of *borealis* from Gaspé, also shown to me by Mr. Bell, appeared to be different from the Bridlington form in being much more transverse. Under these circumstances I have referred the Bridlington shell to the American living species with a doubt, as I am not altogether satisfied on the point.

CARDITA SCALARIS, *Leathes.* Crag Moll., vol. ii, p. 166, Tab. XV, fig. 5.

Localities. Cor. Crag passim. Red Crag, Sutton and Walton. Fluvio-marine Crag, Norwich? Chillesford bed, Chillesford? Middle Glacial, Hopton. Living, North-West America?

This species is exceedingly abundant in the Cor. Crag, and it is not rare in the Red Crag of Suffolk, but many of these latter may be derivatives from the Cor. Crag. One small valve, not quite perfect, was obtained by Mr. Dowson from the Middle Glacial Sands of Hopton and sent to me, and it is given by Dr. Woodward in his Norwich Crag List in White's 'Directory' as from Norwich and Chillesford, but as I have not seen any specimens from either of these places, I have given those localities with a note of interrogation. The species seems to me to survive on the North-West Coast of America, as a shell from there in the British Museum under the name *Cardita ventricosa*, Gould, appears to me to be identical with the Crag form. This species is given as a Bridlington fossil by E. Forbes, 'Mem. Geol. Surv.,' p. 415, 1846, but I have not seen the specimen, and believe it to have been confounded with *borealis* (*analis*). I have not met with it from any of the localities of the Chillesford bed or from the Lower Glacial Sands.

18

CARDITA CORBIS, *Phil.*   Crag Moll., vol. ii, p. 168, Tab. XV, fig. 2 *a, b* (as *Cardita corbis*, var. *nuculina*).

Localities.   Cor. Crag, Sutton.   Red Crag, Walton.   Middle Glacial, Hopton. Living in the Mediterranean.

A single valve is all that has occurred in the Middle Glacial.

CARDITA ANCEPS, *S. Wood.*   Crag Moll., vol. ii, p. 168, Tab. XV, fig. 2 *c, d* (as *Cardita corbis*, var. *exigua*).

VENERECARDRIA ANCEPS, *S. Wood.*   Catalogue, 1840.

Locality.   Cor. Crag, Sutton.

In my 'Catalogue' of 1840 I gave this shell under the above specific designation, but in the 'Crag Mollusca' I placed it as a variety of *C. corbis*, under the name of *exigua*, supposing it to be identical with Dujardin's Touraine shell of that name.   My further examinations have induced me to revert to my views of 1840, and I have accordingly restored it as a species under the name of *anceps*.

C. corbis (*nuculina*) is the only form that I have seen from the Red Crag, and this is from Walton, where it is not very abundant.   *Cardita corbis* is given from the Red Crag at Waldringfield by Mr. Bell and by Mr. Jeffreys, but which of the two forms is referred to I do not know.   *Cardita corbis* is the same as the shell now found living in the Mediterranean.

Dujardin represents his *nuculina* as strongly radiated, and the *exigua* as having somewhat oblique transverse ridges, but of the two Crag forms *corbis*, and *anceps*, it is the latter only which is strongly radiated, from which it would seem that *if* the Touraine and Crag shells are identical it is *anceps* which must be referred to *nuculina*, and *corbis* to *exigua*.   The shape, however, of the shells would lead to the reverse of this reference, and I am, therefore, very doubtful whether there be any identity between the two Crag and the two Touraine shells.   I think, therefore, that I do best and avoid confusion by reverting to my 'Catalogue' name of *anceps* for our present shell, rather than by referring it to either *nuculina* or *exigua*.

Among the synonyms of *C. corbis* I gave *Cardita corbis*, Nyst.   This I now believe to be incorrect, the Belgian Crag shell appearing to be different from either of the forms found in the English Crag and to correspond with the Middle Oligocene shell called *Cardita lævigata* by Dr. Speyer ('Mittel. Oligocän. Söllingen,' p. 60, Tab. III, fig. 7). A shell called *Woodia lævigata* by Von Könen (' Mittel. Oligocän.,' p. 108, Tab. VII, fig. 8, *a—d*) is probably the same, but this I have not seen as an English fossil, though some of the smoother specimens of *anceps* make great approaches to it. Both *corbis* and *anceps* are very abundant in the Coralline Crag.

CARDITA SENILIS, *Lam.* Crag Moll., vol. ii, p. 165, Tab. XV, fig. 1 *a—f.*

Localities. Cor. Crag passim. Red Crag passim except Walton Naze.

The shell which has hitherto gone by the name of *Cardita senilis*, Lam., is one of the most abundant shells in the Coralline Crag, and it is found also in all parts of the Red Crag save Walton, in some, if indeed not all, of which, however, it is probably only present as a derivative. There is some doubt whether this shell be the *Venericardia senilis* of Lam. or the *Cardita rudista* of the same author, under which latter name it is figured by Hörnes, Tab. XXXVI, fig. 2, *a—d*, who associates with it (and I think justly) *Chama rhomboidea*, Broc., Tab. XII, fig. 16. In ' Brit. Conch.,' vol. i, p. xciii, Mr. Jeffreys says, " The *Cardita senilis* of the same beds (Coralline Crag) is the *C. sulcata* of the Mediterranean," and this he has repeated in the list annexed to Mr. Prestwich's Cor. Crag paper. I have, however, carefully compared *sulcata* with the Crag fossil and I cannot coincide in Mr. Jeffreys' opinion. The recent *sulcata* has more rounded ribs, which are nearly smooth, and not imbricated, and the ribs are united at their bases, but in *senilis* they are distant, with a space between them. *Senilis* is exceedingly variable in form, as may be seen by the specimens figured in ' Crag Moll.,' vol. ii, Tab. XV, fig. 1 *a—f*, all of which I believe belong to one and the same species. As so many British authors have referred to this shell for nearly half a century under the name *senilis*, I have not thought it desirable, notwithstanding Dr. Hörnes' identification of it with Lamarck's *rudista*, to make any change in the name.

CARDIUM FASCIATUM, *Mont.* Crag Moll., vol. ii, p. 153, Tab. XIII, fig. 4 (as *C. nodosum*).

Localities. Cor. Crag passim. Red Crag passim. Chillesford bed, Aldeby.

The shell figured in ' Crag Moll.,' under the specific name of *nodosum* may probably be *C. fasciatum*, Mont. (*C. elongatum*, Turt.), as has been said by the author of the ' Brit. Conch.' Although a very common shell in the Cor. Crag, the specimens have (with very rare exceptions) the entire surface removed, and with it the tubercular ornament, and in that state it is of very difficult determination. Mr. Crowfoot sent me several small specimens from Aldeby in which parts of the exterior ornament is well preserved, and these may be referred to this species.

CARDIUM PINNATULUM, *Conrad.*   Crag Moll., vol. ii, p. 154, Tab. XIII, fig. 3 (as *C. nodosulum*).

> CARDIUM PINNATULUM, *Conrad.*   Journ. Acad. Nat. Sc., p. 260, pl. xi, fig. 8, 1831.
> —   —   *Gould.*   Inv. Mass., p. 90, fig. 57, 1841.

Localities.   As in ' Crag Mollusca.'

This in the Crag is rare to my researches.   In the ' Crag Mollusca ' I thought it to be distinct from *C. pinnatulum*, differing in the number of ribs.   I have, however, since seen more of the recent shell, and find that in some specimens they correspond in this respect, so that I have substituted the name of *pinnatulum* for the Crag shell.

CARDIUM STRIGILLIFERUM.   Crag Moll., vol. ii, p. 154, Tab. XIII, fig. 5.

Localities.   Cor. Crag passim.

This Coralline Crag shell is given by Mr. Bell (' Ann. Mag. Nat. Hist.,' 1870) as identical with *C. elegantulum*, Möller, and it is also so referred by Mr. Jeffreys in his list attached to Mr. Prestwich's paper.   I have compared my fossil with the recent shell and think that the two are distinct.   The Crag shell is more tumid than the recent one, and the tubercles on it are narrow and prominent, being, so to speak, perched on the centre of the rib, while in the recent shell the tubercles are more properly imbrices, and are broad and cover the rib.

CARDIUM EDULE, *Linn.*   Crag Moll., vol. ii, p. 155, Tab. XIV, fig. 2.

Localities.   All the Upper Tertiaries of East Anglia.

This shell first appears in the Cor. Crag, where it is very rare, and it is also rare at Walton-Naze, but it is extremely common in the more recent formations.   It is present in the Post-Glacial beds of Kelsey Hill, March, and Hunstanton, also in the Brickearth of the Valley of the Nar.

CARDIUM NODOSUM, *Turt.*   Supplement, Tab. X, fig. 6.

> CARDIUM NODOSUM, *Turt.*   Brit. Biv., p. 186, t. xiii, fig. 8, 1822.
> —   —   *Jeffreys.*   Brit. Conch., vol. xi, p. 283, pl. xxxv, fig. 4.

Locality.   Coralline Crag, Sutton.

There are a number of small specimens in my cabinet which I had imagined to be the young of the shell figured in ' Crag Moll.,' Tab. XIII, fig. 4, as C. *nodosum*.   This shell, however, has since been referred to *C. fasciatum*, Mont., by Mr. Jeffreys, in which reference he is probably correct, and I have, therefore, had figured one of my small specimens which seems to correspond with the *nodosum* of Turton. The recent shells called *nodosum* and *fasciatum* being kept separate by conchologists, I have here, in deference, done the same, although, I confess, not without misgivings. Dr. Lovén ('Ind. Moll. Scandin.,' p. 35), when describing *C. fasciatum* and others, says, "Cardia Europæa misere confusa," and I feel disposed to echo his words.

Mr. Jeffreys, in his list to Mr. Prestwich's paper (p. 138), introduces *C. Norvegicum* as a new Cor. Crag species, but no locality is attached to the name; and in his list of Red Crag shells in the same paper (p. 482) *Card. interruptum* is referred to this species (viz. *C. Norvegicum*). If this be the form upon which the name *Norvegicum* is thus introduced as a Coralline Crag species, it will not require another figure; but I have not yet been able to see a shell like *C. interruptum* from the Coralline Crag, or to refer any of my specimens satisfactorily to *C. Norvegicum*.

CARDIUM PARKINSONI, *J. Sow.* Crag Moll., vol. ii, p. 158, Tab. XIII, fig. 7.

Localities. Red Crag, Walton, Sutton, and Butley.

This species is exceedingly abundant at Walton, but rare in other parts of the Red Crag. Dr. Woodward, in his list in 'White's Directory,' gives it (in fragments) from the Fluvio-marine Crag of Thorpe-by-Norwich, but I have not seen the shell from that Crag, and suspect that the fragments referred to are those of the large individuals of *edule* so common in the Norwich Crag and Lower Glacial sands. *Parkinsoni* much resembles *C. Nuttalli*, Conrad, a shell living in the seas of Upper California ('Journ. Nat. Hist. Soc. Boston,' vol. vii), but I believe our Crag shell to be distinct, as it is less oblique, with fewer and broader ribs, and these are united at their bases, while the ribs in *Nuttalli* have a distinct space between them. In both species the ribs are imbricated.

CARDIUM DECORTICATUM, *S. Wood.* Crag Moll., vol. ii, p. 159, Tab. XIV, fig. 1.

Localities. Cor. Crag passim.

This is abundant in a fragmentary state in the Cor. Crag, but perfect specimens are difficult to obtain; it was a handsome shell, and fragments indicate a length of $3\frac{1}{2}$ inches. *C. venustum* resembles it, and may possibly be the same species; but, as I have before stated, it is a smaller shell and smoother, the ribs in *decorticatum* being prominent and distinct. In the list annexed to Mr. Prestwich's Cor. Crag paper this shell is referred to *C. lævigatum*, Poli, but in the list to the Red Crag paper it is referred to *C. Norvegicum*. I cannot, however, agree in either reference; and have not seen it from the Red Crag, but only the allied form *venustum*.

CARDIUM ISLANDICUM, *Linn.*

Localities. Middle Glacial, Hopton? Upper Glacial, Bridlington?

This species is given by Dr. Woodward in his list of Bridlington fossils ('Geol Mag.,' vol. i, p. 54), but the specimen in the British Museum, upon the authority of which this was done, is too imperfect to be referred with certainty to any species. Umbonal portions of a large *Cardium* similar to the Bridlington species occur in the Middle Glacial sands of Hopton. These, so far as such fragments are reliable, may possibly belong to *C. Islandicum*, or as likely to *C. decorticatum*, and from this uncertainty I have not thought it desirable to figure these fragments.

ERYCINELLA OVALIS, *Conrad.* Crag Moll., vol. ii, p. 171, Tab. XV, fig. 10.

Localities. Cor. Crag passim. Red Crag, Walton (*A. Bell*). Middle Glacial, Hopton.

In assigning the above name to the (once) common Cor. Crag shell I was, as explained in the 'Crag Mollusca,' guided by the report on it kindly brought home for me from Mr. Conrad by Sir Charles Lyell. I have some misgivings about the identity, not having been able to see the American fossil, and if it should hereafter prove to be distinct the specific name of *pygmæa* assigned to it in my Catalogue ('Ann. and Mag. Nat. Hist.,' Dec., 1840) will be applicable to it.

It is singular that, though once very common in the Cor. Crag at Sutton, I have not been able to find a specimen for many years, though I have sifted tons of the Cor. Crag material from the identical spot and horizon at Sutton where I once found it common. Mr. Bell ('Ann. and Mag. Nat. Hist.,' Sept., 1870) gives the species from the Red Crag, Walton, but I have not seen the specimen. Five valves in fair preservation have occurred in the Middle Glacial of Hopton.

CORALLIOPHAGA CYPRINOIDES. Crag Moll., vol. ii, p. 200, Tab. XV, fig. 7 *a—d.*

Locality. Cor. Crag, Sutton and Ramsholt.

My specimens of this are still very rare and very small. It may possibly be the same as *Chama lithophagella*, Broc., which is said to be a common living shell in the Mediterranean, but in the uncertainty I do not think it is necessary to alter the name I had previously given. It cannot be referred to *lithophagella*, Lam. (*Cypricardia*), an externally ornamented West Indian species; my shell is quite smooth.

ASTARTE BOREALIS, *Chemn.* Crag Moll., vol. ii, p. 175, Tab. XVI, fig. 3 *a, b.*

Localities. Fluvio-marine Crag, Bramerton, Thorpe, and Postwick. Chillesford bed, Aldeby, Bramerton, Horstead, and Coltishall. Lower Glacial, Weybourn, Belaugh, and Rackheath. Middle Glacial, Hopton. Upper Glacial, Bridlington. Post Glacial, March.

This is one of the very few shells of the Fluvio-marine Crag, that on the assumption of their being coeval deposits, might have been expected to occur also in the Red, but which have not yet been found in it. Although present in the Fluvio-marine Crag at Bramerton, I am informed that it is rare, while in the Chillesford bed exposed in the same deep section at Bramerton it is common. It occurs in the Chillesford bed at most of the localities except at Chillesford, at which place I have not heard of its occurrence; and it has been found in the Lower Glacial Sands at Belaugh, Rackheath, and Weybourn, the base of those sands at the first of these places being literally a pavement of detached valves of this shell. It is not uncommon at Hopton in a fragmentary state, but only one perfect valve has occurred there. The specimens from March are somewhat peculiar, having fine striations, so that it was inserted by Mr. Seeley in his list of shells in the ' Quart. Journ. Geol. Soc.,' vol. 22, p. 473, as *A. crebricostata.* The March specimens, however, have not the denticulated margin of *crebricostata,* and some of the specimens exceed in magnitude any of *crebricostata* that I have seen. The form called *Withami* (fig. 3 *c—d*) appears to be confined to the Bridlington locality, where it occurs in association with the typical form of *borealis,* and, so far as shape is concerned, seems to bear the same relation to it that *A. elliptica* bears to *A. sulcata;* but there is not the difference of a notched margin such as obtains between *elliptica* and *sulcata.*

ASTARTE BURTINII, *La Jonk.* Crag Moll., vol. ii, p. 188, Tab. XVII, fig. 5 *a—d.*

Localities. Cor. Crag passim. Red Crag, Sutton. Fluvio-marine Crag, Bramerton ? Middle Glacial, Hopton.

This shell still remains to me very rare in the Red Crag. It is given by Dr. Woodward in his list in ' White's Directory ' as in the Norwich Crag, on the authority of a single valve said to have been found there by Mr. Wigham; but I have not been able to hear of its occurrence there from other sources, and its presence in the Fluvio-marine Crag must be received with doubt. I have not met with it from the Lower Glacial sand or from any of the localities of the Chillesford bed. In the Middle Glacial sands of Hopton several young specimens have occured, but all, except one of them, imperfect. I have not met with it from any newer Glacial or from any Post Glacial bed, nor do I know it as living.

ASTARTE COMPRESSA, *Mont.*   Crag Moll., vol. ii, p. 183, Tab. XVI, fig. 8 *a—c.*

Localities.   Red Crag passim.   Fluvio-marine Crag, Bramerton and Thorpe. Chillesford bed, Bramerton, Horstead, Colteshall, and Aldeby.   Lower Glacial, Belaugh and Weybourn.   Middle Glacial, Billockby and Hopton.   Upper Glacial, Bridlington. Post Glacial, Kelsea Hill?

The occurrence of this shell from its first appearance in the oldest part of the Red Crag to the top of the Glacial series seems very uniform.   It is particularly common in the Middle Glacial sands of Hopton, where, although almost always more or less worn, the valves of it occur perfect far more frequently than do those of any other bivalve. Mr. Jeffreys ('Quart. Journ. Geol. Soc.,' vol. xvii, p. 448) gives a fragment of it from the Kelsea Hill gravel, which is the only instance of its occurrence known to me in the Post Glacial deposits of East Anglia.

ASTARTE GALEOTTII? *Nyst.*   Crag Moll., vol. ii, p. 185, Tab. XVII, fig. 3 (as *A. gracilis*).

In the list by Mr. Jeffreys ('Geol. Journ.,' vol. xxvii, p. 138) this Crag shell is considered merely as a var. of *A. compressa*, from which opinion I dissent, believing the two to be specifically distinct; the Cor. Crag shell is not only different in form, but when full grown it has a denticulated margin which is absent from *compressa* in all stages of its existence.

Mr. A. Bell ('Ann. and Mag. Nat. Hist.,' May, 1871) pointed out that *A. gracilis*, so called by me, was not, according to Dr. Weichman, the same species as *A. gracilis*, Munst., adding that the Cor. Crag shell must be referred to *A. Galeotti*, Nyst., which name I had given as a synonym.   Dr. Weichman has obligingly sent some of his German specimens for examination, and I find that they are quite distinct.   The shell called *Galeotti* by Nyst certainly comes near to one of the varieties of my Crag species (fig. 3 *d*), but that is not the general form.   A shell also like this from Thorigny (in the Faluns), given to me by Sir Charles Lyell, equally resembles our small variety, but it does not seem ever to have attained so great a magnitude.   Under these circumstances I have given the shell under the name *Galeotti* with a doubt.

ASTARTE INCRASSATA, *Brocchi.*   Crag. Moll., vol. ii, p. 178, Tab. XVI, fig. 6 *a, b.*

Localities.   Cor. Crag passim.   Red Crag, Sutton.   Middle Glacial, Hopton?
This species is in Mr. Jeffreys' list just referred to and in the 'Brit. Conch.'

regarded as a variety of *A. sulcata*, the ridged or sulcated form being the British and Arctic one, and the smooth form the Mediterranean one. It appears to me, however, that where such distinct forms are characteristic of separate areas it is but a question of words whether for palæontological purposes we call them species or varieties, and it is significant that in the Cor. Crag, with, as I consider, a fauna having its chief affinities with the Mediterranean, we get no trace of *sulcata*, but have *incrassata* in profusion. Some imperfect specimens have occurred in the Middle Glacial at Hopton that seem to belong to this species, but being imperfect I have assigned a Middle Glacial locality to it with a note of interrogation. If, however, better specimens should confirm this, we should, as the undoubted *sulcata* occurs at Hopton, have both the Arctic and Mediterranean forms together in the Middle Glacial deposit. This is probably *Tellina fusca*, Poli.

ASTARTE SULCATA, *Dacosta*. Crag Moll., vol. ii, p. 182, Tab. XVI, fig. 5 *a, b*.

Localities. Red Crag, Sutton. Fluvio-marine Crag, Bramerton and Thorpe? Chillesford bed, Aldeby. Middle Glacial, Hopton. Upper Glacial, Bridlington. Post Glacial, March.

This shell still remains in the Red Crag very rare to my researches. It is given by Dr. Woodward in his Norwich Crag list as occurring both at Bramerton and Thorpe, the young being more frequent, but I have not been able to confirm this through any of my Norwich correspondents. It has been obtained at Aldeby by Messrs. Crowfoot and Dowson. In the Middle Glacial at Hopton two or three perfect valves of young specimens and one full grown have occurred. It is among the specimens in the British Museum from Bridlington, and Mr. Harmer has found it at March.

ASTARTE OMALII, *La Jonk*. Crag Moll., vol. ii, p. 180, Tab. XVII, fig. 1 *a—f*.

Localities. Cor. Crag passim. Red Crag, Sutton. Fluvio-marine, Bramerton? Middle Glacial, Hopton.

The principal portion of a shell, as well as some other fragments, which have occurred at Hopton, enable me to refer this species to the Middle Glacial sands without much hesitation. It is given in Dr. Woodward's list as from Bramerton, but I have not been able to obtain confirmation of this, and have placed a note of interrogation to that locality.

This species is referred by Mr. Jeffreys in his list to *A. undulata*, Gould, but its identity with that shell I fully considered more than twenty years ago, and the reasons for keeping the two species distinct, given by me at p. 180, vol. ii, of 'Crag Mollusca,' appear to me to be still valid. If we were to strain identities in this way a suite of

19

various acknowledged species of this variable genus might be so selected as all to run into one another and the whole of such species accordingly merged in one. In a genus such as this, wherein the specific forms thus graduate into one another, and which, moreover, goes back far into the Mezozoic formations with but little departure from the living types, a more arbitrary line of specific division should be allowed than in the case of species of less variable genera.

ASTARTE FORBESII, *S. Wood.* Crag Moll., vol. ii, p. 192, Tab. XVII, fig. 12 *a, b* (as *A. parva*).

Localities. Cor. Crag passim.

When describing this shell I was not aware that the name of *parva* had been given to another species in this genus; but I find that a fossil, which is quite distinct, has been so called by Dr. Lea ('Contrib. to Geol.,' p. 63, pl. 2, fig. 37), and as this is of prior date (1833), my name, of course, must be suppressed.

In a catalogue of the Mediterranean Mollusca by Mr. Jeffreys, published in the 'Ann. and Mag. Nat. Hist.,' July, 1870, is the name *Astarte parva*, S. Wood; and that gentleman there says, " This may possibly be *A. pusilla* of Forbes; the inside margin is notched in my specimens;" and in the list by the same gentleman accompanying Mr. Prestwich's Cor. Crag paper this identification is inserted without any qualification. I have endeavoured to find out the type-specimens of Forbes' species, but unsuccessfully. It is, however, described by Forbes as " concentricé striata, margine interno denticulato;" but as the markings on my shell are eccentric instead of concentric, and the margin free from denticulations, no greater discordance, so far as description goes, could well occur. In general, where the *full-grown* shell of *Astarte* has a denticulated margin, the *young* has this margin smooth, and as I have now before me one hundred specimens of this Crag species, not one of which has a denticulated margin, I can hardly suppose them all to be immature shells; I therefore cannot with any propriety refer the Crag shell to *pusilla.* It much resembles an Oligocene fossil named and described by Dr. Speyer as *Goodallia Köneni* ('Die. Ober. Oligoc. Test. Detmold,' tab. iv, fig. 6, 1866), which is ornamented with oblique ridges and has a smooth margin, but, judging from the representation, it is not a satisfactory identification; therefore, until specimens can be compared, and as the name *parva* must be abandoned, I have assigned to my shell the name *Forbesii.*

The late Mr. McAndrew gave me some specimens of a small *Astarte* which he obtained in the living state off the Canaries, that much resembles the Cor. Crag shell called *A. parvula;* but I think the recent shell specifically different; it is rather less oblique, the lateral denticles of the recent shell are shorter, and the ventral margin is denticulated. I have also a fossil specimen from Cannes, which is more shaped like Mr. McAndrew's shell, but it has a smooth ventral margin. The following from the Upper Tertiaries of East Anglia I consider as distinct species :

Crenulated in the mature state.	Not crenulated at any stage of growth.
Astarte Basterotii.	Astarte borealis.
,, Burtinii.	,, compressa.
,, crebricostata.	,, elliptica.
,, crebrilirata.	,, Forbesii.
,, incrassata.	,, parvula.
,, Galeottii.	
,, mutabilis.	
,, obliquata.	
,, pygmæa.	
,, Omalii.	
,, sulcata.	
,, triangularis.	

The different formations to which these species belong will be found in my Compendium or General Table.

WOODIA DIGITARIA. Crag Moll., vol. ii, p. 190, Tab. XVII, fig. 8 (as *Astarte digitaria*); Supplement, Tab. X, fig. 8 *a*.

Localities. Coralline Crag passim. Red Crag, Walton, Bentley, and Butley (*Bell*). Middle Glacial, Hopton.

Var. *Hoptonensis,* Supplement, Tab. X, fig. 8 *b*.

Locality. Middle Glacial, Hopton.

The small shell represented in 'Supplement,' Tab. X, fig. 8 *b*, is from the Middle Glacial sand of Hopton, and as it seems to differ materially from the Coralline and Red Crag specimens I think it deserving of a separate figure. This Middle Glacial shell is much less transverse than any of my specimens from either the Coralline[1] or Red Crags, and the elliptical markings much more distant, being scarcely half so numerous as in the Cor. Crag specimens. I have found specimens of the same species at Walton Naze and they are of an intermediate form, with the elliptical lines more distant than those upon the Cor. Crag specimens, but not so much so as upon the Glacial one, while the form of all the Red Crag specimens is generally even more transverse than the Coralline Crag ones.

I had another figure ('Supplement,' Tab. X, fig. 8 *a*) made of the Cor. Crag shell in order to show the ornamentation on the posterior side of the shell, which is not distinctly shown in the figures in Tab. XVII, but the artist has not been fortunate in his representation. These are transverse to the sulcations and form ridges upon them. They

---

[1] Since the above was in type I have found in the Coralline Crag a specimen presenting the elliptical markings nearly as distinct as in the Hopton specimen. This, however, is the only example presenting that feature which I have met with among the very numerous Coralline Crag specimens of this species that have passed through my hands.

are but very faintly perceptible in the Red Crag specimens and not at all in the Glacial one; possibly in both this may be due to wear. One other valve, which is imperfect and which more agrees with the typical form, is all that I have met with from the Middle Glacial sands.

Astarte excurrens ('Crag Moll.,' vol. ii, p. 191, Tab. XVII, fig. 9) seems more probably to belong to the genus *Woodia* than to *Astarte*, and I have so placed it in the tabular list.

Cyprina Islandica. Crag Moll., vol. ii, p. 196, Tab. XVIII, fig. 2.

Localities. Cor. Crag, passim. Red and Fluvio-marine Crags, and Chillesford bed, passim. Lower Glacial, Rackheath, Belaugh, and Weybourne. Middle Glacial, Billockby, Hopton, and Wisset. Upper Glacial, Bridlington, and Dimlington. Post-glacial, March and Kelsea Hill.

This shell is very common in some localities of the Coralline Crag; plentiful in the Red Crag, though seldom perfect; common in the Fluvio-marine Crag and Chillesford beds at all localities, as well as in the Lower Glacial sands of Belaugh, Rackheath, and Weybourne, and in the Middle Glacial sands of Hopton and Billockby. It occurs in the gravel of this deposit also at Wisset (north-west of Halesworth). It is common in the March gravel, and a fragment is mentioned by Mr. Jeffreys at Kelsea Hill. It is the universal shell of nearly all the Upper Tertiaries of East Anglia, but does not appear to have occurred in the Nar Brickearth.

Cyprina rustica, *J. Sow.* Crag Moll., vol. ii, p. 197, Tab. XVIII, fig. 1.

This was at one time abundant in the Cor. Crag at Ramsholt and also near Orford, and I have found it, though rarely, in the Red Crag at Sutton, where probably it may have been a derived shell. The late Mr. Rose informed me that he had found it in the sand of Bradwell (Middle Glacial), but the fragment in his collection to which this name was attached appeared to me to be an imperfect specimen of some species of *Cardium*, with its surface eroded.

Cytherea rudis, *Poli.* Crag Moll., vol. ii, p. 208, Tab. XX, fig. 5 *a—d.*

Localities. Cor. Crag passim. Red Crag passim. Fluvio-marine Crag, Norwich ? Middle Glacial, Hopton.

In his list in White's Directory' Dr. Woodward gives this shell from the Norwich Crag on the authority of a single valve in Mr. Wigham's collection. I have not been able to learn from other sources of its occurrence in that Crag, and have, therefore, placed a note of interrogation to it. In the Middle Glacial sands of Hopton there have occurred many imperfect specimens, composed of the hinge and umbonal region, that seem referable to this species.

VENUS FASCIATA, Crag Moll., vol. ii, p. 211, Tab. XIX, fig. 5.

Localities. Red Crag, Sutton and Walton. Fluvio-marine Crag, Bramerton. Middle Glacial, Billockby and Hopton.

This shell is somewhat rare both in the Red and Fluvio-marine Crags, and I have not met with it from any of the localities of the Chillesford bed or from the Lower Glacial sands. In the Middle Glacial, both at Hopton and at Billockby, it is in extraordinary profusion, but mostly in a fragmentary condition, the hinges being extremely abundant. Very few perfect specimens, and those only of young individuals, have occurred.

VENUS OVATA. Crag Moll., vol. ii, p. 213, Tab. XIX, fig. 4.

Localities. Cor. Crag passim. Red Crag, Sutton, Butley. Chillesford Bed, Aldeby. Middle Glacial, Billockby and Hopton.

This shell, so profuse in the Cor. Crag, is exceedingly rare in the Red and does not appear to have been met with at all in the Fluvio-marine Crag, but it has been found in the Chillesford bed at Aldeby by Messrs. Crowfoot and Dowson. I have not seen it from the Lower Glacial sands, but it is common, though usually in a fragmentary condition, in the Middle Glacial of Hopton and Billockby, only very young specimens occurring perfect.

VENUS VERRUCOSA, *Linn.*

This, in 'Brit. Conch.,' vol. ii, p. 341, is given as a " Cor. Crag species " (S. Wood), but in vol. v, p. 184, it is said " not Coralline Crag." I have not seen it, however, from any of the Tertiaries of East Anglia. It is, I believe, not rare in the Post-glacial deposit at Selsey in Sussex, and it is present, also, in the Clyde beds and Belfast deposit.

VENUS FLUCTUOSA, *Gould*. Supplement, Tab. IX, fig. 8.

> VENUS FLUCTUOSA, *Gould*. Invert. Massach., p. 87, fig. 50, 1841.
> — — *Dekay*. Nat. Hist. New York (Zool.), p. 222, 1843.
> — — *S. P. Woodward*. Geol. Mag., vol. i, p. 54, 1864.
> TAPES — *Binney*. 2nd edit. Gould's Inv. Mass., p. 136, fig. 447.

Spec. char. "Shell moderately small, transversely ovate, lenticular, rather thin. Surface with 20—25 recurved concentrated waves, vanishing at the side ; areola none. Middle tooth in each valve cleft. Epidermis thin, glossy, yellowish, beneath this white. Length 0·8 ; breadth 0·22."

Localities. Middle Glacial, Hopton. Upper Glacial, Bridlington. Recent, Newfoundland and Greenland.

The specimen figured was obtained from the Middle Glacial sand of Hopton Cliff by Mr. Dowson, and it is the only one in fair preservation that has occurred there ; but small specimens somewhat worn, and portions of shells more or less worn, are not uncommon at that locality. This shell has also occurred in the Bridlington bed, but I do not know it from any other formations in Europe than that and the Middle Glacial of East Anglia.

VENUS GALLINA, *Linné*. Supplement, Tab. X, fig. 23.

> VENUS GALLINA, *Linn*. Syst. Nat., edit. xii, p. 1130.
> — — *Phil*. En. Moll. Sic., vol. i, pp. 46, 48.
> — — *Jeffreys*. Brit. Conch., vol. ii, p. 344, var. *gibba?*

Locality. Post Glacial, Kelsey Hill.

In Mr. Prestwich's paper on the Kelsey Hill deposit (' Geol. Journ.,' vol. xvii) is a list of the shells by Mr. Jeffreys, among which is the name of *Venus striatula*, " a single valve," and this specimen Mr. Prestwich has obligingly permitted me to have the use of for the above illustration.

This is of the more rounded form, corresponding in that respect with the Mediterranean form of the species, and it has not the projecting termination on the siphonal side like the British variety. It much resembles the figure in the ' Ency. Meth.,' pl. 286, fig. 3, only that in our shell the concentric ridges are more distant. This may probably be referred to the variety called by Mr. Jeffreys *gibba*, though our shell is not very tumid. Philippi gives this species as recent from the Mediterranean and fossil from the Sicilian beds.

This Kelsey Hill shell has, like the shells of that deposit in general, a very recent aspect, appearing to have retained much of its animal matter. There are vestiges of a finely crenulated margin which it once possessed, but the specimen has been slightly waterworn. I have not met with it from any older deposit in East Anglia, although the name of *Venus gallina* is in the list of shells from Harwich by Mr. Webster before spoken of. As this reference of Mr. Webster's has so long remained unconfirmed he possibly mistook for the present species some specimens of *V. imbricata*.

TAPES PULLASTRA, *Montague.* Supplement, Tab. IX, fig. 1 *a—b.*

> VENUS PULLASTRA, *Mont.* Test. Brit., p. 125.
> TAPES — *Forb. and Hanl.* Brit. Moll., vol. i, p. 382, pl. xxv, fig. 23.
> — — *Jeffreys.* Brit. Conch., vol. ii, p. 355, pl. xxxix, fig. 6.

Localities. Red Crag, Walton? Waldringfield. Middle Glacial, Hopton and Billockby?

The smaller figure represents a perfect specimen (now in the Brit. Mus.) obtained by Mr. Charlesworth from Waldringfield; this is located in a mass of indurated clay. The larger specimen, fig. 1 *a,* was found by myself at Walton, and, I think, belongs to the same species, but the outer surface is gone, either by decortication or abrasion, so that I am not able to tell from its form whether it belongs to this species or to *decussatus.* I have also found an imperfect specimen in the Cor. Crag (almost a facsimile of the specimen figured 1 *a*) with the exterior surface removed. The hinges of either this species or of *virgineus* (or probably of both) are abundant in the Middle Glacial sand of Billockby and Hopton.

TAPES AUREUS, *Gmel.* Crag Moll., vol. ii, p. 202, Tab. XX, fig. 2.

Locality. Fluvio-marine Crag, Bramerton.

This species is, I am informed, abundant near Norwich, but I have not seen it from either the Coralline or the Red Crag. It is, therefore, one of the two or three species occurring in the Fluvio-marine Crag that, having regard to the conditions of the Red Crag deposit, we should have expected to have occurred also in that Crag, on the assumption of its being coeval with the Fluvio-marine.

TAPES DECUSSATUS, *Linn.* Supplement, Tab. X, fig. 4.

> VENUS DECUSSATA, *Linn.* Syst. Natur., p. 1135.

Localities. Post Glacial, Nar Brickearth, Pentney and Bilney.

The specimen figured is from the late Mr. Rose's Nar Brickearth collection, and this deposit is the only one among the Upper Tertiaries of the East of England in which it has for certainty occurred. It is, I believe, abundant in the deposit at Selsey, and in that near Belfast, and in the Clyde beds.

VENERUPIS IRUS, *Linn.* Crag Moll., vol. ii, p. 205, Tab. XIX, fig. 6 *a—b.*

Localities. Cor. Crag, Sutton. Red Crag, Walton.

I have recently found a fragment of this shell in the Cor. Crag of Sutton, having known it previously only from the Red Crag of Walton. *Venerupis Irus* so much resembles the genus *Tapes* in every respect that I believe it belongs to the *Veneridæ* and not to the *Saxicavidæ.*

GASTRANA LAMINOSA, *J. Sow.* Crag. Moll., vol. ii, p. 217, Tab. XXV, fig. 1.

Localities. As in 'Crag. Moll.'

This is (or was) not very rare either in the Cor. Crag of Orford and Sutton or in the Red Crag at Walton Naze and Sutton, but I do not know whether it extended its existence into the Butley Crag. The South African shell, called *Petricola ventricosa*, Krauss, ' Sud. Afrikan. Mollusk,' very much resembles our species, and seems to be its representative in the southern hemisphere in the same manner as many Australian marine shells are said to be identical with those in Europe. *Tellina Guinaca*, Chemn., vol. x, p. 346, tab. 170, figs. 1651-3, a species from Tranquebar, is another shell that can scarcely be removed from our Crag species. *Gastrana laminosa* is given as *Fragilia laminosa* by Mr. Jeffreys in his list accompanying Mr. Prestwich's Cor. Crag paper, p. 139.

DONAX VITTATUS, *Da Costa.* Crag Moll., vol. ii, p. 219, Tab. XXII, fig. 7.

Localities. Fluvio-marine Crag, Bulchamp, Postwick, and Bramerton? Chillesford bed, Horsted and Aldeby. Lower Glacial, Belaugh.

In the ' Crag Mollusca' this shell is given from the Crag of Bramerton, but Mr. Reeve does not appear to have detected it there. It is, however, given by the late Dr. Woodward in his list in White's 'Directory' as from Bulchamp and Postwick as well as Bramerton. If it be thus present in the Fluvio-marine Crag it is another of the few shells that, on the hypothesis of the Red and Fluvio-marine Crags being coeval, ought to occur in the Red Crag, but which has not yet been detected there. I have found this species myself at Horsted and·Belaugh, and it was sent to me from Aldeby by Messrs. Dowson and Crowfoot.

DONAX POLITUS, *Poli*. Crag Moll., vol. ii, p. 220, Tab. XXII, fig. 9.

Localities. Cor. Crag, Sutton and Gedgrave. Red Crag, Walton (*A. Bell*), and Sutton.

Mr. Bell gives this shell ('Ann and Mag. Nat. Hist.,' 1871) from the Red Crag of Walton and Sutton, and I have myself a specimen from the latter locality. At the time of the publication of the 'Crag Mollusca' I had only seen it from the Cor. Crag.

PSAMMOBIA COSTULATA, *Turton*. Supplement Tab. X, fig. 7.

PSAMMOBIA COSTULATA, *Turt*. Conch. Dith., p. 87, tab. vi, fig. 8.

Locality. Coralline Crag, Sutton.

In 'Brit. Conch.,' vol. ii, p 395, the author says, "I observed a specimen (of this species) in Mr. Searles Wood's collection in the Brit. Museum." I have here had the specimen in question figured, and the radiating lines, similar to those which distinguish the recent *costulata*, are very apparent; but the form of the fossil does not accurately agree with that of the recent shell, and I am not satisfied that it is the same species. Should further specimens turn up which present the same difference in form from the recent shell, I should propose for the Crag shell the name *pseudo-costulata*.

PSAMMOBIA TELLINELLA. Crag Moll., vol. ii, p. 223, Tab. XXII, fig. 4.

Locality. Cor. Crag, Sutton.

Perfect specimens of this shell are very rare with me, but I have met with fragments indicating a length of more than an inch, which seems somewhat to exceed that of the British shell of this name, and it is rather more elongated and less tumid. It appears to me more to resemble the Touraine shell called *Psam. affinis*, Dujardin ('Mém. Géol. Soc. Fr.,' tom. ii, pt. ii, p. 257, Pl. XXVIII, fig. 4), but I have not been able to obtain a specimen of that fossil for comparison, and I have left it for the present with its original name. I have seen the Crag shell only from the Cor. Crag of Sutton.

SOLECURTUS STRIGILLATUS, *Linné*. Crag Moll., vol. ii, p. 252, Tab. XXV, fig. 3 (as *Macha strigillata*).

Localities. Cor. Crag, Sutton.

This Mediterranean shell is regarded as distinct from the British one called *candidus*, but I can discover no difference between them except in size and colour, and as stated in

20

the 'Crag Mollusca,' remains of colour in the Crag specimens (which are fragmentary) induced me to refer them to the Mediterranean form. I was misled by Herrmannsen (who gave the date of the name *Macha* as of 1815 instead of 1835) in using that generic name for this shell. *Solecurtus antiquatus* is mentioned in 'Brit. Conch.,' vol. iii, p. 7, and by the author of that work in his list to Mr. Prestwich's Cor. Crag paper, as a Coralline Crag shell; while Mr. Bell, in his paper on the "English Crags" ('Proc. Geol. Association,' 1872) inserts it as a Red Crag species. The only specimens, however (which are all fragmentary), that I have seen from any part of the English Crag belong either to *strigillatus* or *candidus*.

Much difficulty seems still to exist respecting the siphonal side of shells of this family. In 'Brit. Moll.,' vol. i, Pl. I, as I before pointed out ('Crag Moll.,' vol. ii, p. 254), the illustrations for this genus, as well as for other genera in the same plate, show the foot protruded on the siphonal or ligamental side of the shell, and the same misrepresentation is repeated in the generic illustrations in Pl. I, vol. iii, of 'Brit. Conch.' The sinuated mark in the interior of a bivalve, when it exists (as left by the impression of the retractor muscles), is on the side which bears the ligament, and *the siphons* are protruded in that direction, the foot going in the opposite.

CULTELLUS SUTTONENSIS, *S. Wood.* Supplement, Tab. X, fig. 15.

Spec. char. C. *Testa transversa, oblongo-lineari, rectiuscula, lævigata, tenuis, fragilis, antice breviore, rotundato-truncata, postice longiore et latiore; valdè inæqui-lateralis, in valvula dextra bidentatis, in valvula sinistra tridentatis.*

Length $\frac{6}{8}$ of an inch.

Locality. Cor. Crag, Sutton.

Some fragments of this shell have been long in my cabinet, and I had imagined them to belong to the same species as that from the Red Crag of Walton, which I had in the 'Crag Mollusca' figured under the name of *tenuis*, Phil. I now believe these to be specifically different, and a perfect specimen having been obtained by Mr. Robert Bell, I have figured it as above. The shell differs from *pellucida* in the absence of curvature and the broadness of the posterior extremity; it differs also from the Upper Eocene species, *C. Grignonensis*, Desh. ('An. sans. vert. du Bas. de Par.,' Tom. 1, p. 157, Pl. VII, figs. 13—15) in its outline, that shell approaching nearer to *pellucidus* than does our own shell. A fragment in my possession indicates a length of more than three fourths of an inch.

A specimen of *Cultellus* from the Aldeby bed sent me by Messrs. Crowfoot and Dowson is represented in Tab. X, fig. 14. This seems to be intermediate between the Cor. Crag form *Suttonensis* and the recent form *pellucidus*, as the formation from which it comes is correspondently intermediate in time. It is therefore not unlikely that

Suttonensis was the antitype from which the recent *pellucidus* has descended through such modifications as are exhibited in the Aldeby specimen. I have had the recent shell figured beneath these two (Tab. X, fig. 16) for comparison with them.

CULTELLUS CULTELLATUS, *S. Wood.* Crag Moll., vol. ii, p. 258, Tab. XXV, fig. 2 (as *C. tenuis*, Phil.)

CULTELLUS CULTELLATUS, *S. Wood.* Catalogue, Ann. & Mag. Nat. Hist., 1840, p. 245.

Locality. Red Crag, Walton Naze.

In the 'Crag Mollusca' I referred some fragments of a *cultellus* from the Coralline Crag and some perfect valves from the Walton Red Crag to Philippi's species *tenuis*. This I now believe to have been an error. The Red Crag specimens, which are perfect, enable me to refer them to a new species under the name *cultellatus*, which is the name I proposed for both the Coralline and Red Crag forms in my catalogue of 1840.

With respect to the fragments from the Coralline Crag, however, they are insufficient for any satisfactory determination, and may probably be fragments of the before-described *Suttonensis*.

If either *cultellatus* or *Suttonensis* were regarded as identical with *pellucidus*, then most assuredly the Eocene species, *Grignonensis* and *affinis*, would have to be treated as identical also, for the differences between *Suttonensis* and *Grignonensis*, and between *cultellus* and *affinis*, are less than between the Crag shell and *pellucidus*.

In 'Brit. Conch.,' vol. iii, p. 15, *Solen pellucidus* is said to be a Coralline Crag species, but in the list to Mr. Prestwich's Cor. Crag paper *Cultellus tenuis* is mentioned, and not *pellucidus*, while in the Red Crag list *Solen pellucidus* is given as a species from Aldeby.

SOLEN SILIQUA, *Linn.* Crag Moll., vol. ii, p. 255, Tab. XXV, fig. 7.

Localities. Red Crag, Sutton. Fluvio-marine Crag, Bramerton. Chillesford bed, Aldeby? Middle Glacial, Hopton?

This is the only species yet known from the Fluvio-marine Crag in which it is rare. Fragments of a *Solen* have occurred at Aldeby and at Hopton, but neither of them sufficiently characteristic to permit a certain reference to this species. I have therefore given it from the Chillesford Bed and from the Middle Glacial sand with doubt. The species is still unknown to me from the Coralline Crag.

TELLINA PULCHELLA ? *Lamarck*.   Supplement, Tab. X, fig. 5.

Locality.   Cor. Crag, Sutton.

An imperfect valve belonging to the genus *Tellina* was found by myself as above, and it had for some time been in my cabinet.   This specimen, which is the right valve, has only the anterior portion remaining, the siphonal side of the shell, which would best assist in its determination, being unfortunately wanting.   My specimen, however, shows that the hinge possessed two cardinal teeth, one single and the other bifid, and one elongated *lateral* tooth on the broader or anterior side resembling in its dental formula the hinge of *Tellina donacina*, but the present specimen is much larger than any specimen of that species that I have seen, and it is also a flatter shell; like it, however, it is covered with broad and distinct striæ or lines of growth.   I have given it provisionally the above name. Since the specimen was engraved, and while in the hands of the engraver, it has, I regret to say, been destroyed.

TELLINA COMPRESSA, *Broc.*   Crag Moll., vol. ii, p. 234, Tab. XXII, fig. 6 (as *T. donacella*).

Locality.   Cor. Crag, Sutton.

This is an exceedingly rare shell to my researches.   In the ' Crag Moll.' I observed that this might probably be referred to *T. compressa*, Broc., of which a very bad representation is given in ' Conch. Foss. Subap.,' Pl. XII, fig. 9.   It is the same as *T. distorta*, Dubois, ' Volh. Pod.,' Tab. V, figs. 3, 4.   Mr. Jeffreys has obligingly sent me some recent specimens from the Coast of Portugal with the name of *Tellina compressa*, Broc., for comparison, and these correspond with my Crag shell.

TELLINA FABULA, *Gronov.*   Crag Moll., vol. ii, p. 232, Tab. XXI, fig. 3.

Localities.   Fluvio-marine Crag, Bramerton and Bulchamp.   Chillesford bed, Aldeby and Horstead.

In the ' Crag Moll.' this shell was figured from a recent specimen, the only Crag one then known having been lost.   A specimen has lately been obtained from the Fluvio-marine Crag at Bramerton by Mr. Reeve; one was sent to me from Aldeby by Messrs. Crowfoot and Dowson, and two were obtained by myself from Horstead.

TELLINA BALTHICA, *Linné.*   Crag Moll., vol. ii, p. 231, Tab. XXII, fig. 1.

Localities.   Lower Glacial, Weybourne, Belaugh ; Rackheath, Crostwick, Spixworth, and Wroxham.   Middle Glacial, Hopton, Billockby and Clippesby.   Upper Glacial, Bridlington and Dimlington.   Post Glacial, March, Kelsea Hill, Paull Cliff, Hunstanton, and Nar Brickearth.

This shell was erroneously given by me from Bramerton, where it has never occurred. The Weybourne locality for it referred to the Mammaliferous Crag in the ' Crag Mollusca' is the Lower Glacial sand of that place.   Not a vestige of the shell has yet occurred in any bed that can be shown to be as old as the Chillesford Clay.   In all the Glacial and Post-glacial Beds, from the oldest to the newest, it occurs in profusion, which is the character of its occurrence as a living shell.

TELLINA CRASSA, *Penn*.   Crag Moll., vol. ii, p. 226, Tab. XXI, fig. 1.

Localities.   Cor. Crag, Sutton, and near Orford.   Red Crag, Walton, Sutton, and Butley. Fluvio-marine Crag, Thorpe, by Norwich.   Chillesford bed, Chillesford.   Middle Glacial, Hopton and Billockby.

This shell is abundant in every part of the Red Crag except at Walton, and in the Scrobicularia beds or uppermost part of the Red Crag.   In the Fluvio-marine Crag it seems only to have occurred, and that rarely at Thorpe, from the bed of large angular flints, at the base of which section (supposed by some geologists to be of terrestrial origin) Mr. Crowfoot obtained it.   I have not seen it from any locality of the Chillesford Bed except that of Chillesford itself.   I have not met with it from the Lower Glacial Sands, but in the Middle Glacial sands it abounds, though in a more or less fragmentary condition.

TELLINA LATA, *Gmel*.   Crag Moll., vol. ii, p. 228, Tab. XXI, fig. 6.

This shell is unknown at Walton, and rare in the rest of the Red Crag,[1] but very common in the Fluvio-marine Crag and Chillesford bed passim.   It is not uncommon in the Lower Glacial sand of Rackheath and Belaugh, and some fragments from the Middle Glacial sand of both Hopton and Billockby may probably belong to it, but they are not sufficiently perfect for determination.   I observed a perfect specimen in Mr.

---

[1] I erroneously stated in my paper on the " Red Crag," in the ' Quart. Journ. of the Geol. Soc.,' that it was common in the Red Crag of Butley, but such is not the case although I have found it there.

Rose's collection from the Nar Valley Brickearth, and it is mentioned by Mr. Seeley (under the name of *T. proxima*) as occurring in the gravel of March, though Mr. Harmer, who has searched the gravel of that place very assiduously, has not been able to meet with it.

The shell which I have called *lata* was figured by Lister, and called by him *Tellina lata alba*, and the name *lata* was adopted by Gmelin. This is the oldest notice of the shell. The late Edward Forbes, 'Mem. Geol. Sur.,' 1846, p. 412, considered *T. obliqua* as merely a variety of Lister's species from the seas of Norway. Although I have the highest respect for the opinion of the late E. Forbes (whose premature loss all geologists deplore) I must in this instance dissent from it, notwithstanding that it has been adopted by Mr. Jeffreys. Similarly E. Forbes regarded *T. prætenuis* as merely a variety of this shell, and Mr. Jeffreys has followed him. The word variety as distinguished from species is, to my apprehension, too vague to carry any precise meaning with it, but as regards these three forms, *lata*, *obliqua*, and *prætenuis*, there can be no question as to their complete distinction. *T. obliqua* is the only one of the three forms found in the Cor. Crag, but in the Red Crag lying between the Stour and the Alde we get this shell and *T. prætenuis* abounding together, and *T. lata* very rare, while in the Fluvio-marine Crag and Chillesford bed we get all three forms in abundance and well marked. I am sure that all collectors from the Crag will bear me out in saying that the three forms thus occurring in profusion together can be without difficulty selected, and that they do not form merely the terms of an undistinguishable series. They were therefore three distinct forms living in the same sea, and not intermingling; and what else constitutes species? Two of them are, so far as I am aware, not known living, for nothing recent that I have seen can justly be identified with either *obliqua* or *prætenuis*. In the first formation upwards from the Crag, viz. the Lower Glacial sands, *T. prætenuis* becomes rare, while *obliqua* maintains itself there in great profusion. As we ascend in the order of formations we lose the form *prætenuis* altogether, but still meet with *obliqua* in the Middle Glacial sands and at Bridlington, but it has not yet been met with in any Post-glacial bed, or in any of the English Newer Glacial, or in any of the Scotch beds; while *lata* not only occurs in these beds, but survives as an Arctic shell. The history of these most important shells (for it is the most abundant species that are the most important in a geological point of view) is thus clearly traceable in time through the various formations much in the same way as I have traced that of the common *Trophon antiquus*.

ABRA ALBA, *W. Wood.* Crag Moll., vol. ii, p. 237, Tab. XXII, fig. 10.

Localities. Cor. Crag, Sutton. Red Crag, Walton, Sutton, Bawdsey. Fluvio-marine Crag, Thorpe, by Norwich? Bulchamp? Chillesford bed, Chillesford? Aldeby? Post-glacial, Nar Brickearth, Pentney (*Rose*).

The localities of Thorpe, Bulchamp, and Chillesford are on the authority of the late Dr. Woodward's list, and, as I have not been able to confirm them, they are given with doubt. A specimen sent to me by Messrs. Crowfoot and Dowson from Aldeby seems not to be sufficiently perfect to be free from doubt.

ABRA FABALIS, *S. Wood.* Crag Moll., vol. ii, p. 238, Tab. XXII, fig. 12.

Dr. Hörnes has given this Crag shell as a synonym to the fossil from the Vienna basin, which he has called *Syndosmya apelina*, Kien, vol. ii, p. 77, Tab. VIII, fig. 4. This shell, however, judging from the figure he gives, resembles *Ligula profundissima*, Forbes ; but mine is, I think, distinct from either, and I have, therefore, retained its original name.

SCROBICULARIA PLANA, *Da Costa.* Crag Moll., vol. ii, p. 235, Tab. XXII, fig. 14 (as *Trigonella plana*).

Localities. Red Crag, Bawdsey? Chillesford, and Sudbourn. Chillesford bed, Horstead, Chillesford, Tunstall Heath, and Sudbourn Church Walks. Fluvio-marine Crag, Bramerton. Lower Glacial, Belaugh. Middle Glacial, Billockby and Hopton. Post-glacial, March, Hunstanton, Nar Valley, Pentney.

The presence of this shell on the comparison of the Bramerton and Chillesford sections is discussed in the introduction to this Supplement. It occurs abundantly in the newest part of the Red Crag and in the Fluvio-marine Crag at Bramerton. It is not uncommon in the Lower Glacial sands, and in the Middle Glacial sands the hinge portion of specimens are very abundant. In all but the uppermost part of the Red Crag (4''' of the Introduction to this Supplement, page viii), of which it is the characteristic shell, it was unknown to me, but I understand from Mr. A. Bell that he has found it at Bawdsey.

MACTRA OVALIS, *J. Sow.* Crag Moll., vol. ii, p. 246, Tab. XXIII, fig. 1.

Localities. Red Crag passim excepting at Walton. Fluvio-marine Crag passim. Chillesford Bed passim. Lower Glacial, Belaugh, and Middle Glacial, Billockby and Hopton. Post-glacial, Kelsea Hill, Paul Cliff, March, and Hunstanton.

This shell abounds in every part of the Red Crag, excepting at Walton and in the Fluvio-marine Crag in the Chillesford bed and in the Middle Glacial, but is somewhat rare in the Lower.

MACTRA SOLIDA, *Linné.*　Crag Moll., vol. ii, p. 245, Tab. XXIV, fig. 4.

Localities.　Cor. Crag, near Orford.　Red Crag, Sutton, Bentley and Butley. Fluvio-marine Crag, Bramerton.　Chillesford bed, Bramerton and Aldeby ?　Post-glacial, Kelsea Hill, March, and Nar Brickearth.

A specimen from Aldeby, sent me by Messrs. Crowfoot and Dowson, is referable with some doubt to this species.　The specimen in the British Museum, which was the authority on which Dr. Woodward gave it as from Bridlington, does not seem to me to be a fossil of that place.　It is said by Mr. Reeve to be now common at Bramerton.　I have had figured in my Supplement, Tab. X, fig. 10, a peculiar and angular form, from Mr. Rose's collection, and this came from the Cor. Crag, near Orford.

MACTRA SUBTRUNCATA, *Da Costa*, Crag Moll., vol. ii, p. 247, Tab. XXIV, fig. 3.

Localities.　Red Crag, Sutton ?　Fluvio-marine Crag, Bramerton.　Chillesford bed, Aldeby.

MACTRA STULTORUM.　Crag Moll., vol ii, p. 242, Tab. XXIII, fig. 3.

Localities.　Cor. Crag, Sutton.　Red Crag, Sutton and Bawdsey (*Bell*).　Fluvio-marine Crag, Bramerton.　Middle Glacial, Hopton ?

This shell seems to be rare in the Fluvio-marine Crag as well as in the Red, and not to have occurred in either the Chillesford bed, or the Lower Glacial sands; but a fragment comprising nearly the whole of the anterior edge of the shell and the hinge, from the Middle Glacial of Hopton, seems referable to this species.

MACTRA GLAUCA.　Crag Moll., vol. ii, p. 241, Tab. XXIII, fig. 2.

In addition to the locality given in the ' Crag Mollusca' I have obtained several valves of this species from the Red Crag at Bentley, and I have seen others in the collection of Mr. Miller, of Ipswich.　Some fragments that I have from the Cor. Crag of Sutton may probably be referred to this species; but it is difficult to determine satisfactorily from imperfect specimens the difference between *glauca* and *stultorum.*

MACTRA ARCUATA. Crag Moll., vol. ii, p. 243, Tab. XXIII, fig. 5.

Localities. Cor. Crag, Sutton, Ramsholt, and near Orford. Red Crag, Walton, Sutton, Bawdsey. Fluvio-marine Crag, Yarn Hill and Postwick.

This shell has recently been found at Postwick by Mr. Reeve, and some young specimens from Yarn Hill were sent me by Messrs. Crowfoot and Dowson.

In Mr. Jeffreys' list (appended to Mr. Prestwich's paper) *M. arcuata* is given as a var. of *M. glauca*, from which opinion I dissent. There appears to me a greater difference between *glauca* and *arcuata* than there is between *glauca* and *stultorum*, which are both recent forms. *M. arcuata* is only known to me as a fossil.

MACTRA ARTOPTA. Crag Moll., vol. ii, p. 244, Tab. XXIII, fig. 4.

This is at present restricted in England to the Cor. Crag, and there it is rare to my researches. It may probably be referred to *Mactra podolica*, Eichw., in Hörnes' 'Vienna Foss.,' vol. ii, p. 62, Tab. VIII, figs. 1—8, where several varieties are figured. *Mactra deltoides*, Dubois, 'Voth. Pod.,' Tab. IV, figs. 5, 6, may also be the same species; but *semisulcata* is quite distinct. *M. artopta* is not known to me as living species. It is placed by Mr. Jeffreys in his list to Mr. Prestwich's Cor. Crag paper as another variety of *glauca*, but from that I dissent.

LUTRARIA ELLIPTICA, *Lamarck*. Crag Moll., vol. ii, p. 251, Tab. XXIV, fig. 1.
Supplement, Tab. X, fig. 19.

Localities. Cor. Crag, Ramsholt, Sutton, and near Orford. Red Crag, Sutton.

In 'Brit. Conch.,' vol. ii, p. 431, *Lutraria oblonga* is said to occur in the Coralline Crag, and the specimen now figured (which is from my collection in the British Museum) has that name written by Mr. Jeffreys on the back of the tablet, and he has introduced the name in his list to Mr. Prestwich's Cor. Crag paper. I have in consequence had the specimen in question figured. It was included with my other specimens as a variety of *elliptica*, which I still believe it to be, and have in consequence retained it under that name. The specimen in question resembles a figure by Dr. Hörnes, 'Foss. Moll. Wien,' Pl. V, fig. 7 *a—c*, which he gives as a variety of *oblonga*, but which M. Cha. Mayer has considered a distinct species under the name of *Lutraria Hornesii* ('Moll. Tert. du Mus. Zurich,' p. 52, fig. 47, 1867).

21

A fossil in my possession from Bordeaux, with the name of *L. sanna*, which was sent to me many years ago by the Comte du Chastel, seems to correspond with the specimen now figured, but is unlike the figures of *L. sanna* given by Basterot, Pl. 7, fig. 13. I have under these circumstances omitted the name of *L. oblonga* from my tabular list.

THRACIA DISTORTA ? *Mont.*

This species is given by Mr. Jeffreys in his list to Mr. Prestwich's Cor. Crag paper, as occurring in the Coralline Crag. The only specimen that has come to my knowledge, upon which such reference could be made, is that figured by me in the ' Crag Mollusca ' as a distortion of *T. pubescens* (' Crag Moll.,' vol. ii, p. 259, Tab. XXVI, fig. 1 *e, detruncata*), but I do not feel fully justified on the authority of this solitary distorted specimen in adopting the name of *distorta* into the list of Crag shells, and I give it, therefore, with doubt.

THRACIA PAPYRACEA, *Poli.* Crag Moll., vol. ii, p. 260, Tab. XXVI, fig. 2 (as *Thr. phaseolina*).

Localities. Cor. Crag, Sutton, and near Orford. Chillesford bed, Sudbourn Church Walks. Lower Glacial, Belaugh.

Two specimens of this species have been found at Sudbourn, and one by myself in the Lower Glacial sand at Belaugh. This appears now to be generally admitted as *Tellina papyracea* of Poli, and as that has precedence in date to Lamarck's name it is here restored. *Anatina papyracea*, Gould, appears to be different, and to belong to the genus *Cochlodesma*, judging from the representation of the hinge.

THRACIA VENTRICOSA, *Phil.·* Crag Moll., vol. ii, p. 262, Tab. XXVI, fig. 5.

Locality. Cor. Crag, Ramsholt, and near Orford.

This is still with me a very rare shell. In ' Brit. Conch.,' vol. ii, p. 40, as also in the list by Mr. Jeffreys accompanying Mr. Prestwich's paper, this Coralline Crag shell is referred to *Mya convexa*, W. Wood, but I cannot change the name I originally gave to the Crag shell, believing as I do that is distinct from this recent British shell *convexa*. In the ' Crag Mollusca ' I referred the Crag shell to Phillips' species *ventricosa* with doubt, as it did not satisfactorily agree with a Sicilian specimen in my possession, but which specimen, in its turn, did not quite correspond with Phillips' description of this species.

I have not been able to clear up this doubt, but Dr. Hörnes refers one of his Vienna shells to *ventricosa*, placing the Crag species with it as a synonym (Hörnes, ' Foss. Moll.,' vol. ii, p. 48).

PANDORA INEQUIVALVIS, *Linn*. Crag Moll, vol. ii, p. 270, Tab. XXV, fig. 5.

Localities. Cor. Crag, Sutton, and near Orford. Red Crag, Walton Naze, and Waldringfield. Middle Glacial, Hopton.

I expressed my opinion in the ' Crag Mollusca' that *inequivalvis* and *pinna* were not entitled to specific separation, following in this Montague, and I only separated them out of deference to the authors of the ' British Mollusca.' As my view has received the support of the author of the ' Brit. Conchology,' it seems desirable to unite the two forms under the same specific name. In the ' Crag Moll.' I gave *inæquivalvis* as the Coralline Crag form and *pinna* (*obtusa*) as the Red. Since then, however, I have found both in the Cor. Crag of Sutton, and Mr. Bell (' Ann. Mag.,' May, 1871) gives *obtusa* from the Cor. Crag near Orford.

The hinge portion of one specimen has occurred in the Middle Glacial, but to which of the two varieties it belongs cannot be said.

SAXICAVA ARCTICA, *Linné.* Crag Moll., vol. ii, p. 287, Tab. XXIX, fig. 4, and p. 285, Tab. XXIX, fig. 3, as *S. rugosa.*

Localities. Cor. Crag, Sutton. Red Crag, Walton, Sutton, and Butley. Fluvio-marine Crag, Bramerton. Chillesford bed, Bramerton and Aldeby. Lower Glacial, Belaugh. Middle Glacial, Billockby and Hopton. Upper Glacial, Bridlington and Dimlington.

I observed in the ' Crag Mollusca' that, although I had kept *arctica* and *rugosa* as distinct species, I did not believe in any grounds for their distinction. I have, therefore, now united them. The very gigantic *rugosa* form, so characteristic of the Canadian beds, seems confined to the later glacial beds of Britain, as it appears only at Bridlington, Dimlington, and in the yet more recent Clyde deposits. It belongs to the truly arctic fauna that established itself in Britain towards the close of the glacial period, and seems absent from the older glacial beds, in which the small smooth form alone occurs. The rugose form occurs in the Coralline and Red Crags, but of smaller size. The species is common in the Cor. Crag, but not so in the Red. In the Fluvio-marine Crag, in the Chillesford bed, and in the Lower Glacial sands it is rare, but it is extremely common (though always fragmentary) in the Middle Glacial at Hopton, and in all these deposits it is the thin and non-rugose form that occurs.   At Bridlington I believe it is not uncommon, but there and at Dimlington it is the gigantic rugose form that occurs.

SAXICAVA? FRAGILIS, *Nyst.*   Crag Moll., vol. ii, p. 288, Tab. XXIX, fig. 6.

Localities.   Cor. Crag, Sutton, and near Orford.

This shell was described by Montague under the name *Mytilus plicatus*; but *plicatus* of Chemnitz is a different species inhabiting the Nicobar Islands.   I have, therefore, retained Nyst's name for our shell.

There are two or three small bivalves which have been hitherto called by different generic names, and have as yet no proper resting-place, and which I consider ought to have a distinctive generic appellation of their own.   The first of these is the above *fragilis.*   Another is *Saxicava? carinata* ('Crag Moll.,' vol. ii, p. 289, Tab. XXIX, fig. 5), which I believe to be specifically distinct from *fragilis*; while it seems probable that a fossil of the older tertiaries of America called by Lea ('Contrib. to Geol.,' p. 48, pl. i, fig. 16) *Byssomia petricoloides* may prove a third; but the hinge of Lea's specimen is not perfect.

The subjoined list of synonyms shows how the first of these shells, *fragilis,* has been bandied about from genus to genus :

> *Mytilus plicatus*, Mont.   Test. Brit. Suppl., p. 70, 1808.
> *Saxicava plicata*, Turt.   Brit. Biv., p. 22, 1822.
> — ? *fragilis*, Nyst.   Coq. Foss. de Belg., p. 97, Pl. 4, fig. 10 *a, b*, 1843.
> — *rugosa juv.*, Forb. and Hanl.   Brit. Moll., vol. i, pl. 6, figs. 1—3.
> *Sphenia cylindrica*, S. Wood.   Catalogue, 1840.
> *Magdala plicata*, Gray.   List of Brit. Biv., p. 161, 1851.
> *Anatina*   — *id.*   Ann. of Philos., 1825.
> *Psammobia?*— Jeffreys.   Ann. and Mag. Nat. Hist., vol. xix, p. 314.
> *Lyonsia*   — Forb. and Hanl.   Brit. Moll., vol. i, p. 218, 1853.
> *Panopea*   — Jeffreys.   Brit. Conch., vol. iii, p. 75, 1865.
> *Myrina oceanica*, Conti., 1864.   *Fide* Jeffreys, Brit. Conch., vol. v, p. 192.

SPHENIA OVATA?   Crag Moll., vol. ii, p. 276, Tab. XXIX, fig. 7 (as *S. Binghami*).

Locality.   Cor. Crag, Sutton.

The shell figured in the 'Crag Mollusca' was referred with doubt to the recent British species *Binghami.*   Dr. P. Carpenter, however ('Geol. Mag.,' vol. ii, p. 153), says that the Crag shell more resembles *Sph. ovata* from the north-west coast of America.

Although I have not been able to see a specimen of that species to compare with it, yet, as I now think that the shell figured in the ' Crag Moll.' is not *Binghami*, I have assigned it (still with doubt) to the species mentioned by Dr. Carpenter, and on his authority. I have seen no further specimens of the shell.

SPHENIA BINGHAMI, *Turton*.   Supplement, Tab. X, fig. 13.

Localities.   Cor. Crag, Sutton.   Red Crag, Sutton.

The specimen figured in this Supplement was given to me by Mr. Charlesworth, who found it in a fragment of indurated mud from the phosphatic nodule bed at Waldringfield, and this, I have no doubt, is identical with the recent shell *Sphenia Binghami*. I have retained the above generic name as I think the shell is distinct from *Mya* or *Corbula*.   The left valve has a projection on which is placed the cartilaginous connector; this extends along and under the dorsal margin on the siphonal side, and the right valve has a depression for its reception.   The interior of the shell is marked with a shallow sinus.   This shell has been so fully described as a species by the British Conchologists that it is not necessary to repeat the description here.

Genus—LYONSIA.

Tab. X, fig. 17, of Supplement represents the hinge portion of a bivalve from the Coralline Crag of Sutton, which I believe belonged to a species of *Lyonsia*.   This fragment shows the depressed and elongated pit beneath the dorsal margin in which, when alive, was placed the internal connector.   This pit is peculiar in its form, but I cannot find that it has ever been represented; although *Lyonsia Norvegica* has been figured about twenty times and described under eight generic and nine different specific names, the hinge has been mentioned only so far as that it possessed a movable and calcareous ossicle.   My fragment is very thin, fragile, and iridescent.

CORBULA CONTRACTA ? *Say*.   Supplement, Tab. X, fig. 11 *a—c*.

CORBULA CONTRACTA, *Say*.   Journ. Acad. Nat. Sci., ii, 312.
—   —   *Gould*.   Inv. Massach., p. 43, fig. 37.
—   —   Id.   2nd edit., p. 60, fig. 377.

Localities.   Fluvio-marine Crag, Bramerton.   Chillesford bed, Ditchingham and Easton Bavent.   Middle Glacial, Hopton.

Some small specimens of a *Corbula*, which I have referred as above, were sent to me intermixed with *C. striata*, by Mr. Reeve, from Bramerton, and I have also some from Easton Bavent, and from the small patch of fossiliferous pebbles at Ditchingham. Some nearly perfect specimens have also occurred in the Middle Glacial of Hopton. These specimens are smooth, but that is usually the case with all those which can be readily referred to *C. striata;* the principal distinction, as pointed out by the American authors, is the nearly equal size of the valves of the present species, and the more inequilateral form of the valve, the posterior side being elongated. The recent shell is said to be abundant on] the Coast of Georgia and East Florida. The specimen figured is from Bramerton

CORBULA STRIATA, *Walk. and Boys.*　Crag Moll., vol. ii, p. 274, Tab. XXX, fig. 3.

Localities.　Cor. Crag passim.　Red Crag, Walton, Sutton, Bawdsey, and Butley, Fluvio-marine Crag passim.　Chillesford bed passim.　Lower Glacial, Weybourne, Belaugh, and Rackheath.　Middle Glacial, Hopton and Billockby.　Post-glacial, March, Hunstanton, and Kelsea Hill (*Jeffreys*), Nar Brickearth (*Rose*).

This shell is profuse in the Coralline, but somewhat rare in the Red, Crag. In the Chillesford bed it seems to be rare where this is in a Fluvio-marine condition, and commoner as it becomes more marine; but I have not met with it from the most marine development of the bed, viz. that at Chillesford. In the Lower Glacial sands it is common, and in the Middle Glacial very abundant; but I have not seen it from Bridlington. It does not seem common at any of the Post-glacial localities. Fig. 4, Tab. XXX, of ' Crag Moll.,' is the representation of a shell which I referred to *C. rosea*, but which I now believe to be a transverse variety of *striata*.

I have retained the name of *striata* for this shell, conceiving the right of priority to be with Walker and Boys, who figured the recent shell with that name in 1787, whereas the name of *gibba* was given to it by Olivi in 1792. Lamarck at a later date described a Paris basin fossil with the name of *Corbula striata*, and this M. Deshayes has (' An. sans Vert. du Bas. de Paris,' T. 1, p. 221), I think, very properly called *C. Lamarckii,* restoring to the recent shell the name of *striata*. The name *striata* was also adopted by the late Dr. Fleming in his ' Hist. of Brit. Animals,' a very able work at the time it was published, and for which the author has never received the credit he deserved.

CORBULOMYA COMPLANATA, *J. Sow.*　Crag Moll., vol. ii, p. 275, Tab. XXX, fig. 2.

In considering, as I did in the ' Crag Moll.,' the Paris basin shell to be the same as the Crag one, I was influenced by M. Deshayes' having united the Eocene fossil with a shell found in the Touraine Beds. There is, I now think, so great a difference between the two that they ought to be specifically separated; and as Sowerby's name for the Crag shell has priority I so retain it, leaving the Eocene species to find a new one.

NEŒRA OBESA, *Lovén*. Supplement, Tab. X, fig. 9.

> NEŒRA OBESA, *Lovén*. Ind. Moll. Scand., p. 48, 1846.
> — CUSPIDATA, *Jeffreys*. Brit. Conch., vol. v, p. 191.

Locality. Coralline Crag, near Orford.

A single specimen of this species is all that I have seen, which was obtained by Mr. Alfred Bell, and it now belongs to W. Reed, Esq., of York, who has obligingly placed it in my hands for description. Mr. Jeffreys in the Supplement to his 'British Conchology,' p. 192, considered this specimen obtained by Mr. Bell and Lovén's species *obesa* as a variety only of *N. cuspidata*; but in a paper of his in the 'Ann. and Mag. of Nat. Hist.' for June, 1870, on the "Norwegian Mollusca," Mr. Jeffreys says that his reference of *obesa* to *cuspidata* was erroneous, and that he is satisfied that the two species are distinct. Still later, however, in the list to Mr. Prestwich's Coralline Crag paper, he omits *obesa* as a Crag species. I consider it, however, a good species, and have so inserted it.

POROMYA GRANULATA, *Nyst* and *Westendorf*. Crag Moll, vol. ii, p. 268, Tab. XXX, fig. 5 *a—f*.

I have seen this shell from the Cor. Crag only, and there it is by no means abundant. This species was found by Edward Forbes in the Ægean, and although found on the Scandinavian Coast, I consider it properly a southern form. Mr. Jeffreys figured and described a minute bivalve as *Poromya subtrigona*, 'Ann. and Mag. Nat. Hist.,' 3rd series, 1858, p. 42, Pl. II, fig. 1, but in 'Brit. Conch.,' vol. ii, p. 228, he refers this shell to the young of *Kellia cycladia*. I am not aware, therefore, of any other species of this genus besides *granulata*.

PANOPEA NORVEGICA, *Spengler*. Crag Moll., vol. ii, p. 281, Tab. XXIX, fig. 1.

Localities. Red Crag, Sutton and Butley. Fluvio-marine Crag, Bramerton. Chillesford bed, Chillesford. Middle Glacial, Hopton. Upper Glacial, Bridlington.

This shell, though occurring in the Red Crag of Butley, and found in the Chillesford bed at Chillesford with the two valves united, does not appear yet to have occurred at its other localities. In the Middle Glacial sand of Hopton a fragment comprising the hinge and umbonal portion of the shell, showing lines of growth, has occurred. I am not aware whether it be common or not at Bridlington. The genus *Panopea* is certainly not appropriate for this species; but as *Saxicava*, in which other authors place it, appears to me equally inappropriate, I have, until a more suitable genus be erected for it, retained it under the generic name in which it was first figured from the Crag.

PANOPEA FAUJASII, *Menard de la Groye.* Crag Moll., vol. ii, p. 283, Tab. XXVII, fig. 1 *a—f*.

Localities. As in 'Crag Moll.'
In the 'Catal. Syst. et descr. des foss. de Terr. Tert.,' by M. C. Mayer, 1870, three species are given from the English Crag as distinct, viz. *P. Menardi* (*P. gentilis*, Sow.), from the Red Crag, and *P. Rhudolphii* (*P. Ipsviciensis*, Sow.), and *P. Americana*, Conrad, from the Cor. Crag. Having, however, seen and examined a large series of the Crag species of this genus, which present great variation among themselves, I am still of opinion that those shells which I have represented in 'Crag Moll.' all belong to one and the same species.

MYA ARENARIA, *Linn*. Crag Moll., vol. ii, p. 279, Tab. XXVIII, fig. 2.

Localities. Red Crag passim, except Walton. Fluvio-marine Crag passim. Chillesford bed passim, except Chillesford. Lower Glacial, Belaugh, Rackheath, Wroxham, Spixworth, and Crostwick. Middle Glacial, Hopton, Clippesby, and Billockby. Postglacial, March, Hunstanton, and Nar Brickearth, Pentney (*Rose*).
This species I have not yet seen from the Cor. Crag, nor from the oldest part of the Red, viz. that of Walton Naze. I have found it plentifully in the Red Crag at Sutton, where the variety *lata* is also met with, and I have not seen this variety from any other locality. It is common in the Fluvio-marine Crag at Bramerton, where distortions are not uncommon; but at Chillesford, where *truncata*, a less littoral shell abounded, I have not met with it. In the Lower Glacial sands, where they are fossiliferous, it is common. Fragments are numerous in the Middle Glacial sands of Hopton, and it has been found by Mr. Rose in the Nar Brickearth.

MYA TRUNCATA, *Linn.* Crag Moll., vol. ii, p. 277, Tab. XXVIII, fig. 1.

Localities. Cor. Crag passim. Red Crag passim. Fluvio-marine Crag, Bramerton. Chillesford bed, Chillesford and Sudbourn Church Walks. Lower Glacial, Weybourn and Runton. Middle Glacial, Hopton. Upper Glacial, Bridlington. Post-glacial, March, and Kelsea Hill.

This species I have not seen from the Red Crag of Walton Naze, nor does it occur in that of Bentley. It is abundant, though generally small of size, in the Red Crag of Butley, especially in the *Scrobicularia* bed; and I am informed by Mr. Reeve that it is common in the Fluvio-marine Crag at Bramerton. Though once so abundant in the double state in the Chillesford bed at Chillesford, none can now be found there, but a colony of the double shells was found by the Rev. O. Fisher in the same bed at Sudbourn Church Walks. It occurs in the Lower Glacial pebbly sands at Runton Gap, double, and with its syphonal ends erect as it lived, and fragments of it are abundant in the Middle Glacial sand at Hopton. At Bridlington the ordinary form only, so far as I am aware, occurs, but Mr. Leckenby ('Geol. Mag.,' vol. ii, p. 348), mentions the variety *Uddevallensis* as occurring in the Boulder Clay, near Scarborough. *Uddevallensis* does not seem to have appeared in Britain until towards the close of the Glacial period.

PHOLAS CANDIDA, *Linné.* Supplement, Tab. X, fig. 25.

> PHOLAS CANDIDUS, *Linn.* Syst. Nat., ed. xii, p. 1111.
> — PAPYRACEA, *Spengl.* Skriv. Natur. Selsk., vol. ii, pl. i, fig. 4.
> Lister, Hist. Conch., pl. 435, fig. 278.

Length, $1\frac{1}{2}$ inch.

Locality. Fluvio-marine Crag, Bulchamp?

A single specimen of this species was obtained by the Rev. Jno. Gunn some years since, and given to the British Museum, and it has the above locality marked upon the tablet, but upon application to Mr. Gunn for a confirmation of this, he was not able, from lapse of time, to say precisely where the specimen was found.

There is every appearance of its being a Crag fossil, and I have no doubt of its being genuine, and that it may be referred to the above-named species. I have not heard of its having been found elsewhere as a fossil of the Crag or of any other of the Upper Tertiaries of East Anglia.

PHOLAS PARVA, *Pennant*.   Supplement, Tab. X, fig. 26.

> PHOLAS PARVA, *Penn*.   Brit. Zool., vol. iv, p. 77, pl. xl, fig. 13 ?
> —   DACTYLOIDES, *Desh*.   2nd ed., Lam., vol. vi, p. 45.
> —   LIGAMENTINA, Id.   Elem. Conch., pl. iii, figs. 11, 12.
> —   TUBERCULATA, *Turton*.   Brit. Biv., p. 5, pl. i, figs. 7, 8.

Length, ⅝ inch.

Locality.   Red Crag, Waldringfield.

Mr. Alfred Bell and Mr. Charlesworth lately obtained some blocks of clayey material from the nodule pits at Waldringfield, in which were lodged several specimens of this species, with the valves united in their natural position, and I am indebted to both those gentlemen for specimens.

The Crag fossil corresponds with the short variety of the existing species, which is nearly equilateral, but the imbricated radiations are closer and more numerous than upon any recent specimen that I have seen.

PHOLAS BREVIS, *S. Wood*.   Supplement, Tab. X, fig. 24 *a, b*.

Locality.   Cor. Crag, Sutton.

In the ' Crag Moll.,' vol. ii, p. 296, I spoke of a fragment of *Pholas crispata* having been found by myself in the Cor. Crag.   This fragment I now believe does not belong to that species, and I have had it represented as above (Tab. X, fig. 24 *a*).   I have also met with several other large fragments and some small perfect specimens (one of which I have figured 24 *b*) which appear to me to be the young of the same species, although in these young specimens the imbricated radiations approach much nearer to the dorsal margin on the siphonal side than they do on the large fragment.   In the latter, however, they are faintly visible over this part, and may have become obsolete during the growth of the shell.   The small specimens show a highly reflected dorsal edge behind the umbonal region, but this part is broken off in the larger specimens.   In our present species the imbricated rays cover, even in the adult shell, as shown by the large fragment, more or less of the posterior half of the shell, which is not the case with *crispata*.   In other respects the large fragment resembles *crispata*.   The proportions also of both the figured specimens differ from those of *Ph. crispata*.

As the other fragments and imperfect small specimens, of which I have many, uniformly maintain the characters above referred to, the species seems to differ from any other known to me, and I have accordingly proposed for it the above name.

PHOLAS CRISPATA, *Linn.* Crag Moll., vol. ii, p. 296, Tab. XXX, fig. 9.

Localities. Red Crag, Walton, and Sutton. Lower Glacial, Belaugh. Middle Glacial, Hopton. Upper Glacial, Bridlington. Post-glacial, Kelsea Hill and March.

As just explained, the reference by me of this shell to the Cor. Crag was a mistake for the above *Ph. brevis.* Fragments of a large *Pholas* are common in the Middle Glacial sand of Hopton, and, I think that there is little doubt of their belonging to this species; but it cannot be so affirmed with certainty. Mr. Jeffreys gives it (in fragments) from Kelsea Hill, and a fragment probably of this species has occurred at March. None of the specimens that I have seen indicate a magnitude of more than three inches.

Pholas dactylus is mentioned by Mr. Alfred Bell in ' Ann. and Mag. Nat. Hist.,' Sept., 1870, as from the Red Crag of Walton Naze, and Mr. Jeffreys has, on that authority I think, inserted it in his list to Mr. Prestwich's paper. Mr. Bell has not been able to find the specimen.

PHOLAS CYLINDRICA, *J. Sow.* Crag Moll., vol. ii, p. 295, Tab. XXX, fig. 8.

Localities. Red Crag, Sutton and Walton Naze.

I stated in the ' Crag Moll.' that fragments of this shell occurred in the Coralline as well as in the Red Crag; but it now appears to me that these fragments belong to *Ph. brevis* and not to *cylindrica.*

Pholades are spoken of by Sir Charles Lyell in the ' Lond. and Phil. Mag.,' August, 1835, p. 82, as having been found at the depth of six or eight feet below the surface of the Cor. Crag at Sutton where not covered by Red Crag. Mr. Charlesworth, also, speaks of specimens of *Pholas crispata* occurring in the sand of the Cor. Crag at the depth of three feet from the surface; and I have obtained a specimen of the same species from the same locality at nearly four feet from the surface. In the interior of these specimens there was no Coralline Crag, but instead of it fragments of what appear to be Red Crag shells.

If these specimens found buried in the material of the Coralline Crag belong, as I presume they do, to the age of the Red Crag, it is evident that *Pholades* excavate to a depth beyond what they have hitherto been supposed to do, and that either these tubular excavations are by some means kept open to the surface so as to allow of the access of water, or else that the *Pholas* returns upwards to the water by the pressure of the foot, as the *Solen* does. The foot, however, in *Pholas* does not much resemble the same powerful organ in *Solen,* by which that genus is enabled to rise through the sand to the water.

PHOLADIDEA PAPYRACEA, *Solander*.　　Supplement, Tab. X, fig. 27, and Crag Moll., vol. ii,
p. 298, Tab. XXX, fig. 10.

Locality.　Cor. Crag, Sutton.

In the ' Crag Moll.' I figured two disconnected fragments from the Cor. Crag of Sutton of what appeared to me to belong to this species, and I have recently found a small specimen which undoubtedly belongs to it.　This is figured in Tab. X of this Supplement and is from the same locality of Sutton.

In Mr. Jeffreys' list appended to Mr. Prestwich's paper (p. 485) this species is given as occurring in the Red Crag, but as no special locality is mentioned for it and I have not been able to see the specimen, I cannot admit the species as a Red Crag one. The figure I gave in ' Crag Moll.,' Pl. XXXI, fig. 23, is the representation of what I believe to be an extraneous fossil doubtfully belonging to this genus.

BRACHIOPODA.[1]

This division of the Mollusca is composed of animals that are considered to be of lower organisation than the *Bivalvia*.　Although *Brachiopoda* are strictly bivalvia they differ essentially in their internal organisation from the rest of that class, and they have in consequence been separated from that section of the Mollusca.

The animals and shells of this group have been thoroughly examined by several of our ablest comparative anatomists and microscopists, and I must refer the reader to the elaborate work of Mr. Davidson, where the history of the entire Order is most ably given.

Although I defer to the much better knowledge of the subject, and adopt the terms of that gentleman, I cannot but repeat the objection which in my ' Monograph of the Bivalves of the Eocene Mollusca ' I have made to the terms " anterior " and " posterior "

---

[1] The name of the ' Crag Mollusca' given to my Monograph has been objected to, on account of its omitting a portion of the fossils belonging to this class, and, therefore, does not fulfil the conditions required by the title.　It was originally intended that the work should comprise everything belonging to the *Mollusca* which has been found in the Crag, but after its commencement Mr. Davidson undertook to describe for the Palæontographical Society all the fossil *Brachiopoda* of Great Britain, and I thought those of the Crag could not be excluded.　The Crag Brachiopoda being so few, and a desire having been expressed by several persons that the ' Mollusca of the Crag ' should be made complete, the opportunity afforded by the Supplement has been taken to include them.

as applied to those animals, because they are in the first place inapplicable to the character of the organism, and in the next they have never been universally employed in the same sense.

The vent of the animal of the Brachiopoda is said to be situated nearer to the umbo than is the mouth, and on that account the distinctions called "anterior" and "posterior" have been applied. The outer portion of the shell covers the arms and the greater part of the mantle, while the viscera and vital organs are in the umbonal part of the shell; and I think, therefore, that such terms as *visceral region* for the inflated portion and *brachial region* for that which is occupied by the spiral "arm feet" would better define the different positions of the animal, and could never be reversed or misapplied. There is also great confusion, or, at least, want of unanimity, in the use of the distinctive terms "ventral" and "dorsal" valves, the perforated one having sometimes been called ventral and at others dorsal, and *vice versâ*. Professor M'Coy has proposed that the name "entering valve" and "receiving valve" should be given to the two pieces, and Mr. Davidson further suggests that the perforated valve should be called "dental valve," and the imperforated one the "socket valve." However, as this author still distinguishes the two valves as "ventral" and "dorsal" I have followed his example.

In the *Bivalvia* the distinctions are founded upon the hinge furniture, either on the dental character, or upon the position of the *connector*, but in the *Brachiopoda* the distinctions are made dependent upon the apophysary system; viz. on the calcareous internal appendage for the support of their spiral "arm feet." The muscular system of the *Brachiopoda* also differs from that of the *Bivalvia*, inasmuch as the latter open their shells, as well as close them, by means of muscles, of which they have as many as eight, exhibiting a somewhat complex mechanical operation for the opening and closing of the valves; and this arrangement appears to me to be far less simple than that of the elastic connector of the *Bivalvia*, and to effect the object in a far less easy way. Whether this be a proof of inferior or superior organization I am not prepared to suggest, but it does appear from it that the rule that nature adopts the simplest method for attaining a result is not one of universal application.

The valves of the *Brachiopoda* are, with some exceptions, more or less of an ovate form, the longer axis being generally from the umbo to the opposite margin, and the length of the shell is measured in that direction; the breadth, therefore, being at right angles to this line, and the depth having relation only to the tumidity of the valves. This is not in accordance with the measurement of the *Bivalves* as adopted by myself, but it is the mode that appears to be generally accepted, and I have, therefore, measured my shells by that standard.

TEREBRATULA GRANDIS,[1] *Blumenbach.*   Supplement, Tab. XI, fig. 5 *a—g*, from the Cor. Crag ; Supplement, Tab. VIII, fig. 11 *a—c*, from the Red Crag.

TEREBRATULA GRANDIS, *Davidson.*   Brit. Tert. Brach., p. 16, 1852.

Length, 5 inches ; *breadth,* 3¼ inches ; *depth,* 2 inches.

Localities.   Cor. Crag, Ramsholt, Sutton, and near Orford.   Red Crag, Walton Naze, Sutton, and Waldringfield.

Some years ago specimens of this species were procured by myself in great abundance both at Sudbourn and Ramsholt, in the Coralline Crag, but recently they have become somewhat scarce.   This is the largest and most noble of the species that I am acquainted with, and if I had to give it a name it should be called *nobilissima,* but unfortunately it has too many synonyms.

It is only within a few years that I have been able fully to display the internal furniture of this species.   A specimen in the British Museum presented by myself and found at Ramsholt seemed to give fair promise of permitting the removal of the sand from the interior, and I was enabled with care to exhibit *in situ* the short reflected loop of which I have given a figure.   All doubt is by this removed respecting its true generic position, Prof. King having imagined that it was furnished with an internal apparatus much prolonged, like that which he has taken for the type of his genus *Waldheimia,* where the loop is extended to more than two thirds the length of the shell, whereas in *grandis* it proves not to exceed one third.   Mr. Davidson always imagined it to be a true *Terebratula,* but he had not seen the perfection of the loop.   I may further observe that *grandis* possesses a very long *spur* or *crura,* and a very little more extension to this would have united the two parts so as to form a ring like that which characterises the genus *Terebratulina.*   Whether this be so in the young state I have not been able to see.

Specimens of *grandis* have come into my possession that measured nearly five inches in length, and fragments have been found which indicate even larger dimensions.   The difference of size between the fry or infant state and the full-grown shell is so great as to be not often observable in any animal.   I have found what there is every reason to believe is the young (and perfect) state of this species, with its longest diameter not more than the twentieth part of an inch, which will give to the full-grown shell an increase of one hundredfold at least *in linear direction,* and as in these large shells the young state of the longest or perforated valve has entirely disappeared, probably to the extent of one third of an inch, my estimate of these differences between the young and old is rather

[1] For generic and specific descriptions see Davidson, 'Monograph of the British Tertiary Brachiopoda,' published by the Palæontographical Society.

within the mark than otherwise. Although large, this shell is somewhat thin in the brachial region, and the Coralline Crag specimens are generally either cracked or compressed simply by the weight of the matrix in which they are imbedded; but it becomes excessively thick by deposition of calcareous matter in old specimens, in the region of the viscera under the denticles of the larger valve.

Mr. Charlesworth, in the 'Mag. of Nat. Hist.,' 1837, observed that the specimens of *grandis* found in the Red Crag up to that time were always separated valves, and it was inferred in consequence that these specimens were derived from an older formation. Specimens of this shell have, however, lately been turned out by the diggers for Coprolite in the Red Crag at Waldringfield in considerable numbers, many of which are in the possession of Mr. Charlesworth and Mr. Bell. These have the two valves united, and were probably inhabitants of the Red Crag Sea. They differ in outline from the Coralline Crag shell, being generally much more elongated and less regularly ovate or elliptical, having the brachial region more expanded, and the opening for byssus larger; but I believe they are all only a variety of this variable shell, well called *T. variabilis* by Sowerby; and Mr. Davidson is of the same opinion. Specimens of these I have had figured as above referred to. This name is not in the list of "Norwich Crag" shells by Woodward, neither have I seen it from any other locality of the Fluvio-marine Crag, or from the Red Crag of Butley.

TEREBRATULINA CAPUT-SERPENTIS, *Linné*. Supplement, Tab. XI, fig. 3 *a—c.*

TEREBRATULINA CAPUT-SERPENTIS, *Davidson*. Brit. Tert. Brach., p. 12, 1852.

Localities. Coralline Crag, Sutton. Recent, Britain, Mediterranean, &c.

This is a very rare shell to my researches, and the few specimens that I have found are small and very young. They nevertheless present a considerable amount of variation, both in the outward form and also in the number of rays or ribs with which they are ornamented; one specimen, which measures one eighth of an inch in diameter, having only nine coarse large ribs, while another of the same dimensions has fifteen, and one yet smaller only seven. The form also is variable, being elliptical or much elongated in one specimen, while in another it is nearly orbicular. The byssal opening is also variable, but this probably depends upon the position in which the animal had chosen to fix itself. In one specimen the larger valve has a recurved beak like that of *Rhynconella*, while in another this opening is nearly rounded and sloping backwards. In these young shells there is sometimes a sort of incipient shoulder at the hinge-joint like that in the young state of *T. Gervillei*, to which I had in my catalogue doubtfully referred the Crag shell, and the ribs are sometimes nodulous.

This species is one of those which in the recent state is most remarkable for its geographical extension. It is found in the seas of Britain, in the Mediterranean, on the shores of North America, also on the coast of Japan. It (or at least a shell not to be distinguished from it) occurs likewise in the southern hemisphere, and in the Pacific; indeed, it is almost difficult to say where it is not to be found. It has also a bathymetrical range from low-water mark into water of abysmal depth.

This ubiquitous animal has a special claim to our attention, as it seems to offer a defiance to our determinations. Conchologists are perplexed in their examinations, and fail to separate specimens from opposite hemispheres, north and south, east and west, by any distinctive marks usually denominated specific. The late Edward Forbes was inclined to believe that the shell called *T. striata* ('Geol. Rel. of the Existing Fauna and Flora of the Brit. Isles,' p. 73), found in the Chalk and Green Sand presented differences so insignificant from those of the existing *caput-serpentis* that he doubted whether the one was specifically distinct from the other. Mr. Davidson, after having thoroughly examined the fossils called *striata* and *striatula* with *caput-serpentis*, inclines to the opinion that they are all three distinct; but he acknowledges that the differences are but trifling, and even hesitates to give a decided opinion. I have examined carefully these species, and seen the differences pointed out; but unless they are allowed to have more importance than differences of equally slight degree are allowed to have in other families of Mollusca, I much doubt if the three forms be entitled to specific separation. The fossil species from the Upper Secondaries and Lower Tertiaries have generally been considered by palæontologists as specifically distinct; but it may be asked whether the great difference in age of the several deposits in which they have occurred may not in some degree have influenced palæontologists in their determinations. If the Green Sand shell be really identical with *T. caput-serpentis* it has the greatest antiquity of any Mollusc at present known, and, what is also remarkable, the existing shell exhibits no symptom of specific decline, as it flourishes in our own seas at the present day in great profusion; and its world-wide diffusion and great bathymetrical range seems almost confirmatory of its ancient origin.

Argiope cistellula, *S. Wood.*   Supplement, Tab. XI, fig. 4 *a—d.*

Terebratula cistellula, *S. Wood.*   Catal. Ann. and Mag. Nat. Hist., vol. v, p. 253
1840.
Argiope   —   *Davidson.*   Brit. Tert. Brach., p. 10, 1852.

Diam. $\frac{1}{12}$ inch.
Locality. Coralline Crag, Sutton. Recent, Britain.
This does not appear to be abundant anywhere either as a fossil or as a recent species. I have found about twenty specimens in the Coralline Crag, and these show a considerable variation in outline, especially in the elevation of the umbonal portion of the larger valve.

The principal distinction presented by this species, and I believe of the whole genus, is the internal septum or septa of the smaller valve; and the loop is said to connect or to be supported by these thickened internal processes, so that the apophysary system is somewhat variable in different species; but the loop is so delicate in these small shells that it is very rarely preserved, even in the recent specimens, and they are not present in my fossils. The byssal aperture is very large in this species, so also are the hinge-denticles. The dorsal margin of the smaller valve is nearly straight and entire, the inner margin of this valve being crenulated all round up to the dorsal edge, and the valve is divided into two equal parts by a large and prominent septum. Mr. Jeffreys has lately referred the Crag shell with doubt to *lunifera*, a Mediterranean shell (' Rep. Dredg. among the Shetland Islands, 1868 '); but there is great doubt about *lunifera*, so that I have retained my original name of *cistellula*, which, Mr. Davidson tells me, he also intends to do.

RHYNCHONELLA PSITTACEA, *Chemnitz*. Supplement, Tab. XI, fig. 2 *a—c*.

RHYNCHONELLA PSITTACEA, *Davidson*. Brit. Tert. Brach., p. 21, 1852.

Diam. $\frac{7}{8}$ inch.

Localities. Red Crag, Sutton. Fluvio-marine Crag, Bramerton, Thorpe and Postwick. Chillesford bed, Bramerton. Upper Glacial, Bridlington. Post-glacial, March.

This species is not abundant as a British fossil, and the valves are generally separated, although occasionally the two have been found united. It has considerable range as a recent species, but it is confined to the colder regions of the northern hemisphere. A species much resembling it has, however, been found in the southern hemisphere.

There is no loop to this shell, the spiral arms being attached to two small curved processes projecting inwards from the umbo of the imperforated valve; the hinge-teeth of the beaked or larger valve are very strong, and they are supported, as it were, upon a sort of partition extending from the base of the hinge-teeth towards the tumid portion of that valve, by which a strength is given to this shell which I have not observed in other *Brachiopoda*; for what especial purpose, however, this is used I am not able to say. The beak of the larger valve appears to have retained its infant condition, and is not worn away like that of *Terebratula*. Mr. Bell gives the shell from the Red Crag of Sutton (' Ann. and Mag. Nat. Hist.,' Sept., 1870), and Dr. Woodward, in his list in ' White's Directory,' gives it from Thorpe and Postwick. Mr. Reeve has found it, but rarely, in both the beds at Bramerton, and Mr. Harmer found it at March; Dr. Woodward gave it (' Geol. Mag.,' vol. i, p. 53) from Bridlington.

23

DISCINA FALLENS, *S. Wood.*   Supplement, Tab. XI, fig. 6.

> DISCINA NORVEGICA? *S. Wood.*   Ann. and Mag. Nat. Hist., p. 253, 1840.
> ORBICULA LAMELLOSA, *Davidson.*   Brit. Tert. Brach., p. 7, 1852.

Diam. — ?
Locality.   Coralline Crag, Sutton.

The only specimens of this genus that have yet, so far as I know, been found in the Crag are the one figured by Mr. Davidson from my collection in the British Museum, and now figured as above, and another which I have since obtained, but of which the exterior is wholly concealed by *Cellepora.*   The recent shell called *Orbicula lamellosa*, to which this fossil was referred by Mr. Davidson, is a native of the coast of Peru, but he now considers that the reference of the Crag shell to this species was incorrect.

A shell was obtained by Professor King at the depth of 1240 fathoms in Lat. 52° 8′ N., and 15° 30′ W. Long. (about 140 miles from land), which he described under the name of *Discina Atlantica.*   Mr. Jeffreys, in his list of Cor. Crag shells accompanying Mr. Prestwich's paper, has referred my Crag shell to this species of Professor King.

I have had some correspondence with Mr. Davidson respecting this Crag fossil, and he is of opinion with me that it cannot be referred either to *Orbicula lamellosa* or to *Discina Atlantica*, and that, while placing it in the genus *Discina*, it ought, for the present at least, to have a distinct specific appellation.   As Mr. Davidson originally described the Crag shell, I proposed to him that he should give it a new name and point out what he considered were the characters for specific separation, but he has left it to me to do this. The features which, so far as its meagre condition allow me to speak, distinguish the Crag shell from *Atlantica* consist in its less conical form, and in the more excentric position of its apex.   I have, therefore, provisionally called it by the above name, *fallens.*

LINGULA DUMORTIERI, *Nyst.*   Supplement, Tab. XI, fig. 1 *a—c.*

> LINGULA DUMONTIERI, *Nyst.*   In Davidson's Mon. Brit. Tert. Brach., p. 5, 1852.

Length, 1 inch ; *breadth*, ½ inch.
Localities.   Coralline Crag, Sutton, and near Orford.

Fragments of this shell are not particularly rare at Sutton, but I have been unable as yet to obtain a specimen that has not been more or less mutilated.   A similar remark is made by M. Nyst respecting the same species found at Antwerp.

The genus *Lingula* has been generally considered as an inhabitant of tropical or subtropical regions, but the Crag shell is by Mr. Jeffreys referred to a species lately obtained from the seas of Japan, *L. Jaspidea*, A. Adams. I am not able to express any opinion as to the correctness of this identification. My shell differs from *L. tenuis* in being less slender and in having the front margin more obtuse.[1]

The genus *Lingula* occurs at the base of the Silurian system, and has preserved a special uniformity of character throughout the whole geological scale, some of the species (*L. minima* and *L. Symondsi*) differing, so far as their testaceous remains are a guide, in only a trifling degree from our Crag shell. Moreover, so far as these remains indicate, it has never exhibited any large development of individuals, although there is scarcely a formation in which it has not left some trace of its existence, nor do we see any great variation among the individuals in regard to magnitude. As this genus has existed from the oldest Silurian strata, it can scarcely have escaped encountering a great variety of conditions, and if indications of climate are to be inferred from the presence of a genus in any deposit, what are we to infer from the presence of such a genus as *Lingula*?

ADDENDUM.

Several new forms having been discovered since the first part of this Supplement was printed, I find it necessary to add another plate with figures and descriptions of these species, and I take this opportunity to introduce some remarks as to other species which have occurred to me since that portion of this Supplement to which their place properly belonged passed through my hands.

Voluta Lamberti, *J. Sowerby*. Crag Moll., vol. i, p. 20, Tab. II, fig. 3.

Mr. Jeffreys in his list refers this shell (with a note of interrogation) to *Voluta Junonia*, Chemn.

I have again examined and compared my Crag shell with specimens of *V. Junonia* in the British Museum, and I find that it differs from *Junonia* in having a larger pullus, or a more obtuse apex, and when perfect, in the presence of close-set spiral striæ, of which there is no appearance in the recent shell. *Junonia* has also a more expanded outer lip, and is more " emarginate" at the base, and its proportional length of aperture is different from that of the Crag shell. *Junonia* was figured many years ago by Chemnitz, ' Conch. Cab.,' vol. ii, p. 16, pl. 177, figs. 1703 and 1704. Its distribution is somewhat uncertain, but it has recently been found in the Gulf of Mexico.

[1] The muscular impressions as represented in our figure are partly imaginary.

Mr. Sowerby, when originally describing *V. Lamberti* in 1816, spoke of some recent shells which he had seen (five in number) that resembled the fossil; and the Crag shell was treated by Sir Charles Lyell, in one of his early works, as belonging to a living species, on the authority of M. Deshayes. The question was fully examined by Mr. Charlesworth in the 'Magazine of Nat. Hist.' for 1837, pp. 37 and 90, and a figure is there given by him of a specimen of *Lamberti* on parts of which the spiral striæ (sharply cut) to which I have referred are preserved. In most of the Crag specimens these striæ have been removed, but on a small one from the Coralline Crag in the collection presented by me to the British Museum they are present.

COLUMBELLA BORSONI? *Bellardi*. Addendum, Plate, fig. 19.

> COLUMBELLA BORSONI, *Bellardi*. Monog. Columb. Foss., p. 14, t. i, fig. 11.
> —   —   *A. Bell.* English Crags, Proc. Geol. Association, 1872, p. 24.

Locality. Red Crag, Walton Naze.

The specimen figured above, being that upon which this species was introduced into the Crag fauna, has been kindly sent to me by Mr. R. Bell. It appears to me more to resemble the figure of *C. subulata* (*Murex subulatus*, Broc.), but I have not the fossils from Piedmont for comparison. It seems to differ from *C. sulcata* (the common species at Walton) in being more subulate with less convex volutions, in being quite smooth, and in having a more distinct canal. It somewhat resembles the young state of *C. sulcata* ('Crag Moll.,' Tab. 5, fig. 2 *d*), but possesses the denticulations in the mouth which in that species do not appear until the shell is full grown. The apex of our present shell has been destroyed. Under these circumstances I am not satisfied that the specimen is anything more than an aberrant individual of *sulcata*, but have felt that it was desirable to have it figured.

COLUMBELLA MINOR, *Scacchi*. Addendum, Plate, fig. 20.

> COLUMBELLA MINOR (*Scac.* in *Phil.*). Catal., p. 10, No. 12, fig. 11.
> BUCCINUM MINUS, *Phil.* En. Moll. Sic., p. 190, t. xxvii, fig. 12.

Locality. Cor. Crag, near Orford.

A single specimen with this name has also been sent to me by Mr. Robert Bell, and it is, I think, correctly referred. The Crag shell has the surface much eroded, but in its living state it was probably quite smooth.

Columbella abbreviata is given from the Red Crag by Mr. A. Bell in his list ('Proc. Geol. Assoc.'), p. 28, but I have not been able to confirm that reference.

Columbella scripta is given by Mr. Jeffreys in his list, to Mr. Prestwich's Red Crag paper, as from Sutton, Walton, and Waldringfield, but I have not been able to confirm that reference.

LACHESIS, *Risso*.

This generic name has been adopted by all our modern conchologists, and I have here employed it, but upon what special characters it is founded I do not know unless it be upon some peculiarity in the animal.

The name *Lachesis* is unfortunately used for a genus of *Reptilia* and also for a genus of *Arachnidæ*. The little shell *Buccinum minimum*, Montague, now called *Lachesis*, I have not yet seen as a fossil in England.

LACHESIS ANGLICA, *A. Bell*. Addendum, Plate, fig. 7.

Locality. Cor. Crag, near Orford.

Mr. Robert Bell has sent to me a specimen with this name attached. It has six volutions with a deep suture, and the last volution has about a dozen longitudinal ribs or costæ, which are made nodulous by the crossing of three or four large spiral threads or ridges. The mouth is ovate and about one fourth the length of the shell, with the outer lip denticulated within. There are also some coarse striæ below the volutions. It is the only specimen I have seen, and unfortunately the upper portion or apex is destroyed. Judging from the figure this appears to be distinct from *Lachesis (Buccinum) candidissima*, Phil.

BUCCINUM TOMLINEI, *Canham*, m.s. Addendum, Plate, fig. 11.

Locality. Red Crag, Waldringfield.

A single specimen as represented has been sent to me with the above name by Mr. Canham. It was found, he tells me, in the workings for "Coprolite," on the ground of Colonel Tomline, whose name he wishes to commemorate as a friend to science, and a gentleman who has given him every facility for obtaining Crag specimens.

This shell resembles *B. Dalei*, but it differs in several particulars, so as to justify in my opinion a distinct specific position. It has a more elevated spire, with less tumid volutions ; the outer lip is more regularly rounded, and less expanded at the lower part than the lip in *B. Dalei*, and it wants the projecting ridge at the base of the *colummella*.

I have deferred giving this a regular diagnosis, but to leave it for confirmation as a species to the discovery of more specimens, and as it is most probably a shell extraneous to the Red Crag, some of its congeners may make their appearance in an older formation.

NASSA PUSILLINA, *S. Wood.* Supplement, p. 14, Tab. 2, fig. 7. Addendum, Plate, fig. 24.

Since this name was published I have been enabled to examine some recent specimens of that variable shell *Buc. variabilis*, Phil. (*Nassa Cuvieri*, Payr), and one of its varieties appears to correspond with the fossil figured by me under the above name *pusillina*. Mr. A. Bell, in 'Ann. and Mag. Nat. Hist.,' September, 1870, and Mr. Jeffreys in his list to Mr. Prestwich's Cor. Crag paper referred this shell to *N. Cuvieri*, Payr., and I here adopt that name in lieu of *pusillina*.

Very recently Mr. Robert Bell has sent to me a specimen from the Red Crag of Butley, which I have had represented in Addendum Plate, as above referred to. The ribs appear only on some of the volutions, a feature which is exhibited by one of Phillippi's figures of this variable species, and I have accordingly referred our present specimen to *Cuvieri*.

Nassa musiva ? Broc., is inserted in Mr. A. Bell's list as a species from the Red Crag, and a specimen with that name has been sent to me by Mr. R. Bell. It is, however, in bad condition, and appears to me to be one of the varieties of *N. reticosa* much rubbed and worn.

Nassa labiosa, J. Sow., from the Crag, is, I still think, distinct from *B. semistriatum* Broc., but it seems to correspond with a shell called *labiosa* by Beyrich, ('Die. Conch. Nordd. Tertiarb.,' p. 140, Tab. 8, fig. 5 *a—c*). The Red Crag specimens of *N. labiosa* are possibly derivatives from the Coralline Crag.

MUREX INSCULPTUS, *Dujard. ?* Addendum, Plate, fig. 9.

Locality. Red Crag, Waldringfield.

The figure referred to represents a specimen sent to me by Mr. R. Bell with the above name attached to it, and he tells me that it is identical with a fossil from Italy which he received from M. Seguenza with that name. I have, however, been unable to find this name in any publication known to me. M. Dujardin figures and describes a shell as *M. exiguus* in his paper on the fossils of Touraine, 'Geol. Trans. of France,' 1837, p. 296, Plate XIX, fig. 2, but our shell does not well correspond with it. It appears to be intermediate between the figure of *exiguus* and *Buccinum lavatum*, Sowerby. It is, I think, a derived specimen.

Trophon elegans, *Charlesworth*. Crag Moll., vol. i, p. 46, Tab. V, fig. 2; Addendum, Plate, fig. 13.

At p. 98 of this Supplement I expressed my suspicions that this would prove to be a Cor. Crag shell, and to be only derivative in the Red Crag.

I have been confirmed in this by hearing from Mr. Canham that at a temporary recent opening in the Cor. Crag at Sutton the shell shown in the above figure occurred in numbers in association with numerous specimens of *Voluta Lamberti* and *Cassidaria bicatenata*. All of these, however, were in such a decayed state that only the specimen figured above could be extracted, and this is not full grown. Mr. Canham's Red Crag specimen measures $4\frac{1}{2}$ inches in length.

It would appear from this that probably both *T. elegans* and *Cassidaria bicatenata* are present in the Red Crag only as derivatives from the Coralline, for while *V. Lamberti* (which by its occurrence at Yarn Hill is proved to have survived into the later part of the Upper Crag) is common in the only portion of the Red Crag which is not leavened with derivatives, viz. Walton, no trace of either *Cassidaria bicatenata* or our present *Trophon* has ever occurred there.

This shell is, I believe, quite distinct from any variety of *Trophon antiquus*, and is specially distinguished by the apex, which is not obtuse or mammilated, and the apex shown in Tab. VII, fig. 9, of this 'Supplement,' which I had imagined might be that of *Buc. Dalei* now belongs, I have now no doubt, to the present shell.

Fusus despectus, Linné, is given as a Red Crag shell by Mr. Jeffreys, and also by Mr. Bell, as from Sutton, Bramerton, and Thorpe. This I imagine must be the var. of *T. antiquus* with prominent carinæ like the shell figured 'Crag Moll.,' vol. i, Tab. V, fig. 1 *b*, which Edward Forbes ('Mem. Geol. Survey,' vol. i, p. 426, 1864) referred to *Fusus despectus*.

A very imperfect specimen has been sent to me by Mr. Canham with the locality "Red Crag, Waldringfield," which may probably be *Fusus Waelii*, Nyst, spoken of by Sir Charles Lyell, 'Quart. Journ. Geol. Soc.,' vol. viii, pp. 301 and 316, and described by Beyrich, p. 271, Taf. 20, figs. 1—3. This is, no doubt, a derivative.

Trophon Norvegicus? Supplement, p. 21, Tab. V, fig. 14, and Addendum, Plate, fig. 16.

Locality. Red Crag, Sutton.
In Mr. Alfred Bell's paper on the English Crags, 'Proc. Geol. Assoc.,' vol. 2, p. 28,

is the name *Fusus Largillierti*, Fisch, as a species from the Upper Crag. The specimen on the authority of which this name was so introduced has obligingly been sent to me by Mr. Robert Bell (who tells me that Mr. Jeffreys so referred it), and it is figured as above. I have carefully examined and compared it with the figure and description of *F. Largillierti* by M. S. Petit, 'Journ. de Conch.,' vol. 2, p. 255, Plate 7, fig. 6, 1851, and can see no identity. Our shell differs in several respects, more particularly in being strongly striated, while *Largillierti* is smooth and in the form of the volutions and mouth. The specimen appears to me to be a distortion of *Norvegicus*, the position of the canal having been displaced by the same malformation of the animal which imparted the prominent shoulder to the whorl.

PLEUROTOMA CLATHRATA ? *Marcel de Serres.* Addendum, Plate, fig. 8 *a, b.*

> PLEUROTOMA CLATHRATA, *Marc. de Serr.* Geogn. des Ter. Tert. de la Fa., t. xi, figs. 7, 8.
> —   —   *Dujard.* Coq. Foss. Touraine, p. 294, pl. xx, fig. 6.
> —   —   *Hörnes.* Vienna Foss., p. 379, t. xl, fig. 20 *a—c.*

Locality. Cor. Crag, Sutton.

A single specimen in my cabinet which I had considered as a variety of *Pl. Philberti* may, I now think, be referred to the above species. The cancellations are larger, coarser, and fewer, with the knobs more prominent, and the depressions deeper than is the case with *Pl. Philberti;* and, although I have not been able to compare my shell with a Touraine specimen of the species to which I have referred it, the foreign authors seem to me to justify the reference.

PLEUROTOMA TEREOIDES, *S. Wood.* Addendum, Plate, fig. 3 *a, b.*

Spec. desc. Pl. Testa minuta, fusiformi-turrita ; anfractibus convexis, angulatis, supra planulatis, infra convexis ; spiraliter lineatis, lineis incrementi conspicuis ; labro profunde sinuato ; cauda longiuscula aperto ; apertura oblonga, spiram æquante.

Locality. Cor. Crag, Sutton.

A single small specimen has lately rewarded my researches, of which the figure above referred to is a representation. It makes considerable approach to *Pl. teres*, Forbes, but is, I believe, distinct. I have compared specimens of the same size of *Pl. teres*, obtained by Mr. Jeffreys in his deep-sea dredgings, which I believe differ specifically from my Crag shell. Mr. Smith, of the British Museum, has shown to me a small specimen in the national collection from the seas of Japan, which he believes to be at present an undescribed species, and proposes to call *Pl. turritispira*. This makes an approach to the Crag shell, but it differs in having a finely cancellated exterior and a smaller and shallower sinus in the lip. From the affinity which my shell bears to the existing *teres* I have given it the name of *tereoides*.

PLEUROTOMA BERTRANDI? *Payr.* Addendum Plate, fig. 4.

> PLEUROTOMA BERTRANDI, *Payr* (fide *Phil.*). Cat. des Moll. Cors., p. 144, t. vii, figs. 12, 13.
> — — *Phil.* En. Moll. Sic., vol. ii, p. 168.

Localities. Red Crag, Waldringfield, Sutton, and Butley.

This species is given by Mr. A. Bell in his list of the "English Crags," and Mr. Robert Bell has forwarded to me two specimens from Waldringfield with this name attached, and I have myself found a similar one at Butley. There are also some specimens in my collection in the British Museum. I give it with a mark of doubt.

PLEUROTOMA STRIOLATA, *Phil.* Addendum Plate, fig. 2.

> PLEUROTOMA STRIOLATA (*Scac.*), ex *Phil.* Moll. Sic., vol. ii, p. 168, t. xxxi, fig. 7.
> — — *Jeffreys.* Brit. Conch., vol. iv, p. 376, pl. xc, fig. 1.

Locality. Cor. Crag, near Orford.

The shell figured is one of my own finding. Mr. Robert Bell had a very perfect specimen showing the fine striæ, but unfortunately it has been broken or I would have had it figured. My specimen, however, shows traces of the striæ, and I believe it is correctly referred. Mr. Jeffreys in his list to Mr. Prestwich's paper gives this as a Red Crag, but not a Coralline Crag species. I have not, however, been able to confirm the reference of it to the Red Crag.

PLEUROTOMA RUGULOSA, *Phil.*

Mention is made at p. 46 of 'Supplement' of some specimens from the Coralline Crag that might be referred to this species. I have since, however, had reason to doubt whether these may not be the young of some other species.

24

PLEUROTOMA PYGMÆA, *Phil.*

This is given with a note of interrogation by Mr. A. Bell in his paper on the " English Crags," p. 28.　I have not been able to see the specimen.

PLEUROTOMA ICENORUM, *S. Wood.*　Supplement to Crag Moll., p. 35.

This is said by Mr. A. Bell in his list of the " English Crags " to be identical with *Pl. Hosiusii*, Könen, but on comparing my specimen with the German Oligocene shell there appears to be considerable difference.　The Crag shell has much coarser striæ, and a very obtuse apex, while the second volution is smooth.　The knobs on the Crag shell are much larger, and there is a second row of nodules near the suture, a peculiarity which I do not observe on the German shell.　*Icenorum* also has a broader sinus and a shorter canal at the base than has the German shell.

PLEUROTOMA OBLONGA, *Ren*, is given in Mr. Bell's list of Red Crag shells, but with a query.　Mr. Canham has sent to me a specimen from the Red Crag at Waldringfield, and I have also had one from Mr. Bell, both of which resemble in shape the species with this name, but they are in an imperfect condition, with the ornamentation obliterated, so that it would be hazardous to refer them under these circumstances, and as the specimens are, I have no doubt, only derivative in the Red Crag, I have not thought it worth while to have them figured.

APORRHAIS PESPELICANI, *Linné*, var. *Serresianus.*　Supplement to Crag Moll., p. 49, and
　　　　　　　　　　　　　　　　　　　　　　　　　　　Addendum Plate, fig. 6.

CHENOPUS SERRESIANUS? (*Mich.*) *Phil.*　En. Moll. Sic., vol. ii, p. 185, t. 26, fig. 6.

Locality.　Cor. Crag, near Orford.

The above represents one of the specimens previously spoken of at p. 49 of this ' Supplement.'　This appears to me to correspond with the figure and description of the Mediterranean shell as given by Philippi.　Our Cor. Crag specimens are very variable, some having the ornamental nodules, as in this case, small and numerous, while others have little more than half the number, and there are forms intermediate between the two.　Though my specimen has not the digitiform processes perfect, I can see no material difference between it and some of the recent forms of *pespelicani*. *A. MacAndreæ* is probably only a variety. *Chenopus deciscens*, Phil., I do not know, but this is described as "*ultimo quadricarinato carinis nodulosis*," while my shell has only three keels or ridges.

CERITHIUM (TRIFORIS) PERVERSUM? *Linné*, var. *Belli*.   Addendum Plate, fig. 17.

Locality.   Cor. Crag, Sutton.

The above figure represents a small specimen sent to me by Mr. Robert Bell without a name.

The two forms or varieties called *C. perversum* and *C. adversum* (the latter of which is rather more slender than the other) are both found in the Cor. Crag at Sutton; but I have seen nothing so cylindrical as the present specimen. The recent shell, to which I have doubtfully referred it, is described by Mr. Jeffreys as variable, the volutions having three and sometimes four bands ('Brit. Conch.,' vol. iv, p. 261).

As our present shell has four unequal ridges (the lowest one very small, and the next one to it large and nodulous) I have not ventured to consider it as a distinct species on the strength of a solitary specimen, though, should more specimens occur, the question of its specific separation might with more reason be entertained.

CERITHIOPSIS TUBERCULARIS, *Mont*.   Crag Moll., p. 70, Tab. VIII, fig. 5, Supplement, p. 52.

Since the first part of this 'Supplement' was issued Mr. Jeffreys kindly sent me for examination his British recent specimens, from which he formed two distinct species, under the names of *C. Barleei* and *C. pulchella* ('Brit. Conch.,' vol. v, Pl. 81, figs. 2 and 3). It does not appear to me that these shells present sufficient differences from *tubercularis* to entitle them to specific isolation, but they have all their exact representations among the Cor. Crag specimens. The principal differences appear to be a basal ridge or prominent spiral line which is present in some of my Crag specimens. Under these circumstances I have still retained them under the same specific name of *tubercularis*. Var. *subulata*, of 'Crag Moll.,' Tab. VIII, fig. 5 *b*, represents *C. Barleei*, and var. *nana* fig. 5 *c* represents *C. pulchella*, the convexity in the volution of this latter being probably caused by its abbreviated spire.

I have adopted the above generic name in deference to the malacologists, who say that there is a great difference in the animal from that of *Cerithium*, and this is also strengthened (it is said) by the present genus possessing a longer canal.

CANCELLARIA UMBILICARIS? *Brocchi.*　Addendum Plate, fig. 10.

> VOLUTA UMBILICARIS, *Broc.*　Conch. Foss. Subap., vol. ii, p. 312, tab. iii, figs. 10, 11.
> CANCELLARIA —　　　*Bellardi.*　Foss. de Piemont., p. 36, tav. iv, figs. 17, 18.

Spec. Char.　"*C. Testa ventricosa, anfractibus, scalariformibus, canaliculatis, longitudinaliter costata, profunde transversim sulcata, sulcis subimbricatis crispis, umbilico patentissimo, usque ad apicem spiræ pervio.*"—*Broc.*

Locality.　Red Crag, Waldringfield.

The Rev. H. Canham has sent to me for representation a very perfect specimen which he has obtained from the Red Crag at Waldringfield, and this I have referred as above, with some doubt, as it does not strictly conform to any of the figures representing the foreign specimens under the name of *C. umbilicaris.*　Our shell may probably be a dwarf variety of this species, with a depressed spire.

Our present specimen, there can be no doubt, is merely present as a derivative in the Red Crag, but whether it be derived from the Coralline Crag or from some older bed, there are not at present the means of judging.

I have from the Cor. Crag a very imperfect specimen belonging to this genus, which is distinct from any other that I have previously described, but it is too imperfect for representation.　It somewhat resembles *C. elongata*, Nyst.

Cancellaria subspinulosa, Supplement 'Crag Moll.,' Pl. VI, fig. 10, may possibly be the young *C. lyrata*, Broc., but better specimens than I possess will be required for its determination.

PYRAMIDELLA LÆVINSCULA, *S. Wood.*　Crag Moll., vol. i, p. 77, Tab. IX, fig. 2.
Supplement, p. 57.

Mr. Jeffreys has kindly sent to me for examination some recent specimens obtained in the Mediterranean, and from his deep sea dredgings, which he considers identical with my Crag shell.　On comparison I find the recent specimens more conical in shape, and more resembling *P. conulus*, Speyer ('Die Conch. der Cass. Tertiar.,' Tab. XXV, fig. 1. In his letter Mr. Jeffreys says, "There is some difference between the Porcupine and Crag shells, which I consider varietal only, consequent on some alteration in the conditions of habitat.　I make some allowance for the great lapse of time and subsequent change of form."　The lapse of time and alteration of conditions are, I believe, the main causes operating to produce new species, and inasmuch as the more our knowledge of recent and fossil forms extend, the greater will be the perplexity among palæontologists where to draw a line of specific distinction, and the more will they be driven for classification and nomenclature to arbitrary lines of separation, it does not seem to me expedient in the face of the difference pointed out to regard this species as living. There is so much uncertainty respecting the shells that are to be referred to *unisulcata* or *plicosa* that I have retained the name originally given to the Crag species.

SCALARIA SEMICOSTATA, *J. Sowerby*. Addendum Plate, fig. 1.

SCALARIA SEMICOSTATA, *J. Sow*. Min. Conch., tab. 577, fig. 6.

Locality. Red Crag.

A beautiful and perfect specimen of *Scalaria* has been put into my hands by Mr. Charlesworth, but I believe it to be a derivative from one of the older tertiary formations. This specimen came originally from Mr. Whincopp, who had it from one of the diggers for "Coprolite," I presume from Sutton, whence he had most of his specimens.

Two figures are given by Mr. James Sowerby, tab. 577, fig. 5, as *S. reticulata* (*Turbo reticulatus*, Brander), V, fig. 6, as *semicostata*. In the very great accuracy of Mr. James Sowerby's general figures I have perfect faith, and I believe the two to be distinct.

M. Deshayes has figured and described a fossil from the "Sables moyen," 'An. sans Vert. du Bas. de Par.,' vol. ii, p. 343, Pl. 23, figs. 13—16, under the above name, which seems also to correspond with our shell. The specimen figured in 'Min. Con.,' Tab. 577, fig. 6, cannot now be found.

SCALARIA COMMUNIS, *Lam*. Addendum Plate, fig. 5.

SCALARIA COMMUNIS, *Lamark*. An. sans Vert., vol. vi, p. 228.
— — „ *Jeffreys*. Brit. Con., vol. ii, p. 91.

Locality. Post-glacial, March.

An imperfect specimen of this well-known British species was obtained from the March gravel by Mr. Harmer, and sent to me; and as no doubt perfect specimens will hereafter be obtained from that locality, I have had the figure drawn from a recent specimen.

This species is given as from the Red Crag by Mr. A. Bell in his list of the "English Crags," as also by Mr. Jeffreys in his list attached to Mr. Prestwich's Red Crag paper, but I have not been able to confirm these references.

CHEMNITZIA JEFFREYSII ? *Koch* and *Wiechmann*.   Addendum Plate, fig. 14.

> TURBONILLA JEFFREYSII, *Koch & Wiech.* Moll. Faun. Sternb. Gest., p. 103, t. 3,.
> fig. 9 *a, b.*

Locality.   Cor. Crag, Sutton.

I have lately obtained from the Coralline Crag of Sutton an imperfect specimen, which, with the guide of the figures only, I have doubtfully referred to *C. Jeffreysii* of the German authors.   It appears to differ from *Ch. elegantissima* in having the costæ more inclined, and without the bend or flexure present in that species, and from *elegantior* in the shape of the whorls.   The figure in ' Crag Moll.,' vol. i, Tab. X, fig. 5, is not a good representation of the Crag shell *elegantior*, the costæ being straight,. and not wavy, as there represented.

EULIMA STENOSTOMA ?   Supplement, Tab. IV, fig. 25.

Since the engraving of this was made, my solitary specimen has unfortunately been. much injured, and I am not now certain that it has been correctly referred.

ODOSTOMIA ALBELLA, *Lovén.*   Addendum Plate, fig. 15.

> TURBONILLA ALBELLA, *Lovén.*   Ind. Moll. Scand., p. 19, 1846.
> ODOSTOMIA   —   *Jeffreys.*   Brit. Conch., vol. iv, p. 121, pl. 73, fig. 1.

Locality.   Cor. Crag, Sutton.

Mr. Robert Bell has sent to me for representation a specimen from the Coralline Crag, Sutton, with the name of *Odostomia albella*, and this, I think, is correctly referred.   I have myself recently found a rather less perfect specimen from the same locality.   Our shell has a very obtuse apex, smooth and glossy, and probably if we had several specimens quite perfect, one or other might show the sinistral embryonic nucleus spoken of by Mr. Jeffreys at the above reference.   Our shell has a large fold upon the inner side of the aperture, and there is a distinct and somewhat large umbilicus.   The peretreme is sharp and simple, but I am unable to detect any spiral striæ.   *Odost. rissoides*, var. *albella*, Forbes and Hanl., Pl. 96, fig. 5, is probably the same shell.

Litiopa of ' Crag. Moll.,' vol. i, p. 88, Tab. IX, fig. 1, much resembles this species in its obtuse apex, but there is a truncation at the base of that shell which this has not.   I am sorry to say the shell I called *Litiopa* is so scarce that I have not been able to find a specimen for many years.

ODOSTOMA DENTIPLICATA, *S. Wood.*   Addendum Plate, fig. 18.

Locality.   Cor. Crag, Sutton.

A single specimen of this genus has lately been found by myself which, in all respects, except in its denticulated outer lip, corresponds with *Od. plicata*, 'Supplement,' Tab. IV, fig. 22.   Dr. Speyer has figured a shell under the name of *Odontostoma plicatum* ('Tert. Conch. Cassel,' Tab. XXV, fig. 3), which is probably the same as my species. It shows similar denticulations.

There are two or three other Crag shells in this genus, and in the genus *Rissoa* that seem to differ from recent species only in this character.   See my remarks on *Rissoa semicostata*, p. 72 of this 'Supplement.'

MENESTHO BRITANNICA, *A. Bell.*   Addendum Plate, fig. 21.

MENESTHO BRITANNICA, *A. Bell.* "English Crags," Proc. Geol. Assoc., 1872, p. 18.

Locality.   Cor. Crag, Sutton.

The above figure represents the original shell forwarded to me by Mr. Robert Bell upon which the name of *M. Britannica* was introduced by his brother into his list of English Crag shells.   The specimen agrees (especially in the sudden tapering off of the three upper whorls) with the description of *M. albula*, given in the 2nd edit. of Gould's 'Invertebrata of Massachusetts,' p. 333, except in the absence of fine striæ with which the American shell is said to be covered.   Our present fossil is so well preserved that if it had been striated traces of the striæ would most probably be still capable of detection, and as I can detect none I have retained it under the name *Britannica* given to it by Mr. A. Bell.

MENESTHO JEFFREYSII, *A. Bell.*

MENESTHO JEFFREYSII, *A. Bell.*   English Crags, p. 24, Proc. Geol. Assoc., 1872.

Two specimens from the Red Crag of Walton Naze have been sent to me by Mr. Robert Bell with the above name attached, and in his letter is the following remark : " This little shell was first identified by Mr. Jeffreys as a distinct species, and afterwards compared again by himself and my brother, and recognised as identical with some Arctic specimens he possesses.   The name was given with his (Mr. Jeffreys') concurrence and permission."

The name *Menestho Jeffreysii* from the Red Crag is also given by Mr. Jeffreys at page 494 of Mr. Prestwich's Red Crag paper with the remark that the shell was previously known to him as an undescribed Greenlandic species.   I have compared these two Walton specimens with the shorter variety (fig. 12 *b* of Tab. XI of Crag Moll.) of *Rissoa cos-tulata* (*P. Stefanisi* of 'Supplement,' p. 73), and, allowing for the way in which they are worn, I cannot detect any difference between them.

The Walton specimens, although worn, present in places the same ribs and the same cancellation as ornament the Cor. Crag specimen of this variety of *Stefanisi;* and parti-cularly the form, relative dimensions, and position of the slight umbilicus are identical. I have not had the opportunity of examining the recent shells with which our fossils were identified, but if they be thus identical they can only, I think, belong to the shell figured 12 *b* of Tab. XI of the 'Crag Mollusca.' The description given by Mr. A. Bell of his new species *Jeffreysii* (Ann. and Mag. Nat. Hist., May, 1871, p. 10) quite accords with the specimens I examined and with the shell figured by me in Tab. XI of my original work.

FOSSARUS LINEOLATUS, *S. Wood*. Crag Moll., vol. i, p. 121, Tab. VIII, fig. 23 *c—d*, as var. *lineolatus* of *Fossarus sulcatus*.

FOSSARUS LINEOLATUS, *S. Wood*. Catalogue, 1840.

In the 'Crag Mollusca' are figured two varieties of *Fossarus sulcatus*. Mr. Jeffreys has referred one of them (var. *lineolatus*) to *F. Japonicus*, A. Adams. I have compared my Crag fossils with that recent species, and I believe them to be specifically distinct. The recent shell is shorter and more expanded, and it is ornamented with larger, coarser, and fewer ridges.

I treated var. *lineolatus* as a distinct species in my catalogue of 1840; and in this case, as in many others, I am inclined to revert to my views of 1840, and to call this variety a distinct species under my original name of *lineolatus*. M. Weinkauff refers *Fossarus sulcatus* of 'Crag Moll.' to *minutus*, Michaud, 'Bull. Soc. Linn.,' II, t. 122, figs. 7—9; but which of the two Crag varieties he thus refers I cannot make out.

CYCLOSTREMA LÆVIS ? *Phil.* Supplement, Crag Moll., p. 86, Tab. V, fig. 13.

Since my Crag shell was figured Mr. Jeffreys has sent to me for examination a recent specimen which appears precisely to resemble my fossil, and this he considers to be a new species, and proposes for it the name of *basi-striata*. He adds (in Lit.) that *lævis*, Phil., is the same as *serpuloides*, but is different from the Crag shell. I think with him that the Crag shell is distinct from *serpuloides*, and I so considered it at p. 86 of this Supplement, and if it should prove to be the case that the Crag shell is not identical with Philippi's *lævis*, it will require a new specific designation, which may be that of *basi-striata*, which Mr. Jeffreys proposes.

BULLA UTRICULUS ? *Brocchi.* Addendum Plate, fig. 26.

> BULLA UTRICULUS, *A. Bell.* English Crags, p. 19, Proc. Geol. Ass., 1872.

Locality. Cor. Crag, Gedgrave.

The specimen in the above figure was sent to me by Mr. Robert Bell, with the name *B. utriculus,* Broc., attached, being that under which it was inserted by Mr. A. Bell in his paper on the 'English Crags,' p. 19. Upon examination I find this specimen to be quite free from striæ of any kind, while *utriculus* is punctato-striated. It does not correspond with Brocchi's figure of either species.

It resembles, in its freedom from striæ, both Brocchi's and Hörnes' figures of *miliaris;* but the figures of these two authors do not accord with each other in respect to the general form of the shell. Under this uncertainty, although I am inclined to refer the Crag shell to *miliaris* rather than to *utriculus,* I think it likely to give rise to less confusion if I figure it under the name in which it first appeared in Mr. Bell's list of the 'English Crags.'

The shell is umbilicated at both ends, and particularly so at the base.

DENTALIUM ENTALIS, *Linné.* Addendum Plate, fig. 12 *a, b.*

In order to confirm the statements and fragmentary figures already put forward in this Supplement, page 22, and Tab. V, fig. 20, I have had represented a very perfect specimen obtained by Mr. Canham from the Cor. Crag, near Orford, which measures one inch and three quarters in length. The shell is beautifully smooth and glossy, even to its posterior termination, without a vestige of striæ, which, had they ever existed, could not fail to appear on so beautifully preserved a specimen. The terminal slit is *very* narrow, and nearly one fourth the length of the entire shell.

BULIMUS LUBRICUS, *Müller.*

> HELIX LUBRICA, *Müll.* Hist. Verm., pt. ii, p. 104.
> BULIMUS LUBRICUS, *Brug.* Ency. Meth. Vers., vol. i, p. 311.
> ZUA LUBRICA, *Forb. & Hanl.* Brit. Moll., vol. iv, p. 125, pl. cxxv, fig. 8.
> COCHLICOPA LUBRICA, *Jeff.* Brit. Conch., vol. i, p. 292, pl. xviii, fig. 2.

Locality. Red Crag, Butley.

A single specimen, as above referred, was found by Mr. Canham, and it is the only one I have seen from the Crag; it was accompanied with specimens of *Helix hispida,*

25

Succinea putris, *Limnæa palustris*, and what is probably *Limnæa Holbollii* from the same locality. This latter (of which I have myself found several specimens) seems to differ specifically from *palustris* in having a much deeper suture and more convex volutions. *B. lubricus* is common in the Post-glacial Freshwater deposits of this country; but I had not before known it from any deposits so old as the Red Crag. The other species have been previously figured as from the Fluvio-marine Crag, and are only mentioned here as occurring in the Red Crag. I have, at p. 4 of this Supplement, given what appears to me to be the true explanation of the occurrence of land and freshwater shells at this locality of the Red Crag.

CLAUSILIA PLIOCENA, *S. Wood.* Addendum Plate, fig. 22.

I have very recently obtained another land shell from the Cor. Crag of Sutton. This undoubtedly belongs to the above-named genus; but its specific identity is rendered uncertain by the fragmentary condition of the specimen.

Some of the animals of this genus in the living state are of arboreal habits, and my present specimen was probably, like *Helix Suttonensis*, carried into the Crag sea upon some dissevered piece of timber, and there deposited among the marine shells, or it may have been carried to sea on the feet or in the feathers of a bird.

Helix Suttonensis, from the same locality, approaches nearer to a living Madeira species than to any other that I have seen. I had hoped therefore to have been able to identify the present fragment with some species of *Clausilia* from that island, but I have not been successful in so doing; and am compelled therefore to give it provisionally a new name, for it does not agree with any British species known to me.

AVICULA PHALÆNOIDES, *S. Wood.* Supplement, p. 109, Tab. VIII, fig. 12 *a, b,* and Addendum Plate, fig. 23.

I have figured the hinge and umbonal portion of some shells of this genus from the Cor. Crag near Orford. None of the specimens of *A. tarentina* with which I have been able to compare the Crag fragments at all approach in the magnitude and thickness of their hinges these Crag *Aviculæ*, which are about intermediate in this respect between the living British and Mediterranean shell *Tarentina*, and the gigantic form from the Bordeaux beds called *Phalænacea*, Bast. There is such a close resemblance between the largest of these Cor. Crag hinges and those of the Bordeaux fossil that I have assigned to the Crag shell the specific name of *Phalænoides*. The fragments to which I referred in the 'Crag Mollusca,' vol. ii, p. 51, as probably belonging to *A. tarentina*, belong, no doubt, to the same species as the fragments here figured.

SCACCHIA LATA, *S. Wood.*   Addendum Plate, fig. 25.

Locality.   Cor. Crag, Sutton.

The above figure represents a shell recently sent to me from the Cor. Crag of Sutton, by Mr. Canham, which, I believe, is specifically distinct from any other yet described. It is thin, nearly transparent, very inequilateral and tumid, and with a smooth exterior. It slightly resembles *Sc. elliptica ;* but it is more transversely elongate. It is the right valve and is quite free from that sinuosity in the dorsal margin which forms the distinguishing feature of this valve in *elliptica.* My shell is also edentulous in this valve. I have found in the Cor. Crag from the same locality a specimen of what appears to be the left or opposing valve of the same species, and this has one obtuse cardinal tooth, and is similarly inequilateral. The second specimen, which is not quite perfect, has been in my possession these thirty years and more, and I have hitherto been unable to refer it ; but the discovery of the specimen now figured seems to throw light upon it.

The nearest shells to which this species approaches seem to be the older tertiary species *Erycina latens*, Desh., and *E. emarginata*, Desh, ' An. sans vert du Bas de Par.,' vol. i, p. 712, Pl. LI, figs. 24—27, and Pl. LIII, figs. 13—15.

THRACIA INFLATA, *J. Sow.*, var. *dissimilis.*   Addendum Plate, fig. 27.

Locality.   Cor. Crag, near Orford.

A single specimen as above represented has recently rewarded my researches, and from its peculiar form I think it desirable to have it figured. The species to which, as a variety, I have doubtfully referred it is somewhat variable, and I am unwilling, therefore, on the strength of a solitary specimen to describe it as a distinct species. The posterior slope is rugose, but it has not the regular shagreen character of *Thr. pubescens*, nor the convex or protruding ventral margin or sinuation of *convexa.* Should further specimens turn up preserving with integrity the distinguishing characters of this specimen, I should propose to assign it as a species under the name of *Thracia dissimilis.*

Venus dysera, Brocchi, is given in Mr. A Bell's ' English Crags ' as a species from the Cor. Crag, and Mr. Robert Bell has lately sent to me a small specimen from that Crag at Sutton with this name attached. This is, I believe, merely the young state of *Venus imbricata*, and I have several similar specimens in my own cabinet.

CONCLUDING REMARKS.[1]

THE science of Palæontology being one of pure observation, the lapse of a few years may be expected to make considerable alteration in determinations previously arrived at; thus, since the publication of the 'Crag Mollusca' I have been enabled to identify with existing species some of those forms which were considered to be extinct, and to correct errors consequent upon the possession, at the time, of imperfect specimens only, as well as those which the progress of our knowledge of the recent Marine Fauna had served to dispel.

The 'History of the British Mollusca,' by Messrs. Forbes and Hanley, having been published since the completion of my work, has furnished much information, and has supplied many notes for correction, which it had been my intention for some years past to publish with illustrations of new species obtained. I have also had, since then, the still greater benefit of the 'British Conchology,' by J. Gwyn Jeffreys, Esq., F.R.S.

The Crag formations have of late much occupied the attention of geologists and palæontologists, and Mr. Prestwich, in the 'Journal of the Geological Society,' vol. xxvii, 1871, has published three papers upon the Crag, in which he has availed himself of the assistance of Mr. Jeffreys to furnish tabular lists of the Mollusca of the Coralline, and of the Red and Fluvio-marine Crags, and Chillesford bed. As the determinations arrived at by the last-named gentleman differ in so many instances from my own, I have not considered it necessary to review them all in detail, but at the end of this Supplement I have given lists of all the Mollusca which have come under my observation from the Upper Tertiaries of that part of England to which I have confined this Supplement, in the way they appear to me to be specifically separated. Several species were spoken of in my original work as having been obtained by myself in abundance, or, at least, as not at that time being difficult to find, and I fear that this statement may have given disappointment to many collectors. Several species formerly abundant have, I am sorry to say, entirely eluded my search of late years, and this within a few yards distance of, and on the same

---

[1] The name "Coralline Crag" has been employed by myself for the older bed of these different formations, not only from what I conceive to be its special claim in regard to priority of date, but also from its correct and appropriate appellation. This name was given to it by Mr. Charlesworth, in 1835, in consequence of its composition being largely made up of what had been "Corallines," or "little corals," organisms so called by Ellis and others, such as *Cellepores, Retepores, Tubulipores, Flustræ,* &c. &c., but which have since been separated from the true corals. The Red Crag, from its rusty or reddish-brown colour.

vertical horizon as, the place of their former occurrence. On the other hand, some of those previously considered rare have become more plentiful.

In my list will be found many species to which my name is attached. Some of these will hereafter probably require to be altered through having been described by others previously to my work, but I have relinquished my own names wherever, up to the present time, I felt satisfied of the identity of my shell with any other previously described. In every case where I thought it could be justly identified with any other, whether known as living or not, I have not hesitated so to give it, suppressing, whenever occasion required, the name under which it had been previously known from the Crag in favour of any earlier name which the species with which it is thus identified may possess. After all this, however, there remains a wide difference between the view taken by me of the Crag Mollusca, and that taken by the author of the ' British Conchology ;' indeed, while according to that author my tribute of admiration for the persevering industry with which he has so much enlarged our knowledge of the Mollusca of British seas, and of the waters adjoining them, I do not hesitate to point out what, in my judgment, impairs the conclusions he has expressed with reference to Crag species, both in his general work, and in those lists of which he is the author which form part of Mr. Prestwich's papers on the Crag already referred to. It is obvious that this author's leanings are very marked, so as to group together allied Crag forms as varieties only of one species, and especially to make out a Crag shell to be either identical with a living species, or, at most, only a variety of it wherever the slightest presumption can be found for that course. I observe, however, that this reluctance to recognise two distinct, but allied, Crag forms as anything more than a variety gives way when the form has been found living ; and I have been particularly struck with this in the case of *Scalaria subulata*, which in the first, or ' Cor. Crag List,' is put in italics as a variety only of *Scalaria foliacea*, but which in a note to the ' Red Crag List ' (page 496), is restored to specific importance in consequence of its having in the meantime been dredged living by Mr. McAndrew. The degree of difference which is to constitute a species must in a great measure be arbitrary ; but there are many shells of the Crag thus identified with living ones which show quite as little variation from shells of the Older and Middle Tertiaries as they do from their living analogues. If species arise, as I believe they do, by gradual variation from forms previously existing [1] (be the cause of that variation what it may), it is obvious that, if we could get a perfect collection of all the forms that have existed and do exist on the surface of the earth, we should be placed in the dilemma of not being able to draw a specific line anywhere ; and although it must be a long day, if ever, before such a knowledge of animal

---

[1] In a letter to me (April 27th, 1873) the author of the ' British Conchology ' says, " I believe not in evolution, but in descent with modification ;" and I observe the same phrase—" descent with modification " used by his colleague in the " Porcupine" dredging expedition, Professor Thompson, in his late work, ' The Depths of the Sea ' (page 480). I confess that I do not understand the difference between this and evolution, in which I have for very many years been a believer.

life will be acquired, yet every increase of our knowledge of natural history must bring us nearer to this state of things, and proportionately augment our difficulties in the way of specific separation. In the present state of our knowledge, therefore, it seems to me more philosophical, and likely to be more advantageous in working out the history of the past changes of land and water on the globe, if the identification of species be not strained; and that, wherever a form presents differences from any other, though they be but slight, and those differences are fairly maintained in a group of individuals without intermediate forms occurring coevally in the same geographical area, such form should be regarded as a distinct species; in my 'Monograph of the Eocene Bivalves' I have expressed my views on that point more fully. Another reason for not undervaluing even slight differences by which many of the Crag Mollusca are separable from their living analogues, and so reducing them to the inferior importance supposed to be possessed by the term "variety," exists in the discordance between the evidence presented by the Molluscan fauna when thus reduced, and that presented by the other organisms of the Crag period. Thus the evidence of the *Entomostraca*, the *Foraminifera*, the *Polyzoa*, the *Polyparia*, the *Cirripedia*, and the *Echinodermata* of the Crag (all of them studied and described by independent authors) has quite a different bearing from that of the Molluscan fauna, when this last is reduced in the way it has been by the author of the 'British Conchology.'

In the case of the *Entomostraca* the proportion of species not recognised as living is as 13 to 5, of the *Foraminifera* as 47 to 53, of the *Polyzoa* as 65 to 30, of the *Polyparia* 3 to 1, of the *Cirripedia* as 4 to 6, and of the *Echinodermata* as 13 to 3.[1]

Now, although the researches which have been carried on among the living species of these groups of organisms may not have been so extensive as those carried on among the Mollusca, and although we may thus be better acquainted with the living forms of the latter, still, after making very large allowances on this account, we are left with great discrepancies between the evidence afforded by the percentage of forms not known living among the Mollusca, and those among the other groups. These discrepancies are so striking as to suggest caution in accepting the process by which the list of Crag species has been pared down, and so many species eliminated from it in the lists which accompany Mr. Prestwich's Crag papers.

The authors of the 'British Mollusca,' like myself, regard the Molluscan fauna of the Coralline Crag as having its affinities chiefly with that of the Mediterranean; but the

---

[1] A table of the proportions borne of living to extinct forms among these various groups of organism will be found at p. 134 of Mr. Prestwich's paper on the "Cor. Crag." This agrees substantially with the analysis given by me in the text. The number of species of Coralline Crag Mollusca, however, according to the list by Mr. Jeffreys, which accompanies the paper of Mr. Prestwich, is 316, of which he considers 264 to be living, and 52 extinct; thus giving a percentage of 84 recent and 16 extinct. 'Quart. Journ. Geol. Soc.,' vol. xxvii, p. 128.

author of the 'Brit. Conch.' differs from us,[1] and some time ago expressed his opinion that the Cor. Crag sea was the cradle of the British Mollusca. Now, although a number of Cor. Crag species that were not known to live so far north as Britain have lately been discovered living in our seas, this does not appear to me materially to affect the inference of the authors of the 'Brit. Mollusca' and of myself, because Molluscan, like other faunas, overlap each other, and species may yet linger as rareties in areas where they have long ceased to predominate. Similarly several Cor. Crag species have been now found to be denizens of Arctic seas; some few indeed are said by Mr. Jeffreys to be, so far as yet known, exclusively so,[2] while others range from Arctic seas down into British waters. We must remember that since the Coralline Crag period we have had great geological changes, the arctic conditions of the glacial period having fallen upon Britain, and again given way to our present temperate climate. Moreover, at the period of the Coralline Crag the Mediterranean and South European area was probably connected more directly with the Arctic than it now is, but the geographical changes which intervened between this and the Red Crag period appear to have produced an interruption of such direct connection, by giving rise on the east of Britain to a land barrier which shut off the Red Crag sea from the south and left it open to the north; the occurrence of a characteristic Red Crag Molluscan fauna fossil in Iceland, with some traces of it midway about Aberdeen, showing the extension northwards of this sea. By these means, as it seems to me, a part of the Cor. Crag species was induced to retire southwards and died out in Britain, while the other part survived there and some new importations came in. It is as probable that some of our present northern forms originated in and have migrated from seas far to the southward of Britain, as that they did so from the seas to the north, since we find *Panopea Norvegica, Mya truncata, Cyprina Islandica, Lucina borealis, Trophon contrarius,* and perhaps one or two more that are now unknown to the Mediterranean and considered as Arctic-British types, fossil in Sicily in association with forms that are mainly of Mediterranean types. After this change of coast lines, to which the Red Crag was due, had wrought its effect, the arctic conditions of the glacial period supervened, accompanied by a considerable submergence. This submergence destroyed the barrier which shut off the sea of the Red Crag from the southward, and once more brought the sea of the north east of Britain, and through it the Scandinavian and Arctic areas into direct communication with the Lusitanian and Mediterranean seas to which the one part of the Cor. Crag Mollusca had retired, and the conditions favorable to an intermingling, or at least an overlapping of the

[1] 'Brit. Conch.,' pl. lxxxix, Introduction, where it is said, "my investigation of the Crag shells has not led me to form the same conclusion as Messrs. Forbes and Hanley, viz. that most of these ancestors of our living shell-fish are of those forms which we regard as southern types."

[2] There are three species in Mr. Jeffreys' list of Cor. Crag shells to which the letter A alone is attached, signifying that these are exclusively Arctic species, viz. *Velutina virgata* as *V. undata*, Smith; *Cardium strigilliferum* as *C. elegantulum*, Beck; and *Glycimeris angusta* as *G. siliqua*, Chemn. None of these identifications are to my mind satisfactory and I have not adopted them.

faunas were repeated.[1]   After this, and at probably the very last geological period that we recognise in Britain, the two areas were once more divided by an isthmus between Kent and France, which enabled the southern fauna to range up the British Channel without intermingling with or overlapping that of the Arctic province. This was the period of the Selsea mud bed in which a considerable and well-preserved fauna of southern affinities occurs—a fauna which has its living analogue on the southern coast of England, and in which are some few species that now occur only on the Lusitanian or Mediterranean coasts. Besides all this, the exceptional character of the Scandinavian fauna, due, it would seem, to that drift of warm water, which keeps the bays and fiords of Norway free from ice, forms another complicating element in the question ; while the influence of vast stretches of abyssmal water supporting Mollusca (of which some have been hitherto considered characteristic of Mediterranean areas, and others of northern and even Arctic areas) in intermingling distant faunas has yet to be elucidated.

It thus appears to me that if any of these past periods could with propriety be termed " the cradle of the British Mollusca," it would be rather the Red than the Coralline Crag. Strictly speaking, however, it would, owing to the alternate interminglings and separations already referred to, be misleading to refer to any one of these past periods as having formed the cradle of the British Mollusca. In speaking, therefore, of the Coralline Crag fauna having its affinity with that of the Mediterranean, I mean merely to imply that the major part of the living species of the Cor. Crag are of Mediterranean habitat, and that of such among this part as are common to the Mediterranean and British areas, the greater number are more abundant in individuals in the Mediterranean than they are in the British waters ; and that, so far as conditions of temperature can be inferred from a Molluscan fauna, these conditions during the Coralline Crag period would seem to be nearer to those of the seas of southern Europe and of the Azores than to those surrounding the British Islands.

According to the analysis of my Synoptical List (post, page 219), there are 205 species of Gasteropoda and Bivalvia living in the Mediterranean which are identical with Coralline Crag species, among which there are 51 that are not known in the British seas. On the other hand, there are 20 species of British Mollusca in the Coralline Crag that are not living in the Mediterranean. It should not be forgotten that the most abundant, and therefore most characteristic species of the Coralline Crag, such as *Cardita corbis, Cardita senilis, Limopsis pygmæa, Ringicula buccinea*, and others, are southern species unknown to British seas ; and that among the 154 Coralline Crag species occurring both in British and Mediterranean waters there are many which are really characteristic Mediterranean shells, and are only marked British in consequence of some rare occurrences, due to the

[1] The remarkable molluscan fauna, extracted with difficulty from the middle Glacial sands, and still far from complete, throw much light on this history. Several Coralline Crag and Mediterranean species, of which we get no trace in the Red Crag, or at least in the newest part of it, or Chillesford bed, again making their appearance, though the specimens have evidently travelled far along the bottom. At this period the great Glacial submergence fully set in, if, indeed, it did not attain its maximum.

perseverance of our naturalists in dredging the extensive stretch of sea which, extending over 12 degrees of latitude, surrounds the British Isles. In making these comparisons, moreover, it should not be overlooked that we contrast results obtained from a patch of fossil sea bottom only a few yards square in Suffolk,[1] with results obtained from the vast area surrounding the British islands. If instead of this a comparison could be instituted between the Coralline Crag fauna and that which might be obtained from an exhaustive dredging of the bottom of the North Sea off the Suffolk coast, my belief is that the exotic character of the Coralline Crag sea would become much more apparent.

How far during the intercourse which has gone on by ships between Britain and the Mediterranean and Lusitanian coasts for 2000 years past, Mediterranean and Lusitanian Mollusca have been introduced into British waters through the agency of the bottoms, anchors, and especially the ballast of ships, it would be rash to conjecture; but the extent of this cannot have been inconsiderable, especially during the last two centuries. Whatever the degree to which this has extended may be, it has by so much reduced the exclusively Mediterranean and Lusitanian proportion of the Coralline Crag Mollusca below its real amount. Of course the same process must have had similar results in introducing British species into Lusitanian and Mediterranean waters.

The numerous species and profusion of individuals of the genus *Astarte*, the presence of the genus *Cyprina*, and of such shells as *Trichotropis borealis* and *Glycimeris angusta*, represent, on the one hand, what would be urged in support of the arctic and boreal features of the Coralline Crag fauna. On the other hand, the profusion of such species as *Limopsis pygmæa, Cardita senilis, Cardita corbis, Ringicula buccinea, Woodia digitaria*, and *Dentalium dentalis*, and the occurrence of the sixteen genera presently enumerated, represent the Mediterranean features. Besides these we have the tropical forms, *Pyrula* and *Pholadomya*, which were probably present as lingerers from those periods anterior to the Crag which are clearly shown by their fossils to have been more and more tropical as we recede in Tertiary time. Although the genus *Astarte* among recent shells is looked upon as indicating boreal conditions in our present seas, inasmuch as only one species lives in the Mediterranean, yet as we recede in geological time this indication becomes weakened if not negatived altogether, and a proof is afforded that the genus did not in older Tertiary periods originate in icy seas. Thus in the Eocene of England we have no less than four species of this genus in association with such an undoubted tropical Mollusc as the *Nautilus;* and of these, two are from the London clay, the climate of which is proved to have been warm not merely by its *Nautili*, but by the tropical character of the vegetation yielded by the Sheppey deposit. In Mesozoic formations the genus goes back in some abundance of species as far as the Lias in association with gigantic reptilia, through climates indicated by the Purbeck vegetation

---

[1] To any objection that such a dredging would not disclose the contents of the North Sea bed *vertically,* I reply that there are but very few Coralline Crag species which I have not found within a vertical range of three feet in the stackyard pit at Sutton.

to have been warm.   Further, we have the following sixteen Cor. Crag genera not yet known to be living in the seas of Britain, of the North Atlantic, or of the arctic regions, viz. *Panopea*,[1] *Pholadomya*, *Chama*, *Hinnites*, *Erycinella*, *Scintilla*, *Nucinella*, *Lingula* (?), *Sigaretus*, *Pyramidella* (?), *Fossarus* (?), *Cancellaria* (rejecting *Admete*), *Cassidaria*, *Terebra*, *Pyrula*, and *Voluta*, to which might be added a section of *Pleurotoma*, and but for the late dredgings in abyssmal waters, the genus *Verticordia* also.   The presence of these genera in the Coralline Crag seem to me to impart a more southern aspect to its fauna than any analysis of the species themselves would do.

There are a few forms both in the Coralline and in the Red Crag, the living analogues of which survive in seas so remote as to throw no light on the affinities of the faunas of those Crags.   Of these in the Cor. Crag the little *Erato*, which I had identified with the West Indian species *Maugeriæ*, may be instanced. Mr. Jeffreys, however (in a letter), informed me that he thought the specimens in my collection (in the Brit. Mus.) might be stunted forms of *lævis*.   I believe, however, that *E. Maugeriæ* is identical with the shell from the Cor. Crag figured by me under that name.   I have compared many specimens of each without being able to detect a difference that might be called specific, or any more difference than may be observed among the specimens themselves, and I cannot consent to degrade my little Crag shell into the position of a variety, as it and the true *lævis* are found together at the same spot without intermediate forms.

The characteristic shell of the newer part of the Red Crag, of the Fluvio-marine Crag, of the Chillesford bed, and of the Lower, Middle, and Upper Glacial deposits, *Nucula Cobboldiæ*, is another of these instances; the living analogues of this shell, of which there are two or three, being denizens of the North Pacific; and although I have pointed out in the body of this 'Supplement' the characters which seem to me to distinguish *Nucula Cobboldiæ* specifically from all these allied species, yet it is very remarkable that its analogues should be several in number, and all of them confined, as far as yet is known, to so remote a sea.

In 1838 Mr. Conrad published figures and descriptions of some Medial Tertiary fossils of the United States bearing a strong resemblance to Crag forms; and in a paper upon these, published in the 'Proceedings of the Geol. Soc.' for the year 1843, Sir Chas. Lyell gives the names of several species that he considered to be identical with Crag, or, at least, with European fossils.

In the 'Geological Magazine' for April, 1865, vol. ii, is a communication by Dr. P. Carpenter upon the connection between the Crag and the recent Mollusca of the North Pacific, wherein he appears to consider the Crag fauna and the North Pacific fauna to have emanated from the north, the one diverging eastward and the other westward.   He says at p. 153, "Not taking into account *similar* forms, no fewer than twenty-four species

---

[1] *Panopea Norvegica* does not properly belong to the genus *Panopea*, but perhaps to *Panomya*; see p. 161 of this Supplement.

have already been clearly identified on the West[1] Pacific coast; several of these can scarcely have travelled through Behring Straits, not being Boreal forms." The following species, now living on the north-west coast of America, are given by him ('Brit. Assoc. Rep.,' Newcastle, 1863) as identical with Crag forms. Those with an asterisk are also given as Tertiary fossils of Maryland and Virginia by Sir C. Lyell ('Proc. Geol. Soc.,' 1845, p. 555).

Coralline Crag Species.

Mya truncata, *Linn.*
Sphenia ovoidea (?), *Carpenter.*
*Glycimeris generosa (?), *Gould* (Panopea Faujasii).
Saxicava Arctica, *Linn.*
*Tellina donacina, *Linn.*
Astarte triangularis, *Mont.*
* — fluctuatus, *Carpenter.*
Miodon prolongatus (Cardita corbis).
*Lucina borealis, *Linn.*
Cryptodon flexuosus, *Don.*
Verticordia novem-costata (?).
Kellia suborbicularis, *Mont.*
Lascœa rubra (?), *Mont.*
Arca tetragona, *Poli.*

*Arca lactea, *Linn.*
Mytilus edulis, *Linn.*
Modiola modiolus, *Linn.*
Modiolaria marmorata, *Forb.*
Lima subauriculata, *Mont.*
Hinnites giganteus (?), *Gray.*
*Erato columbella, *Carpenter.*
Cerithiopsis tubercularis, *Mont.*
Cerithium adversum, *Mont.*
*Eulima micans, *Carpenter.*
*Solariella peramabilis (?), *Carpenter.*
Margarita Vahlii (?), *Möll.*
*Galerus sinensis, *Linn.*
Cylichna cylindracea, *Pennen.*
— mammillata, *Phil.*

Red Crag Species.

Rhychonella psittacea.
Macoma inquinata.
Astarte compressa.
Cardium Groenlandicum.
Nucula tenuis.
Acila castrensis (?).
Leda minuta.
Yoldia lanceolata.

Bela fidicula.
— excurvata.
Purpura saxicola (?).
Lacuna vincta.
Natica clausa.
Velutina lævigata.
Dentalium Indianorum.

Any inferences which might be drawn from the identifications of Dr. P. Carpenter in the case of the Pacific shells, and of Sir Chas. Lyell in the case of the United States

[1] *East* Pacific coast (*i.e. west* coast of America) is probably intended.

Medial Tertiary shells, depends upon the correctness of the identifications themselves, and this I have not had the means of examining.

A list of extraneous fossils was given by myself in a paper published in the 'Quart. Journ. Geol. Soc.,' 1859, but this would now require to be much added to. These are, however, undoubted derivatives, but I have been greatly embarrassed with specimens that do not belong, so far as I know, to any described species, and to which a strong suspicion attaches that they come from some older Tertiary formation. In such cases, as well as in those of known older Tertiary species, where the mineral condition of the specimen does not of itself indicate that it is derivative, I have figured the specimens, but in the synoptical list which follows I have distinguished all such as clearly derivative. There are besides these many shells in the Red Crag which I am disposed to regard as only present in that formation as derivatives.[1] The number of these is greater than was the case when the 'Crag Mollusca' left my hands in 1856, and I have distinguished them in the list as possibly or probably derivative. In one instance I have been compelled to regard as a genuine fossil of the period what I had before treated as derivative. This is *Voluta Lamberti*. In other instances, such as *Cassidaria bicatenata* and *Trophon elegans*, which were then only known from the Red Crag, I have satisfied myself not only that they are Cor. Crag species, but that their presence in the Red is due only to derivation. The only part of the Red Crag which is genuine and free from derivatives is that of Walton Naze (where *Voluta Lamberti* that I had wrongly regarded as derivative does occur), all the rest of the formation being more or less leavened with derivatives from the Coralline Crag, and from older formations, as well as with shells from older beds of Red Crag age, such as Walton, as explained at page vii of the Introduction to this Supplement. It is unfortunate that we get only this one deposit of Walton with a genuine fauna, since from the change in climate and consequent introduction of many northern forms into the newer parts of the Red Crag (as explained in the Introduction) the Walton fauna, though genuine, does not show what was the true fauna during the *later* stages of the Red Crag formation.

The elimination from the Red Crag fauna of the additional derivatives from the Coralline tends to increase the Palæontological distinction between those deposits, while on the other hand the discovery, since the 'Crag Mollusca' left my hands, of some few Coralline Crag species among the genuine Red Crag fauna pretty well balances this increment. I, however, see no reason to modify the opinion I have always entertained as to the complete separation of the Coralline and Red Crags, although the fauna of the oldest part of the Red Crag has a greater Mediterranean affinity than that of the newer, as pointed out by me in the 22nd volume of the 'Quarterly Journal of the Geological Society,' p. 542. With respect, however, to the Red and Fluvio-marine Crags, and their over-lying Chillesford clay and sand, they can, in my opinion, be regarded as only one deposit, constituting in England the upper Crag, as the Coralline does the lower; and the triple

---

[1] The phosphatic nodule excavation in the Red Crag at Waldringfield is quite a museum of derivatives.

division of the Crag, which has for so many years been assumed, must, in my opinion, be abandoned. The Palæontological difference between the Walton bed and the newest part of the Red Crag, the *Scrobicularia* portion, or even the Butley portion, is far greater than any which exists between these latter and the Fluvio-marine Crag or Chillesford beds; and whether, as discussed at pp. ix and x of the Introduction to this 'Supplement,' the Fluvio-marine Crag of Bramerton (and, of course, the deposits at the other Fluvio-marine Crag localities of Suffolk and Norfolk also) be coeval with the newer part of the Red Crag, or posterior to it, it is sufficiently clear that, from the oldest or Walton Red Crag deposit up to the Chillesford clay itself, all the beds of the English Crag which are posterior to the Coralline are portions of the same geological formation, during the accumulation of which only very slight oscillations in the relative position of the sea and land occurred.

The synoptical list which follows these remarks will show what species of Mollusca have occurred in the Fluvio-marine Crag that have not yet been detected in the Red. With three or four exceptions these are shells that from their minuteness or fragility would only be preserved under exceptional circumstances in so roughly accumulated a deposit as the Red Crag, and their not having been detected in that formation need not surprise us. The most important among these three or four exceptions is the arctic species, *Astarte borealis*, which occurs sparingly in the Fluvio-marine Crag of Bramerton, but is common in the Chillesford bed at the same place, as well as in all the successive glacial deposits. It is, however, significant that while thus common in the Chillesford bed where it overlies the Fluvio-marine Crag, this shell has not been detected in these same Chillesford beds where they overspread the Red Crag; thus indicating, apparently, that its absence from the Red Crag was due to a localisation of the species rather than to any difference in age between the two deposits. The occurrence of a bed of land and freshwater shells in the Red Crag at Butley has, in my opinion, no bearing upon the question of the synchronism between the Red and Fluvio-marine Crags, but is due to the causes referred at p. 4 of this 'Supplement.'

I have not attempted to correlate the British Crag beds with the Crag of Antwerp, or with the Monte Mario, or Sicilian formations, because I feel satisfied that unless I had the means of comparing the specimens themselves from those formations with my own specimens from the English beds, any results that might be published would be more or less illusory. It is not sufficient for reliable identification to look over collections abroad; the specimens from the respective formations must be placed side by side and compared with each other. For this purpose extensive collections from these foreign beds would require to be made; and as I have not the opportunity at my advanced years of forming such, I have thought it best to leave the task of making a satisfactory correlation to other observers. I have therefore confined my tabular lists to the successive beds of the Crag, the Glacial, and Post-glacial series in that portion of England to which I have confined this Supplement, having had the means, except in the few cases expressly mentioned in the work, of comparing all the specimens and of giving them deliberate examination. The Bridlington species have been examined by me either in the collection of Mr. Bean, now in the British Museum, or by means of specimens kindly supplied me for the purpose by Mr. Leckenby, to whom my best thanks are due for his obliging readiness at all times to assist me with specimens. The Burgh, Horstead, and Coltishall specimens, and those from the Lower and Middle Glacial deposits, have all been sifted out by my son under my own eyes from material obtained from time to time from the various localities by Mr. Harmer, and most carefully guarded against intermixture or the possibility of the intrusion of any other material or of specimens from any other place. The March and Hunstanton specimens have all been examined by me in collections made by Mr. Harmer from those places, and the Nar Valley specimens in the collection of the late Mr. Rose. The Kelsea Hill species are principally given on the authority of Mr. J. Gwyn Jeffreys, my son's collection from that place comprising only a portion of the species enumerated by that gentleman. My best thanks are due to Mr. Dowson and Mr. W. Crowfoot for the opportunity of examining the Aldeby specimens, to Mr. Cavell for the sight of specimens from the Cor. Crag near Orford, from Thorpe by Aldborough, and from Easton Cliff, and to Mr. Reeve for specimens from Bramerton. Especially to Messrs. Alfred and Robert Bell for the use of numerous specimens, and to the Rev. H. Canham, of Waldringfield, for his unwearied researches in both the Red and Coralline Crags, are my thanks due. I am also obliged to Dr. Reed, of York, for the use of many specimens which he possesses from the different Crags of Suffolk, as well as to Mr. E. Charlesworth, who has throughout his life taken so active a part in the elucidation of the natural history of the Crags. I must also express my thanks to Mr. Jeffreys for his readiness at all times to assist me with the loan of recent shells in his possession for the purpose of comparison. I have also had the assistance of my son throughout the preparation of this Supplement, and in the re-examination and revision to which I have subjected the determination of all the species given in the 'Crag Mollusca.' In all the instances where not otherwise specified in the text the specimens are from my own collection, and of my own finding.

In conclusion, I would add that I have studiously abstained from recognising the terms "Quaternary" or "Post-Tertiary." Of the terms "Primary," "Secondary," and "Tertiary," adopted by early geologists for great geological divisions, the first has become wholly obsolete. The term "Secondary" becomes yearly more and more vague, and less and less used to define any natural division of geological time; while the term "Tertiary" alone remains convenient in consequence of the yet unbridged chasm which separates the beds of that division from the Cretaceous group. To introduce, however, into geology another division as "Quaternary" or "Post-Tertiary" is not merely to import a term as unmeaning as that of "Primary," which has been universally dropped, but one whose limits cannot be defined by any constant feature, either in physical geology or in palæontology; and it finds a foil when some of our leading naturalists insist, as they have been lately doing, that we are still in the Cretaceous period. If we test these terms " Quaternary," or " Post-Tertiary," by such a standard as that of the reference of those beds in which species not known living occur to the " Tertiary " group, and of those beds which contain none but living species to the " Quaternary," the standard will vary according to what forms of life we select. If we take the Mollusca, we find that a proportion of them not known as living occur in the Lower and Middle Glacial sands, and that, even in the Bridlington Bed, there are two such; so that if the division between Tertiary and Quaternary were based upon Molluscan evidence, we should have to draw the line just above the Bridlington horizon and below that of the Scotch beds, for in the Scotch none but species still living on one side or other of the Atlantic are found. If, instead of this, we take the Mammalia, the line would have to be brought down to that much later period when the *Machairodus* and the extinct species of *Elephant, Rhinoceros,* and *Hippopotamus* died out. On the other hand, if the Quaternary or Post-Tertiary age were attempted to be made co-ordinate with the existence of man, we should not only be placing it in the most uncertain of all positions in consequence of our knowledge of the evidences of man's existence undergoing almost daily extension, but the line so regulated would differ equally from that based on the Molluscan evidence, and from that on the higher animals; the evidence yet obtained of man's existence not going back to those later glacial formations in which the Mollusca belong to species which are all living, but showing him, nevertheless, to have long existed coevally with the extinct Mammalia.

While this difficulty of making any consistent definition of the term is so obvious, we have, on the other hand, from the Older Crag upwards evidence, in the case of the Mollusca at least, of the most gradual transition from a period treated by all as Tertiary, when a considerable proportion of the species consisted of forms not known living, to the most recent beds in which the included remains, to whatever part of the animal kingdom they belong, are all those of living species.

I have therefore referred all formations anterior to the recent, to the Tertiary period,—a period which by long custom has become well known, and must, until our knowledge becomes extended, be retained as convenient; and which, in Europe at least, appears to be sharply defined by unmistakeable physical and palæontological features, though even in this quarter of the world future discoveries may not improbably eventually necessitate its abandonment.

At the end of the synoptical list which follows, the change which has taken place through the formations succeeding the Coralline Crag, from the Mediterranean aspect which the Mollusca of British seas possessed at the period of that Crag, is pointed out.

SYNOPTICAL LIST OF MARINE MOLLUSCA FROM THE UPPER TERTIARIES OF THE EAST OF ENGLAND.

N.B.—The references to the pages of the ' Crag Mollusca' are given in simple figures; those to the Appendix have the letter A prefixed, and those to the Supplement have the letter S prefixed.

PAGE.	GASTEROPODA.	Cor. Crag.	Red Crag, Walton.	Red Crag,* Sutton and Butley.	Scrobicularia Crag.†	Fluvio-marine Crag.	Chillesford beds.	Lower Glacial.	Middle Glacial.	Upper Glacial.	Post Glacial, Kelsey Hill.	Post Glacial, March.	Post Glacial, Hunstanton.	Post Glacial, Nar Brickearth.	Living, Britain.	Living, Mediterranean.	Living, elsewhere.	REMARKS.
14, s 4	Ovula spelta, *Linn.*	×	×	×	...	...	...	...	...	...	...	...	...	...	...	×	×	*Ovula Leathesii,* Crag Moll.
s 4	— — var. brevior	...	×	×	...	...	...	...	...	...	...	...	...	...	...	?	?	
16	Cypræa affinis, *Dujardin*	×	...	...	...	...	...	...	...	...	...	...	...	...	...	...	...	
15	— avellana, *J. Sow.*	×	×	×	...	...	...	...	...	...	...	...	...	...	...	...	...	
16	— Angliæ, *S. Wood*	...	...	×	...	...	...	...	...	...	...	...	...	...	...	...	...	Probably only a variety of *retusa.*
16	— retusa, *J. Sow.*	×	...	×	...	...	...	...	...	...	...	...	...	...	...	...	...	Probably only derivative in Red Crag.
17, s 5	— Europæa, *Mont.*	×	×	×	...	...	...	...	×	...	...	...	...	...	×	×	×	
18	Erato lævis, *Donovan*	×	...	×	...	...	...	...	...	...	...	...	...	...	×	×	×	Probably only derivative in Red Crag. I cannot find anything to justify its insertion as a Fluvio-marine Crag shell.
19	— Maugeriæ, *Gray*	×	...	×	...	...	...	...	...	...	...	...	...	...	...	...	×	One specimen only from Red Crag, and that probably a derivative.
A 310	Mitra ebenus, *Lam.*	×	...	...	...	...	...	...	...	...	...	...	...	...	...	×	×	Probably derivative in Red Crag. Given by Bell (Annals, May, 1871) from Waldringfield.
21, A 311	— — var. plicifera	×	...	...	...	...	...	...	...	...	...	...	...	...	...	×	×	
s 7	— — var. uniplicata	×	...	...	...	...	...	...	...	...	...	...	...	...	...	?	?	
s 8	— fusiformis, *Broc.*	...	...	×	...	...	...	...	...	...	...	...	...	...	...	×	...	Waldringfield only, and derivative.
20, s 7 & 173	Voluta Lamberti, *J. Sow.*	×	×	×	...	×	...	...	...	...	...	...	...	...	...	...	...	
s 7	— (Volutilithes) luctatrix, *Solander*	...	...	×	...	...	...	...	...	...	...	...	...	...	...	...	...	Waldringfield only, and derivative.
s 6	— — nodosa, *J. Sow.*	...	...	×	...	...	...	...	...	...	...	...	...	...	...	...	...	Waldringfield only, and derivative.
s 6	Ancillaria glandiformis, *Lam.*	...	...	×	...	...	...	...	...	...	...	...	...	...	...	...	...	Waldringfield only, and derivative.
26	Terebra canalis, *S. Wood*	×	...	×	...	...	...	...	...	...	...	...	...	...	...	...	...	
s 8	— — var. acuminata	×	...	...	...	...	...	...	...	...	...	...	...	...	...	...	...	*T. exilis?* A. Bell.
26	— inversa, *Nyst*	×	...	×	...	...	...	...	...	...	...	...	...	...	...	...	...	
s 9	Columbella (Astyris) Holbollii, *Moll.*	...	...	...	...	...	...	...	×	...	...	...	...	...	...	...	×	Living in Greenland seas.
s 174	— Borsoni? *Bellardi*	...	×	...	...	...	...	...	...	...	...	...	...	...	...	...	...	A doubtful identification.
23, s 9	— sulcata, *J. Sow.*	×	×	×	?	...	...	...	...	...	...	...	...	...	...	...	...	*Columbella scripta* I have not seen from the Crag, nor *C. abbreviata* either.
s 174	— minor, *Scacchi*	×	...	...	...	...	...	...	...	...	...	...	...	...	...	×	...	
27, s 11	Cassidaria bicatenata, *J. Sow.*	×	...	×	...	...	...	...	...	...	...	...	...	...	...	...	...	Probably only derivative in Red Crag.
s 11	— — var. ecatenata	×	...	...	...	...	...	...	...	...	...	...	...	...	...	...	...	
s 10	Cassis Saburon, *Brug.*	...	...	×	...	...	...	...	...	...	...	...	...	...	...	×	×	Waldringfield only, and derivative in Red Crag.
32, s 15	Nassa conglobata, *Broc.*	...	×	×	...	...	...	...	...	...	...	...	...	...	...	...	?	
31	— consociata, *S. Wood*	×	...	×	...	...	...	...	...	...	...	...	...	...	...	...	...	Probably only derivative in Red Crag.
s 13	— densecostata, *A. Bell*	×	...	...	...	...	...	...	...	...	...	...	...	...	...	...	...	
30	— elegans, *Leathes*	...	×	...	...	...	...	...	...	...	...	...	...	...	...	...	...	
s 11	— granifera, *Dujardin*	×	...	×	...	...	...	...	...	...	...	...	...	...	...	...	...	Probably only derivative in Red Crag.
29, s 12	— granulata, *J. Sow.*	×	×	×	...	...	...	...	×	...	...	...	...	...	...	...	?	Living in Japanese seas. Sec. A. Bell in lit.
29, s 12	— incrassata, *Müll.*	×	...	×	...	...	×	...	×	...	...	...	...	×	×	×	×	
28, s 15 & 176	— labiosa, *J. Sow.*	×	...	×	...	...	...	...	...	...	...	...	...	...	...	...	...	Probably only derivative in Red Crag.
31, s 15 A 315	— Monensis, *Forbes*	...	...	×	...	...	...	...	...	...	...	...	...	...	...	...	...	An uncertain species.

* The column of Sutton and Butley includes all the Red Crag localities except Walton and the Scrobicularia Crag, all such being regarded as newer than the Walton and older than the Scrobicularia beds. The Bentley Crag, however, seems very nearly identical with that of Walton.

† The paucity of species in this column does not arise altogether from the paucity of species in the Scrobicularia Crag, but from its not having been searched in the way that the other beds of the Red Crag have been. I have no doubt that by investigation the names in this column might be much augmented.

PAGE.		Cor. Crag.	Red Crag, Walton.	Red Crag, Sutton and Butley.	Scrobicularia Crag.	Fluvio-marine Crag.	Chillesford beds.	Lower Glacial.	Middle Glacial.	Upper Glacial.	Post Glacial, Kelsey Hill.	Post Glacial, March.	Post Glacial, Hunstanton.	Post Glacial, Nar Brickearth.	Living, Britain.	Living, Mediterranean.	Living, elsewhere.	REMARKS.
32	Nassa prismatica, *Broc.*	×	...	...	...	...	...	...	...	...	...	...	...	...	...	×	×	The true *prismatica* of Brocchi sent to me by Mr. Canham since Supplement in type.
30	— — var. limata	×	×	×	...	...	...	...	...	...	...	...	...	...	...	?	?	
s 13	— propinqua, *J. Sow.*	...	...	×	...	...	...	...	...	...	...	...	...	...	...	...	×	*trivittata* ? Say.
s 14 & 176	— Cuvieri, *Payr.*	...	...	×	...	×	...	...	...	×	...	...	...	...	...	×	×	*N. pusillina* of Supplement to Crag Mollusca.
s 13 A 315	— pulchella, *Andr.*	×	...	×	...	...	...	...	...	...	...	...	...	...	...	...	...	Red Crag, Waldringfield (Bell); probably only derivative in Red Crag.
B 12	— pygmæa, *Lam.*	?	...	?	...	...	...	...	...	...	...	...	...	×	×	×	×	
33 s 15	— reticosa, *J. Sow.*, and numerous varieties.	...	×	×	...	...	×	...	...	...	...	...	...	...	...	...	...	As to *Nassa musiva*, see Sup., p. 176.
s 14 A 315	— reticulata, *Linn.* — — var. nitida	...	...	...	...	...	...	...	...	×	...	×	...	...	×	×	×	
24 s 5 42	Rostellaria lucida, *J. Sow.*	...	...	×	...	...	...	...	...	...	...	...	...	...	...	...	...	*plurimacosta*, Crag Moll.
A 311 s 10	Pyrula reticulata, *Linn.*	×	?	...	...	...	...	...	...	...	...	...	...	...	...	...	?	Probably derivative in the Red Crag.
34 s 16	Buccinum Dalei, *J. Sow.*	×	×	×	...	×	×	...	...	...	...	...	...	...	×	...	×	
s 17	— pseudo Dalei, *S. Wood* ...	×	...	...	...	...	...	...	...	...	...	...	...	...	...	...	...	
s 17	— glaciale, *Linn.*	...	×	?	...	...	...	...	...	...	...	...	...	...	...	...	×	
s 175	— Tomlinei, *Canham*	...	...	×	...	...	...	...	...	...	...	...	...	...	...	...	...	Derivative only.
35 s 18	— undatum, *Linn.*	×	?	×	...	...	...	...	?	×	...	...	×	×	...	...	×	Locality of Cor. Crag omitted accidentally at Supplement, p. 35.
35 s 18	— — var. tenerum	...	×	×	...	×	...	×	...	×	...	×	×	...	...	×	...	Hunstanton since Supplement printed off.
35	— — var. striatum, *Pennant*	...	×	...	...	...	...	...	...	...	...	...	...	...	...	...	...	
s 18	— — var Groenlandicum, *Chemn.*	...	...	...	...	...	...	...	...	...	...	...	...	...	...	...	×	
s 18	— — var. clathratum, *S. Wood*	...	...	×	...	...	...	...	...	...	...	...	...	...	...	...	...	
s 18	— — ? ovulum, *S. Wood* ...	...	...	×	...	...	...	...	...	...	...	...	...	...	...	...	...	Perhaps only a distortion of *undatum*.
36 B 18	Purpura lapillus, *Linn.* (crispata)	...	×	×	×	×	×	×	×	×	×	...	...	...	×	...	×	
36 B 18	— incrassata, *J. Sow.*	...	×	×	...	×	...	...	×	...	...	...	...	...	×	...	×	
38 s 29 ?	— tetragona, *J. Sow.*	...	×	×	...	?	...	...	...	...	...	...	...	...	...	...	...	Given from Norwich in Woodward's list; *sed quæ.*
s 175	Lachesis anglica, *A. Bell*	×	...	...	...	...	...	...	...	...	...	...	...	...	...	...	...	
s 176	Murex insculptus?	...	...	×	...	...	...	...	...	...	...	...	...	...	...	...	...	Derivative only.
s 30	— corallinus, *Scac.*	×	...	...	...	...	...	...	...	...	...	...	...	...	...	...	...	
s 30	— Canhami, *S. Wood*	...	...	×	...	...	...	...	...	...	...	...	...	...	×	×	×	Unique and possibly derivative.
39 s 31	— erinaceus, *Linn.*	...	...	×	...	×	...	...	...	×	...	...	...	...	×	×	×	
40	— tortuosus, *J. Sow.*	×	×	×	...	...	...	...	...	...	...	...	...	...	...	...	...	Fossil in the Vienna beds. Hörnes' Tab. XXV, fig. 12.
41 s 30	Triton connectens, *S. Wood* ...	×	...	...	...	...	...	...	...	...	...	...	...	...	...	...	...	*heptagonus*, Crag Moll.
49	Trophon alveolatus, *J. Sow.*	×	...	×	...	...	...	...	...	...	...	...	...	...	...	...	...	Possibly only derivative in Red Crag.
49	— consocialis, *S. Wood*	×	...	×	...	...	...	...	...	...	...	...	...	...	...	...	...	Possibly only derivative in Red Crag.
s 25	— Actoni, *S. Wood*	...	...	×	...	...	...	...	...	...	...	...	...	...	...	...	...	
44 s 19	— (Neptunea) antiquus, *Linn.*	...	...	×	×	×	×	×	×	...	...	...	...	...	×	...	×	
44 s 19	— — var. striatus	...	...	...	...	...	...	...	...	...	...	...	...	...	...	...	...	
44 s 19	— — var. striatus contrarius	...	×	×	?	×	×	×	?	...	...	...	...	...	×	...	×	
44 s 19	— — var. carinatus	...	...	×	...	?	...	...	...	×	...	...	...	...	...	...	×	
44 s 19	— — var. carinatus contrarius	...	...	×	...	...	...	...	...	×	...	...	...	...	...	...	×	
s 21	— Berniciensis? *King*	...	...	×	...	...	×	...	...	...	...	...	...	...	×	...	×	*Spitzbergensis* of Woodward's list.
48	— costifer, *Nyst.*	×	×	×	...	...	...	...	...	...	...	...	...	...	...	...	?	*Costatus*, J. Sowerby. *Fusus Mitgaui?* Könen.
46 s 22, 98, & 177	— (Atractodon) elegans, *Charlesworth*	×	...	×	...	...	...	...	...	...	...	...	...	...	...	...	...	Probably only derivative in the Red Crag.
47 s 23	— (Tritonofusus) altus, *S. Wood*	...	...	×	...	×	...	...	...	...	...	...	...	...	...	...	...	
46 s 24	— (Sipho) gracilis, *Da Costa*	...	...	×	...	?	...	...	×	...	...	...	...	×	...	×	×	Given in Woodward's list as from Norwich.

PAGE.		Cor. Crag.	Red Crag, Walton.	Red Crag, Sutton and Butley.	Scrobicularia Crag.	Fluvio-marine Crag.	Chillesford beds.	Lower Glacial.	Middle Glacial.	Upper Glacial.	Post Glacial, Kelsey Hill.	Post Glacial, March.	Post Glacial, Hunstanton.	Post Glacial, Nar Brickearth.	Living, Britain.	Living, Mediterranean.	Living, elsewhere.	REMARKS.
s 24	Trophon (Sipho) Jeffreysianus, *Fischer*	...	×	×	...	...	...	...	...	...	...	...	...	...	×	×	×	
A 312 s 21 & 177	— (Strombella) Norvegicus, *Chemn.*	...	...	×	...	?	...	...	...	...	...	...	...	...	×	...	×	Given in Woodward's list as from Norwich.
s 23	— (Sipho) Sabinii, *Gray*	...	...	...	...	...	...	...	...	×	...	...	...	...	...	...	×	
s 25	— Sarsii, *Jeffreys*	...	...	×	...	...	...	...	...	...	...	...	...	...	...	...	×	
A 313 s 24	— — propinquus, *Alder*	×	×	×	...	...	...	...	...	×	...	...	...	...	×	...	×	Given from Bridlington by Woodward.
A 312 s 22	— Turtoni, *Bean*	...	...	×	...	...	...	...	...	...	...	...	...	...	×	...	×	
s 22	— ventricosus? *Gray*	...	...	...	...	...	...	...	...	×	...	...	...	...	...	...	×	
s 27 A 314	— Barvicensis, *Johnston*	...	×	×	...	...	...	...	...	...	...	...	...	...	×	×	×	Walton, *fide* Bell. 310 fathoms off Malta, 'Depths of the Sea,' p. 270.
s 25 A 313	— craticulatus, *Fabr.*	...	...	...	...	...	...	...	...	×	...	...	...	...	...	...	×	*Fabricii?*
s 28	— mediglacialis, *S. Wood*	...	...	...	...	...	...	...	×	...	...	...	...	...	...	...	...	
s 28	— Billockbiensis, *S. Wood*	...	...	...	...	...	...	...	×	...	...	...	...	...	...	...	...	*Murex erinaceus*, juv. ?
48 s 26	— scalariformis, *Gould*	...	...	×	...	×	...	...	×	×	×	×	...	...	...	...	×	Obtained from March since Supplement printed.
s 27	— Gunneri, *Lovén*	...	...	...	...	...	...	...	...	×	×	...	...	...	...	...	×	
50 s 28	— muricatus, *Mont.*	...	×	×	...	...	...	...	...	×	...	...	...	...	×	×	×	
s 26 A 314	— Bamffius, *Mont.*	...	...	...	...	...	...	...	...	×	×	...	...	...	×	...	×	At page 321 of vol. iv of Brit. Con. this shell, under the synonym of *truncatus*, is given from the Norwich and Red Crags, but the author of that work has omitted it from the list in Quart. Journ. Geo. Soc., vol. xxvii, p. 492.
51	— gracilius, *S. Wood*	×	...	...	...	...	...	...	...	...	...	...	...	...	...	...	...	Possibly the young of *Pleurotoma attenuata*.
s 24	— Leckenbyi, *S. Wood*	...	...	...	...	...	...	...	...	×	...	...	...	...	...	...	...	
A 314 s 29	Fusus crispus, *Borson*	...	...	×	...	...	...	...	...	...	...	...	...	...	...	...	...	Derivative in Red Crag.
s 29	— abrasus, *S. Wood*	...	...	×	...	...	...	...	...	...	...	...	...	...	...	...	...	Derivative in Red Crag.
50 s 29	— imperspicuus, *S. Wood*	×	...	...	...	...	...	...	...	...	...	...	...	...	...	...	...	As to *Fusus despectus* and *F. Largillierti*, see Sup., pp. 177-8.
54 s 34	Pleurotoma modiola, *Jan.*	×	×	...	...	...	...	...	...	...	...	...	...	...	?	...	...	*carinata*, Crag Moll. Said to have been dredged off the Irish Coast in 110 fathoms, 'Depths of the Sea,' p. 86.
s 179	— Bertrandi? *Phil.*	...	...	×	...	...	...	...	...	...	...	...	...	...	...	×	...	
s 35	— crispata, *Jan.* — — *var.* papillosa	×	...	...	...	...	...	...	...	...	...	...	...	...	...	×	...	
53 s 32	— intorta, *Broc.*	...	...	×	...	...	...	...	...	...	...	...	...	...	...	...	?	Probably only a derivative in Red Crag.
s 32	— interrupta, *Broc.*	...	...	×	...	...	...	...	...	...	...	...	...	...	...	...	...	Derivative in Red Crag.
55 s 33	— inermis, *Partsch*	×	...	...	...	...	...	...	...	...	...	...	...	...	?	...	?	*nivalis?* Lovén, *porrecta*, Crag Moll.
s 33	— nodifera, *Lam.*	...	...	×	...	...	...	...	...	...	...	...	...	...	...	...	...	Derivative in Red Crag.
s 34	— Tarentini, *Phil.*	×	...	...	...	...	...	...	...	...	...	...	...	...	...	...	...	
53 s 33	— turrifera, *Nyst*	...	...	×	...	...	...	...	...	...	...	...	...	...	...	...	...	Probably only derivative in Red Crag.
54 s 36	— bipunctula, *S. Wood*	×	...	...	...	...	...	...	...	...	...	...	...	...	...	...	...	
54 s 35 & 180	— Icenorum, *S. Wood*	×	...	×	...	...	...	...	...	...	...	...	...	...	...	...	...	Probably derivative in Red Crag.
s 32	— coronata, *Bellardi*	...	...	×	...	...	...	...	...	...	...	...	...	...	...	...	...	Probably only derivative in Red Crag.
s 40	— assimilis, *S. Wood*	...	...	×	...	...	...	...	×	...	...	...	...	...	...	...	...	
s 38	— attenuata, *Mont.*	×	...	...	...	...	...	...	...	...	...	...	...	...	×	×	×	
s 43	— bicarinata, *Couth.*	...	...	×	...	?	...	...	...	...	...	...	...	...	...	...	×	By Mr. Reeve from Bramerton, but I have not seen the specimen.
60	— (Raphitoma) brachystoma, *Phil.*	×	...	...	...	...	...	...	...	...	...	...	...	...	×	×	×	
61	— cancellata, *J. Sow.*	×	×	...	...	...	...	...	...	...	...	...	...	...	×	×	×	*reticulata?* Ren.
57	— castanea, *S. Wood*	×	...	...	...	...	...	...	...	...	...	...	...	...	×	...	...	
s 36	— elegantula, *A. Bell*	×	...	...	...	...	...	...	...	...	...	...	...	...	×	×	×	
58	— costata, *Dacosta*	×	×	×	...	...	...	...	...	...	...	...	...	...	×	×	×	
s 37	— (Conopleura) crassa, *A. Bell*	×	...	...	...	...	...	...	...	...	...	...	...	...	...	...	?	
s 38	— elegantior, *S. Wood*	...	...	...	...	...	...	...	...	×	...	...	...	...	...	...	...	
s 44	— equalis, *S. Wood*	...	...	×	...	...	...	...	...	...	...	...	...	...	...	...	×	*concinnata*, var., Crag Moll. Atlantic, in 358 to 717 fathoms, 'Depths of the Sea,' p. 181.
61 s 42	— hispidula, *Jan.*	×	...	...	...	...	...	...	...	...	...	...	...	...	...	×	...	
s 41	— hystrix, *Jan.*	×	×	...	...	...	...	...	...	...	...	...	...	...	...	×	×	

PAGE.		Cor. Crag.	Red Crag, Walton.	Red Crag, Sutton and Butley.	Scrobicularia Crag.	Fluvio-marine Crag.	Chillesford beds.	Lower Glacial.	Middle Glacial.	Upper Glacial.	Post Glacial, Kelsey Hill.	Post Glacial, March.	Post Glacial, Hunstanton.	Post Glacial, Nar Brickearth.	Living, Britain.	Living, Mediterranean.	Living, elsewhere.	REMARKS.
56 / s 36	Pleurotoma linearis, *Mont.*......	×	...	×	...	...	×	...	×	...	...	...	...	...	×	×	×	
62 / s 42	— lævigata, *Phil.*	?	×	×	...	...	...	...	...	...	...	...	...	...	×	×	×	
63	— Leufroyii, *Michaud*	...	×	×	...	...	...	...	...	...	...	...	...	...	×	×	×	*Boothii*, Crag Moll. Given by Mr. Jeffreys from Sutton, but I only know it from Walton. *Vauquilini*? Payr.
59	— mitrula, *J. Sow.*......	×	×	×	...	...	...	...	...	...	...	...	...	...	...	?	...	
s 39	— Dowsoni, *S. Wood*......	...	...	...	...	...	×	...	?	×	...	...	...	...	...	...	...	
s 40	— robusta, *S. Wood*	...	...	...	...	...	...	...	×	...	...	...	...	...	...	...	...	
s 45	— nebula, *Mont.*......	×	...	...	...	...	...	...	...	...	...	...	...	...	×	×	×	
s 45	— nebulosa, *S. Wood*......	...	...	×	...	...	...	...	...	...	...	...	...	...	...	...	...	
58	— perpulchra, *S. Wood*	×	×	...	...	...	...	...	...	...	...	...	...	...	...	...	...	Given from Walton by Mr. Bell, Ann. and Mag. Nat. Hist. for May, 1871.
s 43	— (Bela) pyramidalis, *Strom.*	...	...	×	...	×	...	...	×	×	×	...	...	...	...	...	×	
64 / s 39	— plicifera, *S. Wood*	...	...	×	...	...	...	...	...	...	...	...	...	...	...	?	?	
s 178	— clathrata, *M. de S.*	×	...	...	...	...	...	...	...	...	...	...	...	...	...	×	...	
57	— Philberti, *Mich.*	×	×	...	...	...	...	...	...	...	...	...	...	...	...	×	...	Given from Walton by Mr. Bell, Ann. and Mag. Nat. Hist. for September, 1870.
s 44	— quadricincta, *S. Wood*...	...	...	×	...	...	...	...	...	...	...	...	...	...	...	...	...	
s 44	— rufa, *Mont.*......	...	...	?	...	?	...	...	...	...	...	...	...	...	...	...	...	
s 39	— scalaris, *Möll.*......	...	...	...	...	...	...	...	×	...	...	...	×	...	...	...	×	*Pl. harpularia*? Couthouy.
s 38	— septangularis, *Mont.*	...	...	...	...	...	...	×	...	...	...	...	...	×	×	×	...	
s 42	— senilis, *S. Wood*	...	...	×	...	...	...	...	...	...	...	...	...	...	...	...	...	Derivative in Red Crag.
62 / s 41	— tenuistriata, *A. Bell*	...	×	...	...	...	...	...	...	...	...	...	...	...	...	...	...	
51 / s 27	— teres, *Forbes*	×	...	...	...	...	...	...	...	...	...	...	...	...	×	×	×	*Trophon paululum* of Crag Moll.
s 178	— tereoides, *S. Wood*......	×	...	...	...	...	...	...	...	...	...	...	...	...	...	...	...	
s 46 & 179	— striolata, *Phil.*	×	...	?	...	...	...	...	...	...	...	...	...	...	×	×	×	It is given by Mr. Jeffreys from the Red Crag, Shottisham, but I have not seen the specimen.
63	— Trevelyana, *Turt.*	...	...	×	...	×	×	...	...	...	...	...	...	...	×	...	×	Chillesford Bed, Horstead.
62 / s 40	— turricula, *Mont.*......	...	×	...	...	×	×	...	×	×	×	×	...	...	×	...	×	As to *P. pygmæa* and *P. oblonga*, see Sup., p. 180.
s 45	— violacea? *Migh. & Ad.*...	...	...	?	...	...	...	...	...	...	...	...	...	...	...	...	×	As to *P. rugulosa*, see Sup., pp. 46 and 179.
s 48	Cancellaria cancellata, *Linn.*	×	...	...	...	...	...	...	...	...	...	...	...	...	...	×	×	
s 46	— contorta, *Bast.*	×	...	...	...	...	...	...	...	...	...	...	...	...	×	×	...	
s 48	— ? Charlesworthii, *S. Wood*	...	...	×	...	...	...	...	...	...	...	...	...	...	...	...	...	
64	— coronata, *Scacchi*	...	×	×	...	...	...	...	...	...	...	...	...	...	...	...	...	Waldringfield only, and derivative.
67 / s 47	— Bellardi, *Mich.*	...	...	×	...	...	...	...	...	...	...	...	...	...	...	...	...	Probably derivative in Red Crag.
67 / s 48	— Bonellii, *Bellardi*	...	...	×	...	...	...	...	...	...	...	...	...	...	...	...	...	Probably derivative in Red Crag.
65	— mitræformis, *Broc.*	×	...	×	...	...	...	...	...	...	...	...	...	...	...	...	?	Probably derivative in Red Crag.
s 49 & 182	— spinulosa? *Broc.*	×	...	...	...	...	...	...	...	...	...	...	...	...	...	...	...	
A 316 / s 182	— scalaroides, *S. Wood*	×	...	...	...	...	...	...	...	...	...	...	...	...	...	...	...	
s 182	— umbilicaris? *Broc.*......	...	...	×	...	...	...	...	...	...	...	...	...	...	...	...	...	Derivative in Red Crag.
s 46	— (Admete) gracilenta, *S. Wood*	×	...	...	...	...	...	...	...	...	...	...	...	...	...	...	...	
66 / s 47	— — subangulosa, *S. Wood*	×	...	...	...	...	...	...	...	...	...	...	...	...	...	...	...	*Admete Reedii*, A. Bell.
66 / s 97	— — viridula, *Fab.*	...	...	×	...	×	...	...	...	...	...	...	...	...	...	...	×	*C. costellifera* of Crag Moll.
s 97	— — *var.* Couthouyi...	...	...	×	...	...	...	...	...	...	...	...	...	...	...	...	×	
s 50	Cerithium (Bittium) aberrans, *S. Wood*	×	...	...	...	...	...	...	...	...	...	...	...	...	...	...	...	
71	— cribrarium, *S. Wood*......	×	...	...	...	...	...	...	...	...	...	...	...	...	...	...	?	
71	— Metaxa, *Delle Chiaje*......	×	...	...	...	...	...	...	...	...	...	...	...	×	×	×	...	
72 / s 50	— perpulchrum, *S. Wood*	×	...	...	...	...	...	...	...	...	...	...	...	...	...	?	...	*C. mammillatum*? Risso.
69 / s 51	— tricinctum, *Broc.*	×	×	×	...	×	×	...	×	...	...	...	...	...	...	...	...	
69	— variculosum, *Nyst*	...	×	...	...	...	...	...	...	...	...	...	...	...	...	...	...	
s 50	— reticulatum, *Da Costa* ...	...	?	...	...	...	...	...	...	...	...	...	×	×	×	×	...	
72 / s 181	— perversum, *Linn.*	×	×	...	...	...	...	...	...	...	...	...	...	×	×	×	×	*C. adversum* of Crag Moll., Red Crag, Walton. Bell, Ann. and Mag., September, 1870.
70	— trilineatum, *Phil.*	×	×	...	...	...	...	...	...	...	...	...	...	...	×	...	...	Walton, according to Bell, Ann. and Mag., September, 1870.
73	— granosum, *S. Wood*......	×	×	...	...	...	...	...	...	...	...	...	...	...	...	...	...	*C. Macandreæ* is said to have the ridges plain, whereas *granosum* is granulated.

PAGE.		Cor. Crag.	Red Crag, Walton.	Red Crag, Sutton and Butley.	Scrobicularia Crag	Fluvio-marine Crag.	Chillesford beds.	Lower Glacial.	Middle Glacial.	Upper Glacial.	Post Glacial, Kelsey Hill.	Post Glacial, March.	Post Glacial, Hunstanton.	Post Glacial, Nar Brickearth.	Living, Britain.	Living, Mediterranean.	Living, elsewhere.	REMARKS.
s 52	Cerithiopsis lactea? *Möller*	×	...	...	...	...	...	...	...	...	...	...	...	...	...	...	×	
70, s 52 & 181	— tubercularis, *Mont.*	×	...	×	...	×	...	...	×	...	...	...	...	...	×	×	×	Derivative in Red Crag.
s 181	— — var. subulatus	×	...	...	...	...	...	...	...	...	...	...	...	...	×	?	?	*C. Barleei,* Jeff.
s 181	— — var. nanus	×	...	...	...	...	...	...	...	...	...	...	...	...	×	?	?	*C. pulchella,* Jeff.
25, s 49	Aporrhais pespelicani, *Linn.*	×	×	×	...	×	...	...	×	...	...	×	...	×	×	×	×	Walton omitted accidentally as a locality in Supplement.
s 180	— — var. Serresianus	×	...	...	...	...	...	...	...	...	...	...	...	...	...	×	...	
67	Trichotropis borealis, *Brod. & Sow.*	×	...	...	...	...	...	...	...	...	...	...	...	...	×	...	×	Tab. VII, fig. 17, of Crag Mollusca.
67	— insignis, *Midd.*	...	...	...	...	...	...	...	×	...	...	...	...	...	...	...	×	Tab. XIX, fig. 11, of Crag Mollusca; living, Behring's Strait.
88	Litiopa? papillosa, *S. Wood*	×	...	...	...	...	...	...	...	...	...	...	...	...	...	...	...	
s 57	Menestho lævigata, *S. Wood*	×	...	...	...	...	...	...	...	...	...	...	...	...	...	...	...	*Menestho albula,* from the Red Crag, I do not know. See page 185 of Supplement as to *M. Jeffreysii.*
s 185	— Britannica, *A. Bell*	×	...	...	...	...	...	...	...	...	...	...	...	...	...	...	...	
75	Turritella imbricataria, *Lam.*	...	...	×	...	...	...	...	...	...	...	...	...	...	...	...	...	Probably derivative in the Crag.
76, s 54	— planispira, *S. Wood*	×	...	...	...	...	...	...	...	...	...	...	...	...	...	?	...	
75, s 52	— incrassata, *J. Sow.*	×	×	×	×	×	...	...	×	...	...	...	...	...	...	×	×	*T. triplicata,* Broc.
75, s 54	— — var. vermicularis	×	...	...	...	...	...	...	...	...	...	...	...	...	...	?	...	
75, s 54	— — var. bicincta	×	...	...	...	...	...	...	...	...	...	...	...	...	...	?	...	
74, s 53	— terebra, *Linn.*	...	...	?	...	×	×	...	×	×	×	×	...	×	×	×	×	*T. communis* of Crag Mollusca.
76, s 53	— (Zaria) erosa, *Couth.*	...	...	...	...	...	...	...	×	...	...	...	...	...	...	...	×	*T. clathratula* of Crag Mollusca.
s 53	— ? penepolaris, *S. Wood*	×	...	...	...	...	...	...	...	...	...	...	...	...	...	...	...	
s 183	Scalaria communis, *Lam.*	...	...	...	...	...	...	...	...	...	...	...	×	...	×	×	×	
94	— clathratula, *Adams*	×	...	...	...	?	...	...	...	...	...	...	...	...	×	×	×	Given in Woodward's list as from Norwich; *sed quæ.*
93	— foliacea, *J. Sow.*	×	...	×	...	?	...	...	...	...	...	...	...	...	...	...	...	Not improbably derivative in Red Crag. Said in Woodward's list to occur in fragments at Norwich; *sed quæ.*
92, s 59	— frondosa, *J. Sow.*	×	...	...	...	...	...	...	...	...	...	...	...	...	...	...	?	*S. soluta?* Tiberi. Mediterranean, 30 to 250 fathoms, 'Depths of the Sea,' p. 192.
91	— fimbriosa, *S. Wood*	×	...	...	...	...	...	...	...	...	...	...	...	...	...	...	...	
92, s 59	— frondicula, *S. Wood*	×	...	...	...	...	...	...	...	...	...	...	...	...	...	...	...	
90, s 59	— Grœnlandica, *Chemn.*	...	...	×	...	×	×	×	×	...	...	...	...	...	×	...	×	
91	— hamulifera, *S. Wood*	×	...	...	...	...	...	...	...	...	...	...	...	...	...	...	...	
90, s 98	— funiculus, *S. Wood*	×	×	...	...	...	...	...	...	...	...	...	...	...	...	...	...	*S. varicosa* of Crag Moll.
93, s 98	— subulata, *J. Sow.*	×	×	...	...	...	...	...	...	...	...	...	...	...	...	...	?	Given by Mr. Bell from Walton, Ann. and Mag., September, 1870.
94, s 58	— Trevelyana, *Leach*	...	...	×	...	×	×	...	×	...	...	...	...	...	×	×	×	
s 58	— Turtonis, *Turton*	...	...	...	...	...	×	...	...	...	...	...	...	...	×	×	×	
95, s 59	— ? cancellata, *Broc.*	×	...	...	...	...	...	...	...	...	...	...	...	...	...	...	...	
95	— ? obtusicostata, *S. Wood*	×	...	...	...	...	...	...	...	...	...	...	...	...	...	...	?	
s 98 & 183	— semicostata, *J. Sow.*	...	...	×	...	...	...	...	...	...	...	...	...	...	...	...	...	Derivative in Red Crag.
77, s 57 & 182	Pyramidella læviuscula, *S. Wood*	×	×	...	...	...	...	...	...	...	...	...	...	...	...	...	...	
80	Chemnitzia (Turbonilla) costaria, *S. Wood*	×	...	×	...	...	...	...	...	...	...	...	...	...	...	?	...	Red Crag, Butley, A. Bell, Ann. and Mag., 1870, but probably only derivative there.
81, s 60	— — elegantior, *S. Wood*	×	×	×	...	...	...	...	...	...	...	...	...	...	...	...	...	*C. elegantissima* of Crag Moll.
82, s 60	— — filosa, *S. Wood*	×	...	...	...	...	...	...	...	...	...	...	...	...	...	?	...	*Parthenia varicosa,* Forbes?
83	— — perexilis, *S. Wood*	×	...	...	...	...	...	...	...	...	...	...	...	...	...	...	...	*C. unica* of Crag Moll., but as it does not seem to belong to Montague's species I have assigned it a new name.
82	— — densecostata? *Phil*	×	...	...	...	...	...	...	...	...	...	...	...	...	×	...	...	
s 184	— — Jeffreysii, *Wiechman*	×	...	...	...	...	...	...	...	...	...	...	...	...	...	...	...	A doubtful identification.

PAGE.		Cor. Crag.	Red Crag, Walton.	Red Crag, Sutton and Butley.	Scrobicularia Crag.	Fluvio-marine Crag.	Chillesford beds.	Lower Glacial.	Middle Glacial.	Upper Glacial.	Post Glacial, Kelsey Hill.	Post Glacial, March.	Post Glacial, Hunstanton.	Post Glacial, Nar Brickearth.	Living, Britain.	Living, Mediterranean.	Living, elsewhere.	REMARKS.
s 59	Chemnitzia (Turbonilla) clathrata, *Jeff.*	×							×						×	×		
81 / s 60	— — internodula, *S. Wood*	×	×	×		×			×							×	×	
79	— — indistincta? *Mont.*	×													×	×	×	
80	— — nitidissima? *Mont.*	×													×	×	×	The striæ are probably absent from wear.
71 / s 60	— — rufa, *Phillipi*	×													×	×	×	
84	— — similis, *S. Wood*	×																{ *C. similis,* Forbes, of Crag Moll., Tab.X, fig. 11 *a;* an error.
84	— — varicula? *S. Wood*	×															?	
s 61	— — rugulosa, *S.Wood*		×															A provisional species.
s 61	— — plicatula? *Broc.*		×	×			×											
84 / s 67	Eulimella acicula, *Phil.*	×	×			×									×	×	×	*Chemnitzia similis* of Crag Moll., fig. 11 *b* & *c.*
85 / s 63	Odostomia conspicua, *Alder*	×													×	×	×	{ Var. β of *O. plicata* of Crag Moll., Tab. IX, fig. 3 *b.* (*O. acuta,* Jeffreys, var.)
85 / A 317 / s 63	— conoidea, *Broc.*	×	?			?		?							×	×	×	Var. α of *O. plicata* of Crag Moll., Tab. IX, fig. 3 *a.*
86 / s 63	— pellucida, *Adams*	×													×	×	×	*O. decussata,* Mont.
s 62	— insculpta, *Mont.*	×													×	×	×	
s 64	— obliqua, *Alder*	×													×	×	×	
A 318	— truncatula, *Jeff.*	×													×			
s 63	— plicata, *Mont.*	×	×												×	×		Not *O. plicata* of p. 85, and Tab. IX, of Crag Moll.
s 185	— dentiplicata, *S. Wood.*	×																{ *O. pupa* of Crag Moll., but as it is neither that shell nor *interstincta,* I propose for it a new name. There is a strong spiral ridge in the lower part of each whorl that is absent in both these species.
86	— chrysalis, *S. Wood*	×																
87 / s 64	— ornata, *S. Wood*	×																*O. similima* of Crag Moll.
s 62	— ? Gulsonæ, *Clark*	×													×		×	*Aclis Gulsonæ* of Jeff.
A 317 / s 64	— unidentata, *Mont.*		×						?						×		×	
s 64 / & 184	— albella, *Lovén*	×													×	×	×	Sardinia sec., Brit. Con.
98 / s 67	Eulima glabella, *S. Wood*	×													×		×	
96 / s 67	— intermedia, *Contraine*	×	×												×	×	×	{ Var. *vulgaris* of *E. polita* in Crag Moll., Tab. XIX, fig. 1 *a.*
96 / s 65	— polita, *Linn.*	×	×												×	×	×	Fig. 16 of Tab. XIX of Crag Moll.
s 65	— similis? *D'Orb.*		×												?	?	?	(*distorta,* Phil.), a doubtful identification.
s 65 / & 184	— stenostoma? *Jeffreys*	×													×		×	
97 / s 66	— subulata, *Donov.*	×													×	×	×	
s 66	— bilineata? *Alder*	×													×	×	×	A doubtful identification.
99 / s 55	Alvania supranitida, *S. Wood*	×													×	×	×	{ *Alvania ascaris* of Crag Moll. *Aclis ascaris* I do not recognise in the Crag.
99 / s 56	— — albella, *Leach*	×													×	×	×	*Aclis Walleri,* Jeff.
109 / s 65	Eulimene pendula, *S. Wood*		×															
109 / s 65	— terebellata, *Nyst.*		×	×														Derivative?
109 / s 71	Hydrobia ulvæ, *Pennant*					×	×								×	×	×	Also in older Post Glacial bed, Gedgrave.
s 97	Rissoa abyssicola, *Forbes*	×													×	×	×	
104	— confinis, *S. Wood*	×				?												{ I do not know this shell from Bramerton, and as it has no distinguishing mark in the list to Mᵗ. Prestwich's Red Crag paper (p. 490) it has probably got in there by some oversight.
107	— — *var.* supracostata	×																{ *R. supracostata* of Crag Moll. I have treated this as a variety of *confinis,* because the absence of costæ from the lower whorl, which is its only distinction, is exhibited by more than one species of living *Rissoæ.*
106 / s 73 / & 185	— Stefanisi, *Jeff.*	×	×													?	?	{ *R. costulata* of Crag Moll. It is said to be living. *Menestho Jeffreysi,* A. Bell.

PAGE.		Cor. Crag.	Red Crag, Walton.	Red Crag, Sutton and Butley.	Scrobicularia Crag.	Fluvio-marine Crag.	Chillesford beds.	Lower Glacial.	Middle Glacial.	Upper Glacial.	Post Glacial, Kelsey Hill.	Post Glacial, March.	Post Glacial, Hunstanton.	Post Glacial, Nar Brickearth.	Living, Britain.	Living, Mediterranean.	Living, elsewhere.	REMARKS.
106	Rissoa crassi-striata, *S. Wood*...	×	...	...	...	...	...	...	...	...	...	...	...	...	...	...	...	
102, s 72, & 104	— curticostata, *S. Wood*	...	...	×	...	×	...	...	×	...	...	...	...	...	...	...	...	R. semicostata, Woodward (*non* Mont.), and R. pulchella of Crag Moll.
s 72, 105	— eximia? *Jeffreys*	×	...	...	...	...	...	...	...	...	...	...	...	...	×	...	...	
105	— obsoleta, *S. Wood*	×	...	...	...	...	...	...	...	...	...	...	...	...	...	...	...	
s 71	— proxima, *Alder*	×	...	...	...	...	...	...	...	...	...	...	...	...	×	×	×	
103	— concinna, *S. Wood*	×	...	...	...	...	...	...	...	...	...	...	...	...	...	...	...	R. punctura of Crag Moll. I there gave the reference to punctura, Mont., with a doubt; and as that doubt is confirmed by others, I have reverted to the name concinna, given in my Catalogue of 1842.
103, s 73	— reticulata? *Mont.*	×	...	...	...	...	...	...	...	...	...	...	...	...	×	×	×	
100	— striata, *Mont.*	×	...	...	...	...	...	...	...	...	...	...	...	...	×	×	×	I do not know this shell from any other than the Cor. Crag. It is given in the Red Crag list of Mr. Prestwich's paper (p. 491), but as no locality is inserted it has probably got in there by some oversight.
A 318	— soluta, *Phil.*	×	...	...	...	...	...	...	...	...	...	...	...	...	?	×	?	
102	— vitrea, *Mont.*	×	...	...	...	...	...	...	...	...	...	...	...	...	×	×	×	
101	— Zetlandica, *Mont.*	×	×	×	...	...	...	...	...	...	...	...	...	...	×	×	×	
s 73	— senecta, *S. Wood*	×	...	...	...	...	...	...	...	...	...	...	...	...	...	...	...	A doubtful species.
117, s 87	Cæcum glabrum, *Mont.*	×	...	...	...	...	...	...	...	...	...	...	...	...	×	×	×	
s 87	— liratum, *Carpenter*	×	...	...	...	...	...	...	...	...	...	...	...	...	...	...	...	
116, s 87	— mammillatum, *S. Wood*...	×	...	×	...	...	...	...	...	...	...	...	...	...	...	...	?	Probably only derivative in Red Crag.
115, s 87	— trachea, *C. M.*	×	...	...	...	...	...	...	...	...	...	...	...	...	×	×	?	C. tumidum, Carp.
121, A 317	Fossarus sulcatus, *S. Wood*...	×	...	...	...	...	...	...	...	...	...	...	...	...	...	...	...	
121, s 186	— lineolatus, *S. Wood*	×	...	...	...	...	...	...	...	...	...	...	...	...	...	...	...	Var lineolatus of F. sulcatus in Crag Moll.
316, s 80	Lacuna vincta, *Mont.*	...	...	...	...	×	...	...	...	×	×	...	...	...	×	...	×	L. divaricata.
s 80	— crassior	...	...	...	...	?	...	...	×	...	...	...	...	...	×	...	×	I do not know the species from the Fluvio-marine Crag, but it is given from Bramerton in Mr. Prestwich's Red Crag paper.
122, s 79	— reticulata, *S. Wood*	×	...	...	...	...	...	...	...	...	...	...	...	...	...	...	...	Genus Macromphalus, S. Wood.
120, s 80	— suboperta, *J. Sow.*	...	×	×	...	...	...	...	...	...	...	...	...	...	...	...	...	
118, s 79	Littorina littorea, *Linné*	...	...	×	...	×	×	×	×	×	×	×	...	×	×	...	×	
118, s 79	— rudis, *Maton*	...	...	...	...	×	×	×	?	...	...	...	...	...	×	...	×	
s 78	Amaura candida, *Möll.*	...	...	×	...	...	...	...	...	...	...	...	...	...	...	...	×	Since Supplement was printed this has been found also by Mr. Crowfoot in the Red Crag of Boyton.
s 74	Natica Alderi, *Forbes*	×	...	...	...	×	...	...	×	...	×	×	...	×	×	×	×	It is given in Mr. Prestwich's Red Crag paper as from the Red Crag, but I have not yet seen it from there unless it be the shell called N. Guillemini, infrà.
142, s 76	— catena, *Da Costa*	...	...	×	...	×	×	×	×	...	...	...	...	...	×	×	×	
141, s 77	— catenoides, *S. Wood*	...	×	×	...	...	?	...	...	...	...	...	...	...	...	...	...	
147, s 75	— clausa, *Brod. & Sow.*	...	...	×	...	×	×	×	×	×	...	...	...	...	×	...	×	
145, s 76	— cirriformis, *J. Sow.*	×	...	...	...	...	...	...	...	...	...	...	...	...	...	...	...	Not N. heros, Say.
146, s 75	— Grœnlandica, *Beck*	...	...	×	...	×	×	...	×	×	...	...	...	...	×	...	×	
142, s 74	— Guillimini, *Phil.*	...	...	×	...	...	...	...	...	...	...	...	...	...	?	?	?	Possibly N. Alderi or the young of catenoides.
145, s 78	— (Amauropsis) helicoides, *Johnson*	...	...	×	...	×	×	×	×	×	×	...	×	...	×	...	×	
144, s 75	— hemiclausa, *J. Sow.*	...	×	×	...	?	×	...	...	...	...	...	...	...	...	...	...	It is given in Mr. Prestwich's list from the Fluvio-marine Crag, Bulchamp, but I have not seen it.
s 74	— helicina, *Broc.*	×	×	×	...	...	...	...	...	...	...	...	...	...	...	...	...	Red Crag of Bentley and Walton only.

PAGE.		Cor. Crag.	Red Crag, Walton.	Red Crag, Sutton and Butley.	Scrobicularia Crag.	Fluvio-marine Crag.	Chillesford beds.	Lower Glacial.	Middle Glacial.	Upper Glacial.	Post Glacial, Kelsey Hill.	Post Glacial, March.	Post Glacial, Hunstanton.	Post Glacial, Nar Brickearth.	Living, Britain.	Living, Mediterranean.	Living, elsewhere.	REMARKS.
148 } s 76	Natica multipunctata, *S. Wood*	×	×	×	...	...	?	...	...	...	...	...	...	...	...	...	...	Not *millepunctata*.
143	— proxima, *S. Wood*	×	...	×	...	...	...	...	...	...	...	...	...	...	?	?	?	*N. sordida?* Phil.
s 77	— pusilla, *Say*	×	...	...	...	...	...	...	...	...	...	...	...	...	...	...	×	
146 } s 76	— occlusa, *S. Wood*	...	...	×	...	...	×	...	...	×	...	...	...	...	...	...	×	
s 78	— Montacuti, *Forbes*	...	...	...	...	...	...	...	×	...	...	...	...	...	×	...	×	
143	— varians, *Dujardin*	...	×	...	...	...	...	...	×	...	...	...	...	...	×	...	×	Probably derivative in Red Crag.
149	Sigaretus excavatus, *S. Wood*	×	...	...	...	...	...	...	...	...	...	...	...	...	...	...	?	
151	Marsenia tentaculata, *Linn.*	×	...	...	...	...	...	...	...	...	...	...	...	...	×	×	×	
152	Velutina lævigata, *Pennant*	...	...	...	×	...	...	...	...	...	...	...	...	...	×	...	×	
153	— undata, *J. Smith*	...	...	...	×	...	...	...	...	...	...	...	...	...	...	...	×	
153	— virgata, *S. Wood*	×	...	...	...	...	...	...	...	...	...	...	...	...	×	...	...	
113	Vermetus intortus, *Lam.*	×	...	×	...	...	...	...	...	...	...	...	...	...	×	...		{ *V. subcancellatus*, Phil., T. IX, fig. 20. Probably only derivative in Red Crag.
114	— Bognoriensis	...	...	×	...	...	...	...	...	...	...	...	...	...	...	...		Probably derivative from Eocene.
	— ? triqueter, *Bivonæ*	?	...	?	...	...	...	...	...	...	...	...	...	...	?	...		{ It is doubtful whether the specimens be not Annelids of the genus *Vermilia*.
129	Trochus Adansoni, *Payr.*	×	...	×	...	...	...	...	...	...	...	...	...	...	×	...		Probably only derivative in Red Crag.
128	— villacus? *Phil.*	×	...	×	...	...	...	...	...	...	...	...	...	...	×	...		Probably only derivative in Red Crag.
130	— — Kicksii? *Nyst*	×	...	×	...	...	...	...	...	...	...	...	...	...	...	...		{ Probably only a var. of *Adansoni*, and derivative in Red Crag.
s 82	— bullatus, *Phil.*	×	×	...	...	...	...	...	...	...	...	...	...	...	...	...		
131 } s 81	— cinerarius, *Linn.*	...	×	×	...	...	...	...	×	...	...	×	...	...	×	×	×	
131	— cineroides, *S. Wood*	...	×	×	...	...	...	...	...	...	...	...	...	...	...	...		
125	— conulus, *Linn.*	×	...	...	...	...	...	...	...	...	...	...	...	×	×	?		
133	— ditropis, *S. Wood*	×	...	...	...	...	...	...	...	...	...	...	...	...	...	...		
125	— formosus, *Forbes*	×	...	×	...	...	...	...	...	...	...	...	...	...	×	...	×	*T. alibastrum*, Beck; *T. occidentalis*, Migh.
127	— millegranus, *Phil.*	×	×	...	...	...	...	...	...	...	...	...	...	×	×	×		{ Red Crag, Walton, *fide* Bell, Ann. and Mag. Nat. Hist., 1871.
127	— multigranus, *S. Wood*	×	...	×	...	...	...	...	...	...	...	...	...	...	...	?		Cor. Crag, *fide* Bell, Ann. and Mag., 1871.
129	— Montacuti, *W. Wood*	×	...	×	...	...	...	...	...	...	...	...	...	×	×	×		Possibly only derivative in Red Crag.
126 } s 81	— noduliferens, *S. Wood*	...	×	×	...	×	...	?	...	...	...	...	...	...	...	...		*T. papillosus* of Crag Mollusca.
133	— obconicus, *S. Wood*	×	...	...	...	...	...	...	...	...	...	...	...	...	...	...		
s 81	— granulatus, *Born*	...	×	...	...	...	...	...	...	...	...	...	...	×	×	×		
126	— sub-excavatus, *S. Wood*	...	×	×	...	...	...	...	...	...	...	...	...	...	...	...		Possibly a variety of *noduliferens*.
132	— tricariniferus, *S. Wood*	×	...	...	...	...	...	...	...	...	...	...	...	...	...	...		
s 80	— turgidulus? *Broc.*	×	...	...	...	...	...	...	...	...	...	...	...	...	...	...		
130	— tumidus, *Mont.*	...	...	×	...	×	×	...	...	...	...	...	...	×	×	×		
124 } s 81	— zizyphinus, *Linn*	×	...	...	?	...	...	?	...	...	...	...	...	×	×	×		
123 } A 321	— crenularis	...	...	...	...	...	...	...	...	...	...	...	...	...	...	...		Spurious.
134	Margarita elegantissima, *Bean*	...	...	...	...	...	...	...	×	...	...	...	...	...	...	×		Living, Greenland and Spitzbergen.
s 84	— argentata, *Gould*	...	...	...	...	...	×	...	...	...	...	...	...	...	...	×		
s 83	— Groenlandica, *Chemn*	...	...	...	...	×	...	?	...	...	...	...	×	...	×			
135 } s 83	— maculata, *S. Wood*	×	×	...	...	...	...	...	...	...	...	...	...	...	...	...		{ One small specimen from the Red Crag at Walton is in my cabinet.
136 } s 84	— trochoidea, *S. Wood*	×	...	...	...	...	...	...	...	...	...	...	...	...	...	...		
139 } s 84	Adeorbis pulchralis, *S. Wood*	×	...	...	...	...	...	...	...	...	...	...	...	...	...	...		
137 } s 84	— striatus, *Phil.*	×	...	...	...	...	...	...	...	...	...	...	...	...	...	...		
139	— subcarinatus, *Mont.*	×	×	×	...	...	...	...	...	...	...	...	...	×	×	×		
137 } s 84	— supranitidus, *S. Wood*	×	...	...	...	...	...	...	...	...	...	...	...	...	...	...		
138 } s 84	— tricarinatus, *S. Wood*	×	...	...	...	...	...	...	...	...	...	...	...	...	...	...		
s 85	Solarium vagum, *S. Wood*	...	...	...	...	...	...	...	...	...	...	...	...	...	...	...		Derivative.
s 86 & 186	Cyclostrema lævis, *Phil.*	×	...	...	...	...	...	...	...	...	...	...	...	×	×	?		*C. basi-striatum?*
122 } s 86	— ? sphæroidea, *S. Wood*	×	...	...	...	...	...	...	...	...	...	...	...	...	×	...		*Turbo sphæroidea* of Crag Moll.
s 86	Homalogyra atomus, *Phil.*	×	...	...	...	...	...	...	...	...	...	...	...	×	×	×		
163	Scissurella crispata, *Flem.*	×	...	...	...	...	...	...	...	...	...	...	...	×	×	×		
168	Fissurella græca, *Linn.*	×	×	×	...	...	...	...	...	...	...	...	...	×	×	×		
s 90	— costaria, *Grateloup*	×	...	×	...	...	...	...	...	...	...	...	...	...	×	...		Probably derivative in Red Crag.

PAGE		Cor. Crag.	Red Crag, Walton.	Red Crag, Sutton and Butley.	Scrobicularia Crag.	Fluvio-marine Crag.	Chillesford beds.	Lower Glacial.	Middle Glacial.	Upper Glacial.	Post Glacial, Kelsey Hill.	Post Glacial, March.	Post Glacial, Hunstanton.	Post Glacial, Nar Brickearth.	Living, Britain.	Living, Mediterranean.	Living, elsewhere.	REMARKS.
166, s 91	Cemoria (Puncturella) Noachina, *Linn.*	...	...	...	...	...	...	...	...	×	...	...	...	...	×	×	×	Recently found in Mediterranean, Brit. Con., vol. v, p. 200.
165, s 90	Emarginula crassa, *J. Sow.*	...	...	×	...	...	...	...	...	...	...	...	...	...	×	×	×	Recently found in Mediterranean, Brit. Con., vol. v, p. 200.
164	— — var. crassalta	×	...	...	...	...	...	...	...	...	...	...	...	...	...	...	...	
164, s 89	— fissura, *Linn.*	×	×	×	...	...	...	...	...	?	...	...	...	...	×	×	×	
s 89	— — var. rosea, *Bell.*	...	×	...	...	...	...	...	...	...	...	...	...	...	×	×	×	
s 89	— — var. elongata	×	...	...	...	...	...	...	...	...	...	...	...	...	...	×	...	
159, s 89	Calyptræa Chinensis, *Linn.*	×	×	×	...	×	×	...	×	...	...	...	...	...	×	×	×	
155, s 88	Capulus Ungaricus, *Linn.*	×	×	×	...	?	...	...	×	...	...	...	...	...	×	×	×	
156, s 88	— (Brocchia) partim sinuosus, *S. Wood*	...	...	×	...	...	...	...	...	...	...	...	...	...	...	...	...	Doubtful species. Figs. 3 a, b, of Tab. XVII of Crag Mollusca.
s 88	— — sinuosus, *Broc.*	×	...	...	...	...	...	...	...	...	...	...	...	...	...	...	...	Doubtful species.
156, s 88	— recurvatus, *S. Wood*	×	×	×	...	...	...	...	...	...	...	...	...	...	×	×	...	Fig 3 f, of Tab. XVII, of Crag Moll.
156	— obliquus, *S. Wood*	×	×	×	...	...	...	...	...	...	...	...	...	...	...	...	...	Cor. Crag, *fide* Bell, Ann. and Mag., 1870.
157	— fallax, *S. Wood*	×	...	...	...	...	...	...	...	...	...	...	...	...	...	...	...	
161	Tectura virginea, *Müller*	×	...	×	...	...	...	...	...	...	...	...	...	...	×	×	×	Found by myself in Cor. Crag at Sutton since Crag Mollusca published.
161, s 91	— fulva, *Müller*	×	?	...	...	...	...	...	×	...	...	...	...	...	×	×	...	
162, s 91	— parvula, *S. Woodward*	...	...	...	...	×	...	...	...	...	...	...	...	...	...	...	...	A doubtful species, possibly the young of *Patella vulgata.*
183	Patella vulgata, *Linn.*	...	...	×	...	...	...	...	...	...	...	...	...	...	×	×	×	The Fluvio-marine Crag shell is that called *T. parvula.*
189, s 93	Dentalium abyssorum, *Sars*	...	...	...	...	...	...	...	×	...	...	...	...	...	×	×	×	Fig. 2 of Tab. XX of Crag. Moll., as *D. entale* from Bridlington.
188, s 92	— dentalis, *Linn.*	×	...	×	...	...	...	...	×	...	...	...	...	...	...	×	×	Probably derivative in Red Crag.
s 92 & 187	— entalis, *Linn.*	×	...	...	...	...	...	...	...	×	...	...	...	...	×	×	×	Not the *entale* of Crag Moll., which is *D. abyssorum.*
s 92	— rectum, *Linn.*	...	...	×	...	...	...	...	...	...	...	...	...	...	...	?	...	Probably derivative in Red Crag.
190	— bifissum, *S. Wood*	×	...	...	...	...	...	...	...	...	...	...	...	...	...	?	?	*G. Dischides,* Jeff.
s 95	Chiton discrepans? *Brown*	×	...	...	...	...	...	...	...	...	...	...	...	...	×	×	×	
185	— fascicularis, *Linn.*	×	...	...	...	...	...	...	...	...	...	...	...	...	×	×	×	
186, s 95	— Hanleyi, *Bean*	×	...	...	...	...	...	...	...	...	...	...	...	...	×	×	×	*C. strigillatus* of Crag Moll.
186, s 95	— Rissoi, *Payr*	×	...	...	...	...	...	...	...	...	...	...	...	...	...	×	...	
22, s 96	Ringicula buccinea, *Broc.*	×	...	×	...	...	...	...	...	...	...	...	...	...	...	×	×	Derivative in Red Crag.
22, s 97	— ventricosa, *J. Sow.*	×	...	×	...	×	×	...	...	...	...	...	...	...	...	×	×	
169, s 93	Actæon Noæ, *J. Sow.*	...	×	×	...	?	...	...	...	...	...	...	...	...	...	...	...	
171, s 94	— levidensis, *S. Wood*	×	...	...	...	...	...	...	...	...	...	...	...	...	...	...	...	
170, s 94	— subulatus, *S. Wood*	×	...	×	...	...	...	...	...	...	...	...	...	...	...	...	...	
170, s 93	— tornatilis, *Linn.*	×	...	×	...	×	×	...	...	...	...	...	...	...	×	×	×	
s 94	— ? Etheridgii, *A. Bell*	...	×	...	...	...	...	...	...	...	...	...	...	...	...	...	?	
s 187	Bulla(Cylichna)utriculus, *Brocc.*	×	...	...	...	...	...	...	...	...	...	...	...	...	×	×	×	
176	— alba, *Brown*	...	...	...	...	×	...	...	...	...	...	...	...	...	×	...	×	*B. cylindracea,* var., Tab. XXI, fig. 1 b, of Crag Mollusca. *Volvaria alba,* Brown.
174, A 322	— (Volvula)acuminata, *Brug.*	×	...	...	...	...	...	...	...	...	...	...	...	...	×	×	×	
173, A 322	— (Cylichna)conuloïdea, *S. Wood*	×	...	...	...	...	...	...	...	...	...	...	...	...	...	...	?	*B. conulus* of Crag Moll., Vigo Bay, 380 to 994 fathoms, 'Depths of the Sea,' p. 184.
176	— — concinna, *S. Wood*	×	...	...	...	...	...	...	...	...	...	...	...	...	...	...	...	
175	— — cylindracea, *Pennant*	×	...	×	...	...	...	...	...	...	...	...	...	...	×	×	×	The Norwich specimen mentioned in Woodward's list is that inserted in the Fluvio-marine Crag column under the name of *C. alba.* Probably only derivative in Red Crag.
178	— (Utriculus) Lajonkaireana, *Bast.*	×	...	...	...	...	...	...	...	...	...	...	...	...	...	?	...	*B. mamillata?* Phil.
177	— — Regulbiensis, *Adams*	...	...	...	...	×	×	...	...	...	...	...	...	...	×	×	×	*B. obtusus,* Mont.
176	— — truncata, *Adams*	×	...	...	...	...	×	...	...	...	...	...	...	...	×	×	×	

PAGE.		Cor. Crag.	Red Crag, Walton.	Red Crag, Sutton and Butley.	Scrobicularia Crag.	Fluvio-marine Crag.	Chillesford beds.	Lower Glacial.	Middle Glacial.	Upper Glacial.	Post Glacial, Kelsey Hill.	Post Glacial, March.	Post Glacial, Hunstanton.	Post Glacial, Nar Brickearth.	Living, Britain.	Living, Mediterranean.	Living, elsewhere.	REMARKS.
178	Bulla (Utriculus) nana, *S. Wood*	×	...	...	...	...	...	...	...	...	...	...	...	...	...	?	...	
180	Bullæa catena, *Mont.*	×	...	...	...	...	...	...	...	...	...	...	...	...	×	×	×	*B. sculpta* of Crag Moll.
179	— quadrata, *S. Wood*	×	...	...	...	...	...	...	...	...	...	...	...	...	×	×	×	310 fathoms, off Malta, 'Depths of the Sea,' p. 270.
181	— scabra, *Müller*	×	...	...	...	...	...	...	...	...	...	...	...	...	×	×	×	*B. dilatata*, Crag Moll.
182 / s 96	— ventrosa, *S. Wood*	×	...	...	...	...	...	...	...	...	...	...	...	...	...	?	...	
173	Scaphander lignarius, *Linn.*	×	...	...	...	...	...	...	...	...	...	...	...	...	×	×	×	*Bulla lignaria* of Crag Moll.
s 96	— librarius? *Lovén*	×	...	...	...	...	...	...	...	...	...	...	...	...	×	...	×	
11	Melampus pyramidalis, *J. Sow*	...	×	×	...	?	×	...	...	...	...	...	...	...	...	...	...	*Conovulus pyramidalis* of Crag Moll.
12 / s 3	— fusiformis, *S. Wood*	...	...	×	...	×	...	...	...	...	...	...	...	...	...	...	...	*C. myosotis* of Crag Moll.

PTEROPODA.

PAGE.		Cor. Crag.	Red Crag, Walton.	Red Crag, Sutton and Butley.	Scrobicularia Crag.	Fluvio-marine Crag.	Chillesford beds.	Lower Glacial.	Middle Glacial.	Upper Glacial.	Post Glacial, Kelsey Hill.	Post Glacial, March.	Post Glacial, Hunstanton.	Post Glacial, Nar Brickearth.	Living, Britain.	Living, Mediterranean.	Living, elsewhere.	REMARKS.
191 / s 99	Cleodora infundibulum, *S. Wood*	×	...	...	...	...	...	...	...	...	...	...	...	...	?	...	?	

BIVALVIA.

PAGE.		Cor. Crag.	Red Crag, Walton.	Red Crag, Sutton and Butley.	Scrobicularia Crag.	Fluvio-marine Crag.	Chillesford beds.	Lower Glacial.	Middle Glacial.	Upper Glacial.	Post Glacial, Kelsey Hill.	Post Glacial, March.	Post Glacial, Hunstanton.	Post Glacial, Nar Brickearth.	Living, Britain.	Living, Mediterranean.	Living, elsewhere.	REMARKS.
8 / s 100	Anomia ephippium, *Linn.*	×	...	×	...	×	×	...	×	×	...	...	...	...	×	×	×	
—	— — var. aculeata	×	...	...	...	×	...	...	×	...	...	...	...	...	×	×	×	
11 / s 100	— striata, *Lovén*	×	...	×	...	×	...	...	×	...	...	...	...	...	×	×	×	
10 / s 101	— patelliformis, *Linn.*	×	×	×	...	×	...	...	×	...	...	...	...	...	×	×	×	
13 / s 101	Ostrea edulis, *Linn.*	×	×	×	...	×	...	...	...	...	×	×	×	×	×	×	×	
s 101	— cochlear	×	...	×	...	...	...	...	...	...	...	...	...	...	×	×	×	*O. edulis*, var. *spectrum*, of Crag Moll. Derivative in Red Crag.
s 102	— plicatula, *Gmelin*	×	...	×	...	...	...	...	...	...	...	...	...	...	...	×	...	Derivative in Red Crag.
17	— princeps, *S. Wood*	×	...	×	...	...	...	...	...	...	...	...	...	...	...	...	...	Derivative in Red Crag.
s 102	— flabellula, *Lamk.*	...	...	×	...	...	...	...	...	...	...	...	...	...	...	...	...	Derivative in Red Crag.
19 / s 102	Hinnites Cortesyi, *De France*	×	...	×	...	?	...	...	...	...	...	...	...	...	...	...	...	Derivative in Red Crag.
29	Pecten Bruei, *Payr*	×	...	...	...	...	...	...	...	...	...	...	...	...	×	×	×	358 to 717 fathoms, in Atlantic, as *P. aratus*, 'Depths of the Sea,' p. 181.
38	— dubius, *Broc.*	×	×	×	...	...	...	...	...	...	...	...	...	...	...	...	...	
30 / s 106	— septemradiatus, *Chemn.*	...	...	×	...	...	...	...	...	...	...	...	...	...	×	×	×	*P. Danicus* of Crag Moll.
24 / s 104	— (Pleuronectia) Gerardii, *Nyst*	×	...	...	...	...	...	...	...	...	...	...	...	...	...	...	...	
40 / s 103	— Islandicus, *Müller*	...	...	...	...	...	...	...	×	...	...	...	...	...	...	...	×	Upper Glacial, Bridlington, *fide* Woodward, Geol. Mag., vol. i, p. 53.
22	— (Janira) maximus, *Linn.*	×	...	×	...	...	...	...	...	...	...	...	...	...	×	×	×	
23	— — var. grandis, *J. Sow.*	×	...	×	...	...	...	...	...	...	...	...	...	...	...	...	...	Probably derivative in Red Crag.
35 / s 105	— opercularis, *Linn.*	×	×	×	×	×	×	...	×	...	...	...	...	...	×	×	×	
37	— var. gracilis, *J. Sow.*	...	...	×	...	...	...	...	...	...	...	...	...	...	...	...	...	*P. gracilis*, Sow., of Crag Moll.
32 / s 103	— princeps, *J. Sow*	×	...	...	...	...	...	...	...	...	...	...	...	...	...	...	...	
s 103	— var. pseudo-princeps	...	...	...	...	×	...	...	...	...	...	...	...	...	...	...	...	
33 / s 105	— pusio, *Pennant*	×	×	×	...	...	...	...	×	?	...	...	...	...	×	×	×	
25	— (Pleuronectia) similis, *Laskey*	×	...	...	...	...	...	...	...	...	...	...	...	...	×	×	×	
27	— tigrinus, *Müller*	×	×	×	...	?	×	...	...	...	...	...	...	...	×	×	×	Recently found in Mediterranean, Brit. Con., vol. v, p. 167.
41 / s 104	— varius, *Linn.*	?	...	...	...	...	...	...	×	...	...	...	×	...	×	×	×	
s 104	— niveus ? *McGil.*	?	...	...	...	...	...	...	...	...	...	...	...	×	...	...	...	
A 323 / s 106	— (Janira) Westendorpianus, *Nyst*	...	...	×	...	...	...	...	×	...	...	...	...	...	...	...	...	Probably a Cor. Crag species, and only derivative in Red Crag. Fossil Belgian Crag.
43 / s 108	Lima exilis, *S. Wood*	×	×	×	...	...	...	...	...	...	...	...	...	...	...	...	...	
44 / s 108	— hians, *Gmelin*	×	...	...	...	...	...	...	?	...	...	...	...	...	×	×	×	
45	— Loscombii, *G. B. Sow.*	×	×	...	...	...	...	...	...	...	...	...	...	...	×	×	×	
48 / s 108	— ovata, *S. Wood*	×	...	...	...	...	...	...	...	...	...	...	...	...	...	...	...	
46 / s 109	— plicatula, *S. Wood*	×	...	...	...	...	...	...	...	...	...	...	...	...	...	?	...	Possibly young of *squamosa*.

PAGE.		Cor. Crag.	Red Crag, Walton.	Red Crag, Sutton and Butley.	Scrobicularia Crag.	Fluvio-marine Crag.	Chillesford beds.	Lower Glacial.	Middle Glacial.	Upper Glacial.	Post Glacial, Kelsey Hill.	Post Glacial, March.	Post Glacial, Hunstanton.	Post Glacial, Nar Brickearth.	Living, Britain.	Living, Mediterranean.	Living, elsewhere.	REMARKS.
47 / s 107	Lima subauriculata, *Mont.*	×	...	...	...	...	...	...	...	...	...	...	...	...	×	×	×	
47 / s 107	— elongata, *Forbes*	×	...	...	...	...	...	...	...	...	...	...	...	...	×	×	×	Var. *elongata* of *L. subauriculata* in Crag Moll.
s 109	— squamosa, *Lamk.*	×	...	...	...	...	...	...	...	...	...	...	...	...	...	×	...	
51 / s 109 / & 188	Avicula phalænoides, *S. Wood*	×	...	...	...	...	...	...	...	...	...	...	...	...	...	...	...	*A. Tarentina* of Crag Moll.
50 / s 110	Pinna pectinata, *Mont.*	×	?	×	...	...	...	?	...	...	...	...	...	...	×	×	×	
s 110	— rudis, *Linn.*	×	...	...	...	...	...	...	...	...	...	...	...	...	...	×	×	
52	Mytilus edulis, *Linn.*	...	...	×	×	×	×	×	×	×	×	×	×	×	×	×	×	
52	— var. hesperianus, *Lam.*	×	...	...	...	...	...	...	...	...	...	...	...	...	...	...	...	
s 110	— giganteus, *S. Wood*	...	...	×	...	...	...	...	...	...	...	...	...	...	...	...	...	Derivative.
58	Modiola barbata, *Linn.*	...	×	...	...	...	...	...	...	...	...	...	...	...	×	×	×	
60	— costulata, *Risso*	×	×	...	...	...	...	...	...	...	...	...	...	...	×	×	×	
63 / s 111	— discors, *Linn.*	?	...	...	...	...	×	...	...	...	...	...	...	...	×	×	×	
62	— marmorata, *Forbes*	×	×	...	...	...	...	...	...	...	...	...	...	...	×	×	×	
57	— modiolus, *Linn.*	?	...	×	...	×	...	...	×	...	...	...	...	...	×	...	×	{ Upper Glacial, Bridlington, *fide* Woodward, Geol. Mag., vol. ii, p. 53 (*M. vulgaris*). Doubtful whether the fragments in Cor. Crag belong to this species.
60 / s 111	— Petagnæ, *Scac.*	×	×	...	...	...	...	...	...	...	...	...	...	...	...	×	×	
59 / s 111	— phaseolina, *Phil.*	×	...	...	...	...	...	...	...	...	...	...	...	...	×	×	×	
64	— rhombea, *Berkley*	×	...	...	...	...	...	...	...	...	...	...	...	...	×	×	×	
61	— sericea, *Bronn.*	×	...	...	...	...	...	...	...	...	...	...	...	...	...	...	...	
77	Arca lactea, *Linn.*	×	×	×	...	...	...	...	...	...	...	...	...	...	×	×	×	
79	— pectunculoïdes, *Scac.*	×	...	...	...	...	...	...	...	...	...	...	...	...	×	×	×	
76 / s 116	— tetragona, *Poli*	×	...	×	...	...	...	...	...	...	...	...	...	...	×	×	×	
66 / s 116	Pectunculus glycimeris, *Linn.*	×	...	×	×	×	×	...	×	...	...	...	...	...	×	×	×	
66 / s 116	— pilosus, *Linn.*	×	...	×	...	...	...	...	...	...	...	...	...	...	...	×	...	Probably derivative in the Red Crag.
66 / s 116	— subobliquus, *S. Wood*	...	×	×	...	...	...	...	...	...	...	...	...	...	...	...	...	
70 / s 117	Limopsis aurita, *Broc.*	×	...	×	...	...	...	...	...	...	...	...	...	...	×	×	×	Probably derivative in Red Crag.
71 / A 324 / s 117	— pygmæa, *Phil.*	×	×	×	...	...	...	×	...	...	...	...	...	...	...	×	×	{ The statement in App., p. 324, that this had been dredged in Arctic seas is erroneous, the species dredged being different.
73	Nucinella miliaris, *Desh.*	×	...	...	...	...	...	...	...	...	...	...	...	...	...	...	...	
82 / s 111	Nucula Cobboldiæ, *J. Sow.*	...	...	×	...	×	×	×	×	×	...	...	...	...	...	...	...	
81 / s 113	— lævigata, *J. Sow.*	...	×	×	...	...	...	...	...	...	...	...	...	...	...	...	...	
s 113	— — var. calva, *S. Wood*	×	...	...	...	...	...	...	...	...	...	...	...	...	...	...	...	
85 / s 114	— nucleus, *Linn.*	×	...	×	...	...	...	?	...	...	...	...	...	...	×	×	×	
87	— — var. radiata	?	...	×	...	...	...	...	...	...	...	...	...	...	×	×	...	Supplement, Tab. X, fig. 3.
s 113	— nitida, *G. Sow.*	×	...	...	...	...	...	...	...	...	...	...	...	...	×	×	×	
84 / s 114	— tenuis, *Mont.*	?	...	×	...	×	×	...	?	×	...	...	...	...	×	×	×	
86 / s 113	— proxima, *Say*	×	...	...	...	...	...	...	...	...	...	...	...	...	...	...	×	North America only. *N. trigonula* of Crag Moll.
92 / s 115	Leda caudata, *Donov.*	...	...	...	...	...	...	...	×	...	...	...	...	...	...	...	×	
88 / s 115	— lanceolata, *J. Sow.*	...	...	×	...	×	×	...	×	...	...	...	...	...	...	...	×	Arctic seas only.
92 / s 115	— minuta, *Mont.*	...	...	×	...	...	...	...	...	...	...	...	×	...	...	...	×	
90 / s 114	— oblongoides, *S. Wood*	...	...	×	...	×	×	×	×	×	...	...	...	...	...	...	?	*Leda myalis* of Crag Moll.
93	— pernula, *Müll.*	...	...	...	...	...	...	...	×	...	...	...	...	...	...	...	×	{ Arctic seas, 358 to 717 fathoms in Atlantic, 'Depths of the Sea,' p. 181.
95	— pygmæa, *Münst.*	×	×	...	...	...	...	...	...	...	...	...	...	...	×	×	×	Walton, *fide* A. Bell, Ann. and Mag. Hist., Sept., 1870.
s 115	— myalis, *Couth.*	...	...	...	...	×	...	×	...	...	...	...	...	...	...	...	×	North America only.
91	— semistriata, *S. Wood*	×	...	...	...	...	...	...	...	...	...	...	...	...	...	...	...	

PAGE.		Cor. Crag.	Red Crag, Walton.	Red Crag, Sutton and Buxley.	Scrobicularia Crag.	Fluvio-marine Crag.	Chillesford beds.	Lower Glacial.	Middle Glacial.	Upper Glacial.	Post Glacial, Kelsey Hill.	Post Glacial, March.	Post Glacial, Hunstanton.	Post Glacial, Nar Brickearth.	Living, Britain.	Living, Mediterranean.	Living, elsewhere.	REMARKS.
115 / s 121-2	Lepton deltoideum, *S. Wood*	×	…	×	…	…	…	…	…	…	…	…	…	…	…	…	…	
116 / s 121-2	— depressum, *Nyst*	×	…	…	…	…	…	…	…	…	…	…	…	…	…	…	…	
116 / s 121-2	— nitidum, *Turton*	×	…	…	…	×	×	…	…	…	…	…	…	…	×	×	×	
114 / s 121	— squamosum, *Mont.*	×	…	…	…	…	…	…	…	…	…	…	…	…	×	×	×	
s 121-2	Lasæa Clarkiæ, *Clarke*	×	…	…	…	…	…	…	…	…	…	…	…	…	×	…	…	
125 / s 121	— rubra, *Mont.*	×	.:	…	…	…	…	…	…	…	…	…	…	…	×	×	×	*Kellia rubra* of Crag Moll.
124 / s 121	— pumila, *S. Wood*	×	…	…	…	…	…	…	…	…	…	…	…	…	…	…	?	*Kellia pumila* of Crag Moll.
s 121-3	— intermedia, *S. Wood*	…	…	…	…	×	…	×	…	…	…	…	…	…	…	…	…	
s 121-3	Bornia ovalis, *S. Wood*	×	…	…	…	…	…	…	…	…	…	…	…	…	…	…	…	
126 / s 121-5	Montacuta bidentata, *Mont*	×	×	…	…	×	…	×	…	…	…	…	…	×	×	×	×	
129 / s 121-6	— ferruginosa, *Mont.*	×	…	…	…	×	…	…	…	…	…	…	…	…	×	×	×′	
127 / s 121	— truncata, *S. Wood*	×	…	…	…	…	…	…	…	…	…	…	…	…	…	…	…	Fossil in Vienna beds.
s 121-6	— elliptica, *S. Wood*	×	…	…	…	…	…	…	…	…	…	…	…	…	…	…	…	
131 / s 121-6	Sphenalia donacina, *S. Wood*	×	…	…	…	…	…	…	…	…	…	…	…	…	×	…	…	*Montacuta donacina* of Crag Moll.
128 / s 121-7	— substriata, *Mont*	×	…	…	…	…	×	…	…	…	…	…	…	…	×	×	×	*Montacuta substriata* of Crag Moll.
118 / s 121-4	Kellia suborbicularis, *Mont*	×	×	×	…	…	…	…	…	…	…	…	…	…	×	×	×	
120 / s 121-5	Scintilla ambigua, *Nyst*	×	×	×	…	…	×	…	…	…	…	…	…	…	…	…	…	*Kellia ambigua* of Crag Moll.
123 / s 121	— compressa, *Phil.*	×	…	…	…	…	…	…	…	…	…	…	…	…	…	…	…	*Kellia coarctata* of Crag Moll.
122 / s 121-4	Scacchia cycladia, *S. Wood*	×	…	…	…	…	…	…	…	…	…	…	…	…	…	…	…	*Kellia cycladia* of Crag Moll.
121 / s 121	— elliptica, *Phil.*	×	…	…	…	…	…	…	…	…	…	…	…	…	…	×	×	*Kellia elliptica* of Crag Moll.
s 189	— lata, *S. Wood*	×	…	…	…	…	…	…	…	…	…	…	…	…	…	…	…	
120 / s 121-4	— orbicularis, *S. Wood*	×	…	…	…	…	…	…	…	…	…	…	…	…	×	×	…	*Kellia orbicularis* of Crag Moll.
135 / s 121-7	Cryptodon rotundatum, *S. Wood*	×	…	…	…	…	…	…	…	…	…	…	…	…	…	…	…	*Cryptodon ferruginosum* of Crag Moll.
134 / A 324 / s 121	— sinuosum, *Donovan*	×	…	…	…	×	…	…	…	…	…	…	…	…	×	×	×	
137 / s 127	Loripes divaricatus, *Linn.*	…	…	×	…	×	…	…	×	…	…	…	…	…	×	×	×	
139 / s 128	Lucina borealis, *Linn.*	×	×	×	…	×	×	×	×	…	…	…	…	…	×	×	×	
140	— crenulata, *S. Wood*	×	…	…	…	…	…	…	…	…	…	…	…	…	…	…	…	*L. exigua* ? Hörnes.
141	— decorata, *S. Wood*	×	…	…	…	…	…	…	…	…	…	…	…	…	…	…	…	*L. trigonula* ? Bronn.
146 / s 129	Diplodonta Astartea, *Nyst*	×	…	×	…	×	…	…	…	…	…	…	…	…	…	?	?	
145 / s 128	— dilatata, *S. Wood*	×	…	×	…	…	…	…	…	…	…	…	…	…	…	…	…	Possibly derivative in Red Crag.
144 / s 128	— rotundata, *Mont.*	×	…	×	…	…	…	…	…	…	…	…	…	…	×	×	×	
148 / s 130	Lucinopsis Lajonkairii, *Payr*	×	…	×	…	…	…	…	…	…	…	…	…	…	…	…	…	Probably derivative in the Red Cräg.
s 129	— undata, *Pennant*	…	…	…	…	…	×	…	…	…	…	…	…	…	×	×	×	
150 / s 130	Verticordia cardiiformis, *S. Wood*	×	…	…	…	…	…	…	…	…	…	…	…	…	…	…	…	*Hippagus Verticordius* of Crag Moll.
162 / s 130	Chama gryphoides, *Linn*	×	…	×	…	…	…	…	…	…	…	…	…	…	…	×	×	Probably derivative in Red Crag.
168 / s 130	Cardita borealis, *Conrad*	…	…	…	…	…	…	…	…	×	…	…	…	…	…	…	×	*C. analis*, Crag Moll. Living in N. E. American seas.
167	— Chamæformis, *Leathes*	×	?	×	…	…	…	…	…	…	…	…	…	…	…	…	…	Probably derivative in Red Crag.
168 / s 132	— corbis, *Phil.*	×	×	×	…	…	…	…	×	…	…	…	…	…	×	×	…	
168 / s 132	— anceps, *S. Wood*	×	?	?	…	…	…	…	…	…	…	…	…	…	…	…	…	If in Red Crag it is probably so only as a derivative.
167	— orbicularis, *Leathes*	×	…	?	…	…	…	…	…	…	…	…	…	…	…	…	…	If in Red Crag it is probably so only as a derivative.
166 / s 131	— scalaris, *Leathes*	×	×	×	…	?	?	…	×	…	…	…	…	…	…	…	?	*C. ventricosa* ? Gould, a Pacific shell.

PAGE.		Cor. Crag.	Red Crag, Walton.	Red Crag, Sutton and Butley.	Scrobicularia Crag.	Fluvio-marine Crag.	Chillesford beds.	Lower Glacial.	Middle Glacial.	Upper Glacial.	Post Glacial, Kelsey Hill.	Post Glacial, March.	Post Glacial, Hunstanton.	Post Glacial, Nar Brickearth.	Living, Britain.	Living, Mediterranean.	Living, elsewhere.	REMARKS.
165 / s 133	Cardita senilis, *Lam.*	×		×												×		Probably derivative in Red Crag.
157	Cardium angustatum, *J. Sow.*			×		×												Fluvio-marine Crag (Thorpe by Aldbro'), *fide* A. Bell, Ann. and Mag. Nat. Hist., 1870.
152	— echinatum, *Linn.*			×										×	×	×	×	
155 / s 134	— edule, *Linn.*	×	×	×	×	×	×	×	×	×	×	×	×	×	×	×	×	
153 / s 133	— fasciatum, *Mont.*	×		×			×								×	×	×	*C. nodosum* of Crag Moll.
s 134	— nodosum, *Turt.*	×		?											×	×	×	
154 / s 134	— pinnatulum, *Con.*			×													×	*C. nodosulum* of Crag Moll.
s 137	— Islandicum? *Linn*							?	?								×	
158 / s 135	— Parkinsoni, *J. Sow.*		×	×		?												Not *C. Nuttalli.*
154 / s 134	— strigilliferum, *S. Wood*	×																
160	— (Aphrodita) Groenlandicum, *Chemn.*			×		×	×										×	
159	— interruptum, *S. Wood*			×														
159 / s 135	— decorticatum, *S. Wood*	×																
160	— venustum, *S. Wood*		×	×														Var. of *decorticatum*?
171 / s 136	Erycinella ovalis? *Conrad*	×	×						×									
177	Astarte Basterotii, *De la Jonk*	×	?	×														
188 / s 137	— Burtinii, *De la Jonk*	×		×		?			×									Probably derivative in Red Crag.
175 / s 137	— (Triodonta) borealis, *Chemn.*					×	×	×	×	×		×					×	
175 / s 137	— — var. Withami								×									
183 / s 138	— (Nicania) compressa, *Mont.*		×	×		×	×	×	×	×	×			×			×	
186	— crebricostata, *Forbes*								×								×	
184	— crebrilirata, *S. Wood*		×	×														Neither *A. depressa* of Brown nor *A. crebricostata* of Forbes.
185 / s 138	— Galeottii? *Nyst*	×		×														*A. gracilis* of Crag Moll. Probably derivative in Red Crag.
185	— — var. incerta	×																
178 / s 138	— incrassata, *Broc.*	×		×					?							×	×	Probably derivative in Red Crag.
179	— mutabilis, *S. Wood*	×		×														Fig. 1 e, f, Tab. XVI, of Crag Moll., from Bridlington, is probably a variety only of *borealis*. Probably derivative in Red Crag.
189	— obliquata, *J. Sow.*		×	×														
180 / s 139	— Omalii, *De la Jonk*	×		×		?			×									Probably only a derivative in Red Crag.
192 / s 140	— Forbesii, *S. Wood*	×																*A. parva* of Crag Moll.
175 / s 140	— parvula, *S. Wood*	×																
187	— pygmæa, *Munst.*	×																
182 / s 139	— sulcata, *Da Costa*			×		?	×		×		×			×			×	
181	— elliptica, *Brown*					?				×					×	×	×	
173	— triangularis, *Mont.*	×	×												×	×	×	
190 / s 141	Woodia digitaria, *Linn.*	×	×						×						?	×	×	*Astarte digitaria* of Crag Moll.
191 / s 142	— excurrens, *S. Wood*	×																*Astarte excurrens* of Crag Moll.
196 / s 142	Cyprina Islandica, *Linn.*	×	?	×	×		×	×	×	×	×	×	×		×		×	Fossil in Mediterranean.
197 / s 142	— rustica, *J. Sow.*	×		×														Probably derivative in Red Crag.
193	Isocardia cor, *Linn.*	×		×											×	×	×	Probably derivative in Red Crag.
210	Venus casina, *Linn.*	×		×											×	×	×	Probably derivative in Red Crag.
212	— imbricata, *J. Sow.*		×	×														The var. *gibberosa* is probably derivative in the Red Crag. As to *V. dysera*, see Sup., p. 189.
212	— — var. gibberosa	×		×														
211 / s 143	— fasciata, *Da Costa*		×	×		×			×						×	×	×	
s 144	— fluctuosa, *Gould*								×	×							×	North-east American and Arctic seas only.

PAGE.		Cor. Crag.	Red Crag, Walton.	Red Crag, Sutton and Butley.	Scrobicularia Crag.	Fluvio-marine Crag.	Chillesford beds.	Lower Glacial.	Middle Glacial.	Upper Glacial.	Post Glacial, Kelsey Hill.	Post Glacial, March.	Post Glacial, Hunstanton.	Post Glacial, Nar Brickearth.	Living, Britain.	Living, Mediterranean.	Living, elsewhere.	REMARKS.
213 s 143	Venus ovata, *Pennant*	×	...	×	...	...	×	...	×	...	...	...	...	...	×	×	×	Probably derivative in the Red Crag, where it is very rare.
A 326 s 144	— gallina, *Linn.*	...	...	...	...	...	...	...	...	...	×	...	...	...	×	×	×	
207	Cytherea Chione, *Linn.*	×	...	×	...	...	...	...	...	...	...	...	...	...	×	×	×	In the Red Crag it has only occurred at Waldringfield (Bell, Ann. and Mag., 1870), and is probably derivative.
208 s 142	— rudis, *Poli*	×	×	×	...	?	...	...	×	...	...	...	...	...	...	×	×	
198	Circe minima, *Mont.*	×	...	×	...	...	...	...	...	...	...	...	...	...	×	×	×	Probably derivative in the Red Crag.
215	Artemis lincta, *Pult.*	×	×	...	...	...	...	...	...	...	...	...	...	...	×	×	×	
215	— lentiformis, *J. Sow.*	?	×	...	...	...	...	...	...	...	...	...	...	...	...	...	...	The only Cor. Crag specimen is fig. 7 *c*, of Tab. XX, of Crag Moll., and this may possibly be a deformity of *A. lincta.*
202 s 145	Tapes aureus, *Gmel.*	...	...	...	...	×	...	...	...	...	...	...	...	...	×	×	×	
A 327 s 145	— decussatus, *Linn.*	...	...	...	...	...	...	...	...	...	...	...	×	×	×	×	×	
203	— perovalis, *S. Wood*	×	...	...	...	...	...	...	...	...	...	...	...	...	...	...	...	
A 327 s 145	— pullastra, *W. Wood*	?	?	×	...	...	...	?	...	...	...	...	...	...	×	×	×	Only unique specimens from Coralline Crag and from Walton, which are in bad preservation.
204	— texturatus, *Lam.*	...	×	×	...	...	...	...	...	...	...	...	...	...	×	×	×	Red Crag of Waldringfield, *fide* A. Bell.
201 s 145	— virgineus, *Linn.*	×	×	×	...	×	...	...	?	...	...	...	...	...	×	×	×	
200 s 136	Coralliophaga cyprinoides, *S. Wood*	×	...	...	...	...	...	...	...	...	...	...	...	...	...	?	...	
205 s 146	Venerupis Irus, *Linn.*	×	×	...	...	...	...	...	...	...	...	...	...	...	×	×	×	
217 s 146	Gastrana laminosa, *J. Sow.*	×	×	×	...	...	...	...	...	...	...	...	...	...	...	...	?	Closely resembles a South African shell.
220 s 147	Donax politus, *Poli*	×	×	×	...	...	...	...	...	...	...	...	...	...	×	×	×	
219	— trunculus, *Linn.*	...	...	×	...	...	...	...	...	...	...	...	...	...	×	×	×	
219 s 146	— vittatus, *Da Costa*	...	...	?	...	×	...	...	...	...	...	...	...	...	×	×	×	Given by Mr. Jeffreys from the Red Crag of Sutton, but I cannot learn on what authority.
221 s 147	Psammobia Ferroensis, *Chemn.*	×	...	...	...	...	...	...	...	...	...	...	...	...	×	×	×	
	— costulata, *Turt.*	×	...	...	...	...	...	...	...	...	...	...	...	...	×	×	×	
223 s 147	— tellinella, *Linn.*	×	...	...	...	...	...	...	...	...	...	...	...	...	×	...	×	
222	— vespertina, *Chemn.*	×	...	...	...	...	...	...	...	...	...	...	...	...	×	×	×	
227	Tellina balaustina, *Linn.*	×	...	...	...	...	...	...	...	...	...	...	...	...	×	×	×	
231 s 151	— Balthica, *Linn.*	...	...	...	...	...	...	×	×	×	×	×	×	×	×	×	×	
230	— Benedenii, *Nyst et Westend*	...	...	×	...	...	...	...	...	...	...	...	...	...	...	...	...	Only a derivative in the Red Crag, and may probably turn up in the Coralline.
226 s 151	— crassa, *Gmel.*	×	×	×	...	×	×	...	×	...	...	...	...	...	×	×	×	
234 s 150	— compressa, *Broc.*	×	...	...	...	...	...	...	...	...	...	...	...	...	...	×	×	*Tellina donacilla* of Crag Moll.
233	— donacina, *Linn.*	×	...	×	...	...	...	...	...	...	...	...	...	...	×	×	×	Mr. Bell writes me that he has obtained a specimen from the Red Crag of Boyton. This is the only instance known to me from the Red Crag, and is doubtless due to derivation from the Coralline.
232 s 150	— fabula, *Gronov*	...	...	...	...	×	×	...	...	...	...	...	...	...	×	×	×	
228 s 151	— lata, *Gmel.*	...	...	×	...	×	×	×	×	×	...	?	...	×	...	...	×	Arctic only.
228 s 151	— obliqua, *J. Sow.*	×	?	×	×	×	×	×	×	...	...	...	...	...	...	...	×	One specimen found at Walton by Mr. Norton, but this is of doubtful origin.
230 s 151	— prætenuis, *Leathes*	...	?	×	×	×	×	×	...	...	...	...	...	...	...	...	...	One specimen in my cabinet is labelled Walton, but it is of doubtful origin.
s 150	— pulchella? *Lam.*	×	...	...	...	...	...	...	...	...	...	...	...	...	...	...	...	
235 s 153	Scrobicularia plana, *Da Costa*	...	...	×	×	×	×	...	...	...	...	...	×	×	×	×	×	*Trigonella plana* of Crag Moll. *S. piperata*, auctorum.
237 s 152	Abra alba, *W. Wood*	×	×	×	...	?	?	...	...	...	...	...	...	...	×	×	×	
238 s 153	— fabalis, *S. Wood*	...	×	...	...	...	...	...	...	...	...	...	...	...	...	...	...	
240	— obovalis, *S. Wood*	...	...	×	...	×	...	...	...	...	...	...	...	...	...	?	...	*Erycina ovata?*
239	— prismatica, *Mont.*	×	...	...	...	...	?	...	...	...	...	...	...	...	×	×	×	Given by Mr. Jeffreys from Aldeby, but I have not been able to learn on what authority.

PAGE.		Cor. Crag.	Red Crag, Walton.	Red Crag, Sutton and Butley.	Scrobicularia Crag.	Fluvio-marine Crag.	Chillesford beds.	Lower Glacial.	Middle Glacial.	Upper Glacial.	Post Glacial, Kelsey Hill.	Post Glacial, March.	Post Glacial, Hunstanton.	Post Glacial, Nar Brickearth.	Living, Britain.	Living, Mediterranean.	Living, elsewhere.	REMARKS.
243 / s 155	Mactra arcuata, *J. Sow.*	×	×	×	...	×	×	...	...	...	...	...	...	...	...	...	...	
244 / s 155	— artopta, *S. Wood*	×	...	...	...	...	...	...	...	...	...	...	...	...	...	...	...	
249	— (Mesodesma) deaurata, *Turt.*	...	×	×	...	...	...	...	...	...	...	...	...	...	...	...	×	North American seas only, the British specimens being spurious. See Forb. & H., vol. i, p. 346.
241 / s 154	— glauca, *Born.*	...	...	×	...	...	...	...	...	...	...	...	...	...	×	×	×	
246 / s 153	— ovalis, *J. Sow.*	...	...	×	×	×	×	×	×	...	×	×	×	...	...	×	×	The Mediterranean analogue is *M. triangula.*
249	— var. constricta, *S. Wood*	...	×	...	...	...	...	...	...	...	...	...	...	...	...	...	...	
244	— procrassa, *S. Wood*	...	...	×	...	...	...	...	...	...	...	...	...	...	...	...	...	
245 / s 154	— solida, *Linn.*	×	×	×	...	×	×	...	...	×	×	...	...	×	×	×	×	
242 / s 154	— stultorum, *Linn.*	×	?	×	×	...	...	...	?	...	...	...	...	...	×	×	×	
247 / s 154	— subtruncata, *Da Costa*	...	...	?	×	×	×	...	...	...	...	...	...	...	×	×	...	
248	— — var. obtruncata	×	...	×	×	...	...	...	...	...	...	...	...	...	...	...	...	
A 325	— triangula, *Renier*	×	...	...	...	...	...	...	...	...	...	...	...	...	...	×	...	*M. triangulata* of Crag Moll.
245	— truncata, *Mont.*	...	×	×	...	...	...	...	...	...	...	...	...	×	...	...	...	
251 / s 155	Lutraria elliptica, *Lam.*	×	...	×	...	...	...	...	...	...	...	...	...	...	×	×	×	Fragments only in Red Crag, and these probably only derivative. As to *L. oblonga*, see p. 155 of Supplement.
252 / s 147	Solecurtus strigillatus, *Linn.*	×	...	?	...	...	...	...	...	...	...	...	...	...	...	×	×	*Macha strigillata* of Crag Moll.
s 148	Cultellus Suttonensis, *S. Wood*	×	...	...	...	...	...	...	...	...	...	...	...	...	...	...	...	
258 / s 149	— cultellatus, *S. Wood*	...	×	...	...	...	...	...	...	...	...	...	...	...	...	...	...	*C. tenuis* of Crag Moll.
s 148	— pellucidus ? *Penn*	...	...	...	...	?	...	...	...	...	...	...	...	...	×	×	×	Tab. X, fig. 14, of Supplement.
254	Solen gladiolus, *Gray*	...	×	×	...	...	...	...	...	...	...	...	...	...	...	...	×	
255 / s 149	— siliqua, *Linn.*	...	...	×	...	×	?	...	?	...	...	...	...	...	×	×	×	
256	— ensis, *Linn.*	×	×	×	...	...	?	...	?	...	...	...	...	...	×	×	×	See Supplement, p. 149, as to occurrence at Aldeby and Hopton.
264	Cochlodesma complanatum, *S. Wood*	...	×	...	...	...	...	...	...	...	...	...	...	...	...	...	...	
264	— praetenerum, *S. Wood*	×	...	...	...	...	...	...	...	...	...	...	...	...	...	...	...	
264	— praetenue, *Pult.*	×	...	...	...	...	...	...	...	...	...	...	...	...	×	×	×	Red Crag, Brightwell, with doubt, in Mr. Jeffreys list. Neither *Conradi* nor *Corbuloides.* See C. M., p. 261. If in Red Crag, derivative.
261 / s 189	Thracia inflata, *J. Sow.*	×	...	?	...	...	...	...	...	...	...	...	...	...	...	...	...	
s 156	— distorta ? *Mont.*	?	...	...	...	...	...	...	...	...	...	...	...	...	×	×	×	
259 / s 156	— papyracea, *Poli*	×	...	...	...	×	×	...	...	...	...	...	...	...	×	×	×	*T. phaseolina* of Crag Moll.
259	— pubescens, *Pult.*	×	...	...	...	...	...	...	...	...	...	...	...	...	×	×	×	
262 / s 156	— ventricosa, *Phil.*	×	...	...	...	...	...	...	...	...	...	...	...	...	×	×	...	
s 159	Lyonsia — ?	×	...	...	...	...	...	...	...	...	...	...	...	...	...	...	...	Species undeterminable.
270 / s 157	Pandora inequivalvis, *Linn.*	×	×	×	...	...	...	...	×	...	...	...	...	...	×	×	×	
273 / s 161	Necœra cuspidata, *Oliver*	×	...	...	...	...	...	...	...	...	...	...	...	...	×	×	×	
s 161	— obesa, *Lovén*	×	...	...	...	...	...	...	...	...	...	...	...	...	...	...	×	Scandinavian seas, 380 to 994 fathoms in Vigo Bay and 2435 fathoms in Atlantic, 'Depths of the Sea,' pp. 184 and 96.
272	— jugosa, *S. Wood*	×	...	...	...	...	...	...	...	...	...	...	...	...	...	...	?	I have not had the opportunity of comparing this with *N. lamellosa*, Sars. 380 to 994 fathoms in Vigo Bay, 'Depths of the Sea,' p. 184.
268 / s 161	Poromya granulata, *Nyst*	×	...	...	...	...	...	...	...	...	...	...	...	...	×	×	×	
274 / s 160	Corbula striata, *Walker & Boys*	×	×	×	...	×	×	×	×	...	×	×	×	×	×	×	×	
s 159	— contracta ? *Say*	...	...	...	×	...	...	...	...	...	...	...	...	...	...	...	×	North American seas.
275 / s 160	Corbulomya complanata, *J. Sow.*	...	×	×	...	...	...	...	...	...	...	...	...	...	...	...	...	*Corbula complanata* of Crag Moll.
287 / s 157	Saxicava arctica, *Linn.*	×	×	×	...	×	×	×	...	...	...	...	...	...	×	×	×	
285 / s 157	— var. rugosa, *Linn.*	×	×	×	...	...	...	...	×	...	...	...	...	...	×	×	×	Only the gigantic form in Upper Glacial.
288 / s 158	— fragilis, *Nyst*	×	...	...	...	...	...	...	...	...	...	...	...	...	×	×	×	*Panopea plicata.* Algeria, sec. Weinkauff.

PAGE.		Cor. Crag.	Red Crag, Walton.	Red Crag, Sutton and Butley.	Scrobicularia Crag.	Fluvio-marine Crag.	Chillesford beds.	Lower Glacial.	Middle Glacial.	Upper Glacial.	Post Glacial, Kelsey Hill.	Post Glacial, March.	Post Glacial, Hunstanton.	Post Glacial, Nar Brickearth.	Living, Britain.	Living, Mediterranean.	Living, elsewhere.	REMARKS.
289 } s 158	Saxicava carinata, *S. Wood*......	×	...	...	...	...	...	...	...	...	...	...	...	...	...	...	..	
283 } s 162	Panopea Faujasii, *Men de la Groye* }	×	...	×	...	...	...	...	...	...	...	...	...	...	...	×	×	{ *P. glycimeris*, Born. *P. Aldrovandi*, Phil. Derivative in Red Crag.
281 } s 161	— ? Norvegica, *Spengler* ...	...	...	×	...	×	×	...	×	×	...	...	...	...	×	...	×	Northern seas only.
266	Pholadomya hesterna, *J. Sow*....	×	...	...	...	...	...	...	...	...	...	...	...	...	...	...	...	
276 } s 159	Sphenia ovata ?......	×	...	...	...	...	...	...	...	...	...	...	...	...	...	...	?	{ *Sphenia Binghami* of Crag Moll. *S. ovata* is a Pacific shell.
s 159	— Binghami, *Turt*	...	×	×	...	...	...	...	...	...	...	...	...	...	×	×	×	
279 } s 162	Mya arenaria, *Linn*........	...	...	×	×	×	×	×	×	...	...	×	×	×	×	...	×	
277 } s 163	— truncata, *Linn*.	×	...	×	×	×	×	×	×	×	×	×	...	×	...	×		
291	Glycimeris angusta, *Nyst*........	×	...	×	...	...	...	...	...	...	...	...	...	...	...	...	...	{ Only fragments in Red Crag, and these probably derivative. Not *G. siliqua*.
298 } s 166	Pholadidea papyracea, *Solander*	×	...	...	...	...	...	...	...	...	...	...	...	×	...	×		{ The specimen of *Pholadidea* in Crag Moll., Tab. XXXI, fig. 23, is some derivative, and its species being doubtful, it is excluded from the list.
s 163	Pholas candida, *Linn*.	...	...	...	...	×	...	...	...	...	...	...	...	×	×	×		
296 } s 165	— crispata, *Linn*...............	...	×	×	...	...	×	×	×	×	×	...	×	...	×			
s 164	— brevis, *S. Wood*............	×	...	...	...	...	...	...	...	...	...	...	...	...	...	...		
295 } s 165	— cylindrica, *J. Sow*.	...	×	×	...	...	...	...	...	...	...	...	...	...	...	...		
s 165	— dactylus ? *Linn*...........	...	?	...	...	...	...	...	...	...	...	...	...	×	×	×		
s 164	— parva, *Penn*...............	...	...	×	...	...	...	...	...	...	...	...	...	×	×	×		
292	Gastrochæna dubia, *Penn*.........	×	...	×	...	...	...	...	...	...	...	...	...	×	×	×	} The tubes only of both these species are found in the Red Crag, and it is probable that they are there only as derivatives.
300	Teredo Norvegica, *Spengler*......	×	...	×	...	...	...	...	...	...	...	...	...	×	×	×	
	BRACHIOPODA.																	
s 172	Lingula Dumortieri, *Nyst*	×	...	...	...	...	...	...	...	...	...	...	...	...	...	...	?	
s 171	Discina fallens, *S. Wood*.........	×	...	...	...	...	...	...	...	...	...	...	...	...	...	...	...	
s 170	Argiope cistellula, *S. Wood*......	×	...	...	...	...	...	...	...	...	...	...	×	?	?			
s 169	Terebratulina caput-serpentis, *Linn*. }	×	...	...	...	...	...	...	...	...	...	...	...	×	×	×		
s 168	Terebratula grandis, *Blumenb*...	×	×	×	...	...	...	...	...	...	...	...	...	...	...	...	...	
s 171	Rhynchonella psittacea, *Chemn*	...	...	×	...	×	×	...	...	×	...	×	...	...	×	...	×	

After rejecting species which are mentioned *in it* as doubtful or undeterminable, and also every repetition of a species in the form of a variety, and rejecting from the Red Crag all species marked in the list as derivative, or as possibly or probably so, and (with the exceptions mentioned in the footnote[1]) treating all ?'s as × 's, the result of the foregoing list is, as regards the Gasteropoda and Bivalvia, given on the following page :

I have made no analysis of the Scrobicularia Crag, as the fauna obtained from it is too meagre to make it worth while.

[1] The following species occurring in the columns of the list are rejected from the summary, viz. In the Cor. Crag, *Pecten varius*, *Pecten niveus*, *Lima plicatula*, *Modiola modiolus*, *Artemis lentiformis*, *Tapes pullastra*, and *Thracia distorta*. *Pyramidella læviuscula* is also treated as not living. In the Walton Red Crag, *Tapes pullastra*, *Cyprina Islandica*, *Tellina obliqua*, and *Tellina prætenuis*. In the rest of the Red Crag, *Solecurtus strigillatus*. In the Fluvio-marine Crag, *Scalaria clathratula*, *Scalaria foliacea*, *Rissoa confinis*, *Hinnites Cortesyi*, *Astarte Burtinii*, and *Astarte Omalii*.

FORMATION.	British and not Mediterranean.	British and Mediterranean.	Mediterranean and not British.	Neither British nor Mediterranean.	Not known living.	Total.	REMARKS.
Coralline Crag	20	154	51	24	142	391	Of the 24 species not known either as British or Mediterranean there is but one (and that a doubtful identity), *Cerithiopsis lactea*, which is Arctic. The remainder inhabit seas of lower latitude than Britain. See Note A.
Walton (or Older Red) Crag	13	61	14	10	50	148	Of the 10 species not British or Mediterranean two are Arctic and the rest inhabit seas of lower latitude than Britain. See Note B.
Rest of Red Crag (except Scrobicularia beds)	30	78	14	22	55	199	Of the 22 species not British or Mediterranean thirteen are Arctic, and the rest inhabit seas of lower latitude than Britain. See Note C.
Fluvio-marine Crag	36	38	7	12	18	111	Of the 12 species not British or Mediterranean three inhabit seas of lower latitude than Britain, and the rest are Arctic. See Note D.
Chillesford beds..............	19	44	1	9	14	87	Of the 9 species not British or Mediterranean two inhabit seas of lower latitude than Britain, and the rest are Arctic. See Note E.
Lower Glacial	13	9	...	4	3	29	The 4 species not British or Mediterranean are all Arctic. See Note F.
Middle Glacial	21	43	8	10	12	94	Of the 10 species not British or Mediterranean three inhabit seas of lower latitude than Britain, and the rest are Arctic. See Note G.
Upper Glacial	21	10	...	23	5	59	The 23 species which are neither British nor Mediterranean are all Arctic. See Note H.
Post Glacial	19	26	...	4	...	49	The 4 species which are neither British nor Mediterranean are all Arctic. See Note I.

NOTE A.—The twenty-four species in the Coralline Crag are the following, viz. *Erato mangeriæ*, which is West Indian; *Nassa granulata*, which is Japanese; *Pyrula reticulata*, a tropical form; *Trophon Costifer*, *Pleurotoma hispidula, Cerithium cribrarium, Chemnitzia varicula, Sigaretus excavatus, Bullæa ventrosa*, and *Lasæa pumila*, which are reported by Mr. Jeffreys to have occurred in the abyss of the Atlantic; *Cancellaria mitræformis*, and *Scalaria obtusicostata*, which he reports as West Atlantic; *Trochus multigranus, Neæra obesa*, and *N. jugosa*, which he reports from Atlantic and Lusitanian abysses (and the two first of which are also Scandinavian); *Scalaria subulata*, which he reports from the Canaries; *Bulla conuloidea*, which he reports from the Lusitanian Coast; *Cæcum mammillatum* and *Skænia ovata*, identified by Dr. P. Carpenter with Pacific shells; *Cardita scalaris*, another Pacific shell; *Gastrana laminosa*, a South African shell; *Natica pusilla* and *Nucula proxima*, species inhabiting the coast of the United States; *Cerithiopsis lactea*, which, if identical with the Crag shell, (of which I am doubtful) is the only exclusively Arctic species among the twenty-five.

29

Note B.—The ten species from Walton are the following, viz. *Nassa conglobata*, African; *N. granulata*, Japanese; *Actæon Etheridgii*, which Mr. Jeffreys speaks of as identical with a species (*exilis*) from the Atlantic abyss; *Trophon costifer*, from the same abyss; *Scalara subulata*, which Mr. Jeffreys mentions from Teneriffe; *Gastrana laminosa*, South African; *Cardita scalaris*, Pacific; *Mactra deaurata*, which seems to me to be *M. Jauresii* of the United States Coast; and *Buccinum glaciale* and *Solen gladiolus*, which are Arctic.

Note C.—The twenty-two species from the Red Crag other than Walton are the following, viz. *Nassa granulata*, Japanese; *Gastrana laminosa*, South African; *Trochus multigranus*, reported by Mr. Jeffreys from the Atlantic and from the Scandinavian Coast; *Trophon costifer*, from the Atlantic abyss; *Cardita scalaris*, Pacific; *Nassa propinqua* (*trivittata?* Say), *Pleurotoma bicarinata*, *Cardium pinnatulum*, and *Mactra deaurata* (*M. Jauresii*), from the coast of the United States; and *Buccinum glaciale*, *Trophon Sarsii*, *T. scalariformis*, *Pleurotoma pyramidalis*, *P. violacea*, *Cancellaria viridula*, *Amaura candida*, *Natica occlusa*, *Leda lanceolata*, *L. oblongoides* (*L. Arctica?* Gray), *Cardium Grœnlandicum*, *Tellina lata*, and *Solen gladiolus*, which are Arctic.

Note D.—The twelve species from the Fluvio-marine Crag are the following, viz. *Cardita scalaris*, Pacific; *Pleurotoma bicarinata*, United States Coast; *Trophon Costifer*, from the Atlantic abyss; and *Pleurotoma pyramidalis*, *Cancellaria viridula*, *Velutina undata*, *Leda lanceolata*, *L. oblongoides* (*L. Arctica?* Gray), *L. myalis*, *Cardium Grœnlandicum*, *Astarte borealis*, and *Tellina lata*, which are Arctic.

Note E.—The nine species from the Chillesford beds are the following:—*Cardita scalaris*, Pacific; *Corbula contracta*, United States Coast; and *Natica occlusa*, *Margarita argentata*, *Leda lanceolata*, *L. oblongoides* (*L. Arctica?* Gray), *Cardium Grœnlandicum*, *Astarte borealis*, and *Tellina lata*, which are Arctic.

Note F.—The four Lower Glacial species are—*Leda oblongoides* (*L. Arctica?* Gray), *L. myalis*, *Astarte borealis*, and *Tellina lata*, all of them Arctic.

Note G.—The ten Middle Glacial species are the following, viz. *Nassa granulata*, Japanese; *Cardita scalaris*, Pacific; *Corbula contracta*, United States Coast; and *Trophon scalariformis*, *Leda lanceolata*, *L. oblongoides* (*L. Arctica?* Gray), *Cardium Islandicum*, *Astarte borealis*, *Venus fluctuosa*, and *Tellina lata*, which are Arctic.

Note H.—The twenty-three Upper Glacial species are, *Columbella Holbollii*, *Trophon Sabini*, *T. ventricosus*, *T. craticulatus*, *T. scalariformis*, *T. Gunneri*, *Pleurotoma elegantior* (*P. elegans?* Moll.), *P. pyramidalis*, *P. scalaris*, *Trichotropis insignis*, *Turritella erosa*, *Natica occlusa*, *Margarita elegantissima*, *Pecten Islandicus*, *Leda caudata*, *L. pernula*, *L. oblongoides* (*L. Arctica?* Gray), *Cardita borealis*, *Cardium Islandicum*, *Astarte borealis*, *A. crebicostata*, *Venus fluctuosa*, and *Tellina lata*, all of them Arctic.

Note I.—The four Post-glacial species are, *Trophon scalariformis*, *Pleurotoma pyramidalis*, *Astarte borealis*, and *Tellina lata*, all of them Arctic.

The results of the table indicate an almost identical percentage of forms not known as living in the case of the older Red Crag and the Coralline, which is in conflict with the geological break, which I still believe to exist between the two formations.　　With the exception of the Middle Glacial column of it, however, the table shows very forcibly the diminishing Mediterranean aspect of the fauna as we ascend in the geological scale.

In the Coralline Crag there are fifty-two Mediterranean forms not living in the British seas, and only twenty of the converse character; and of these twenty, two, viz. *Odostomia Gulsonæ* and *Psammobia tellinella*, are Lusitanian. In the Walton Red Crag the respective numbers are fourteen and thirteen, but in the rest of the Red Crag the British species not living in the Mediterranean are in number more than double those of the converse character; while in the Fluvio-marine Crag these proportions are increased five-fold, and in the Chillesford beds nineteen-fold. In the Lower Glacial there occur thirteen, in the Upper Glacial twenty-one, and in the Post Glacial nineteen British species unknown in the Mediterranean, but in none of these three deposits does there occur a single species of the converse character.

Simultaneously with these features we find (as shown in notes A to I) a proportional increase of the Arctic species as we ascend through the Crag and Glacial series; and that even in the Post Glacial deposits no less than four out of a total of forty-nine are Arctic shells of the preceding Glacial Period which have since receded from the British Coasts.

The Middle Glacial fauna stands out in some discord with the above, since in it not only do several Mediterranean species unknown to British seas reappear, but the proportion which these bear to the number of British species not known in the Mediterranean is as eight to twenty-one—a much larger proportion than exists in the Fluvio-marine Crag, and altogether beyond the proportions exhibited by the intervening formations. The explanation is probably to be found in the Molluscan remains of this deposit having travelled from some distance, as mentioned in the introduction to this 'Supplement' (p. xxiii). Altogether this Middle Glacial assemblage is a very interesting one, and the most important of any of the formations succeeding the Crag. Several of the species which occur in it seem to have disappeared from the British Coast during the earlier part of the Red Crag; and while some of these are not known living, others are confined to the Mediterranean or other southern waters.

I have only to add that I am equally convinced with the authors of the introduction to this work that the Molluscan remains of the Middle Glacial Sand (fragmentary and worn as they occur in it) are not derived from any older deposit, but are contemporaneous with the sand which contains them.

INDEX

OF

SUPPLEMENT TO THE CRAG MOLLUSCA.

PLATE I.

(This Plate was engraved December, 1867.)

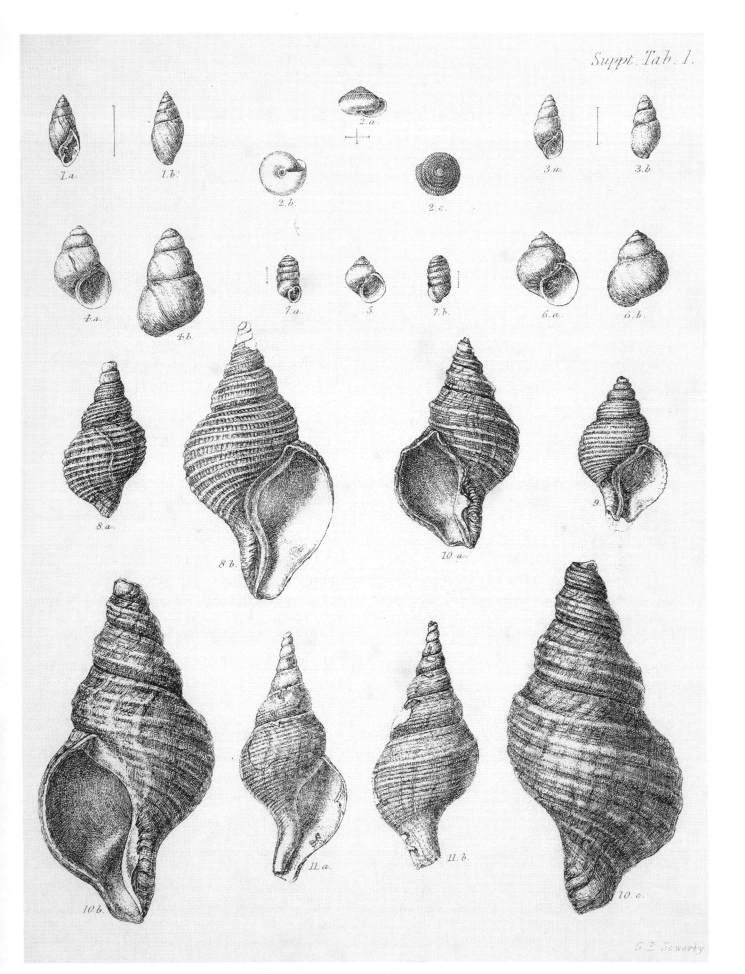

G. E. Sowerby

PLATE II.

(This Plate was engraved in 1869.)

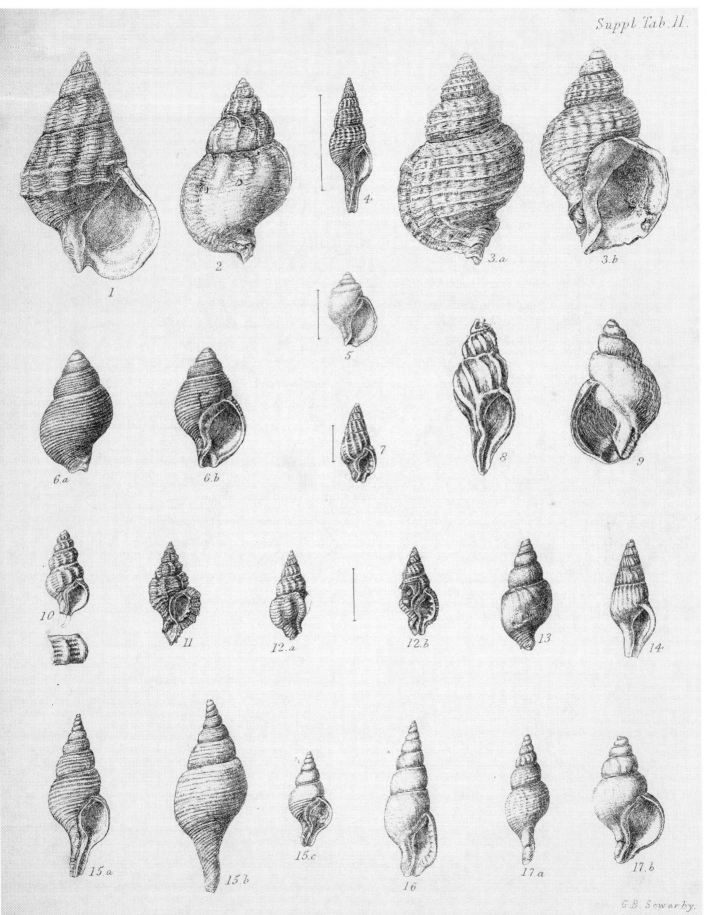

G.B. Sowerby.

PLATE III.

(This Plate was engraved in the beginning of the year 1870.)

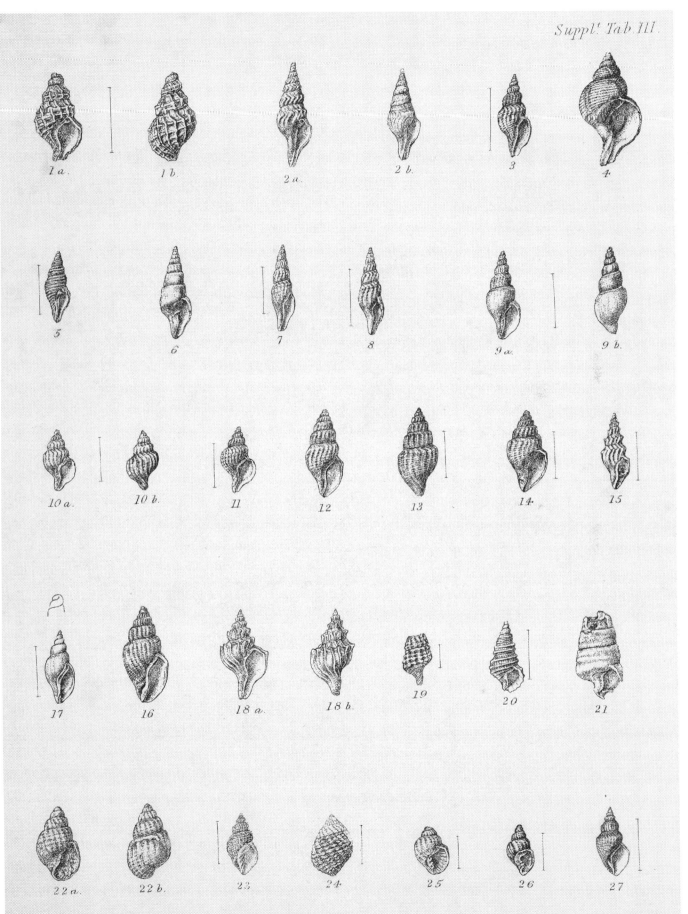

PLATE IV.

(This Plate was engraved in 1870.)

G. B. Sowerby

PLATE V.

(This Plate was engraved in 1870.)

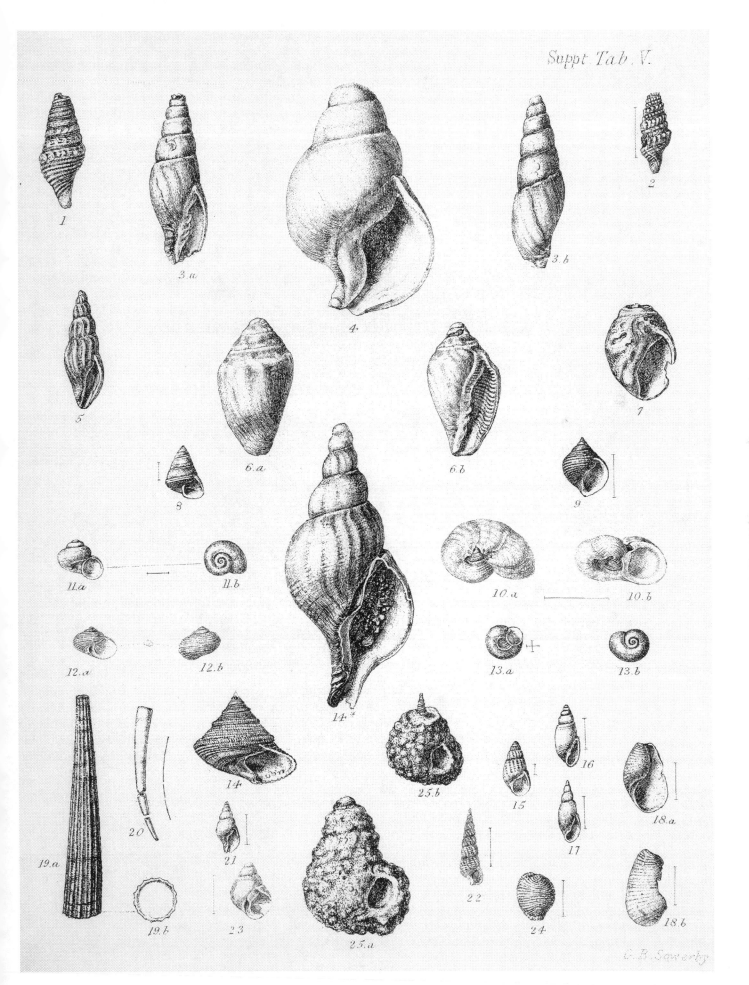

PLATE VI.

(This Plate was engraved in 1871.)

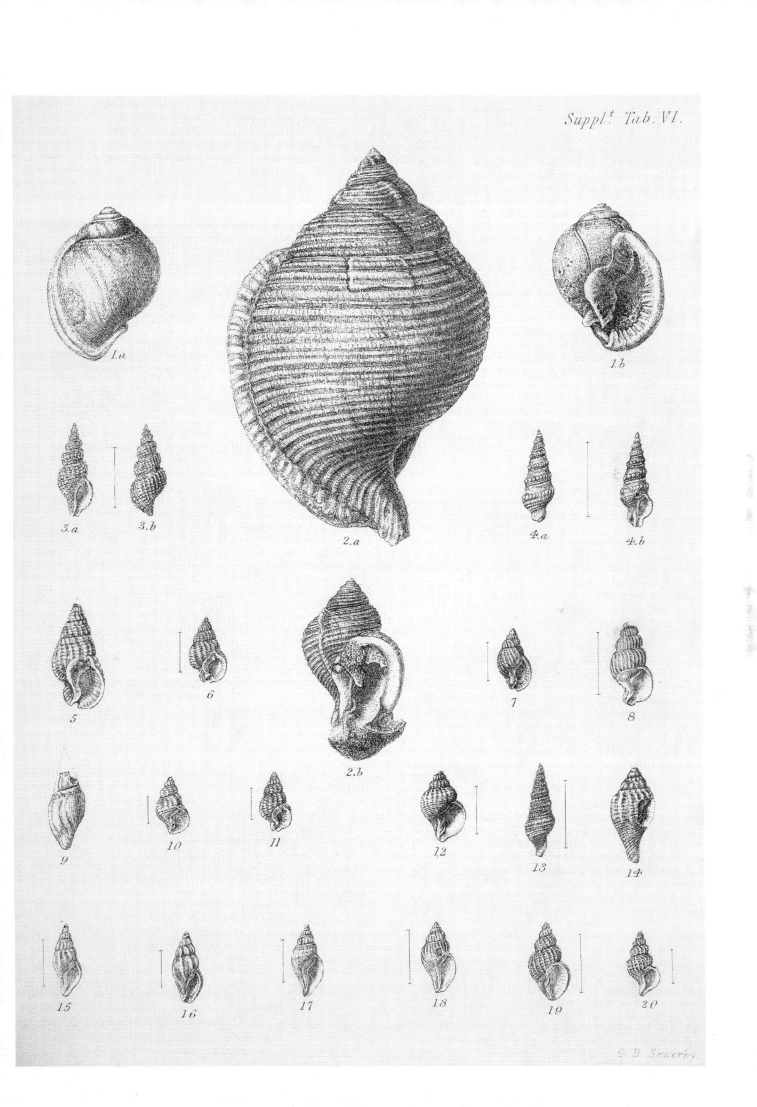

PLATE VII.

(This Plate was engraved in 1871.)

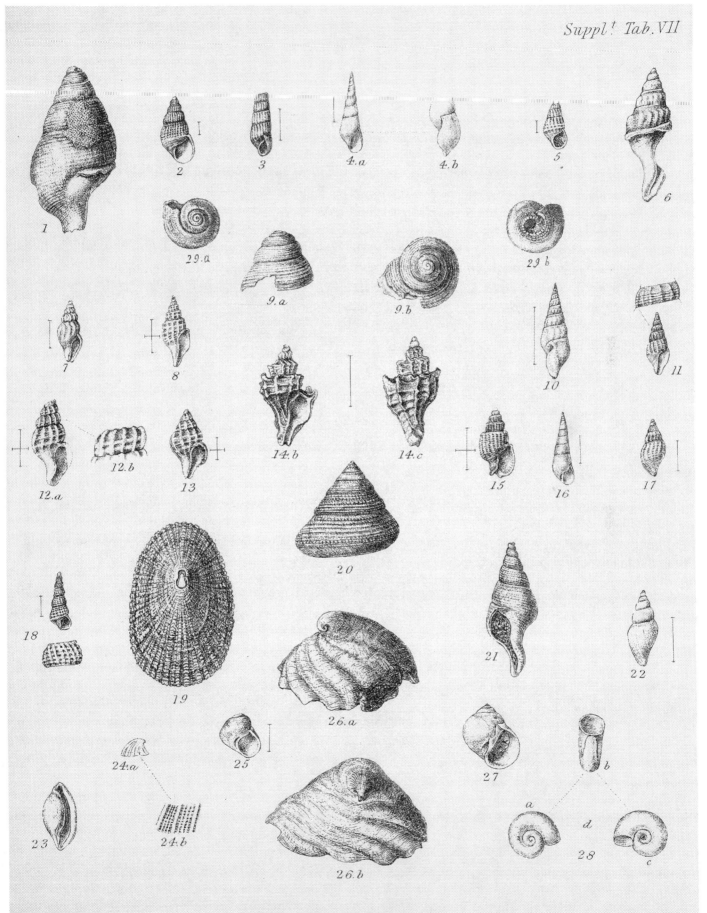

G.B.Sowerby

PLATE VIII.

(This Plate was engraved in 1870.)

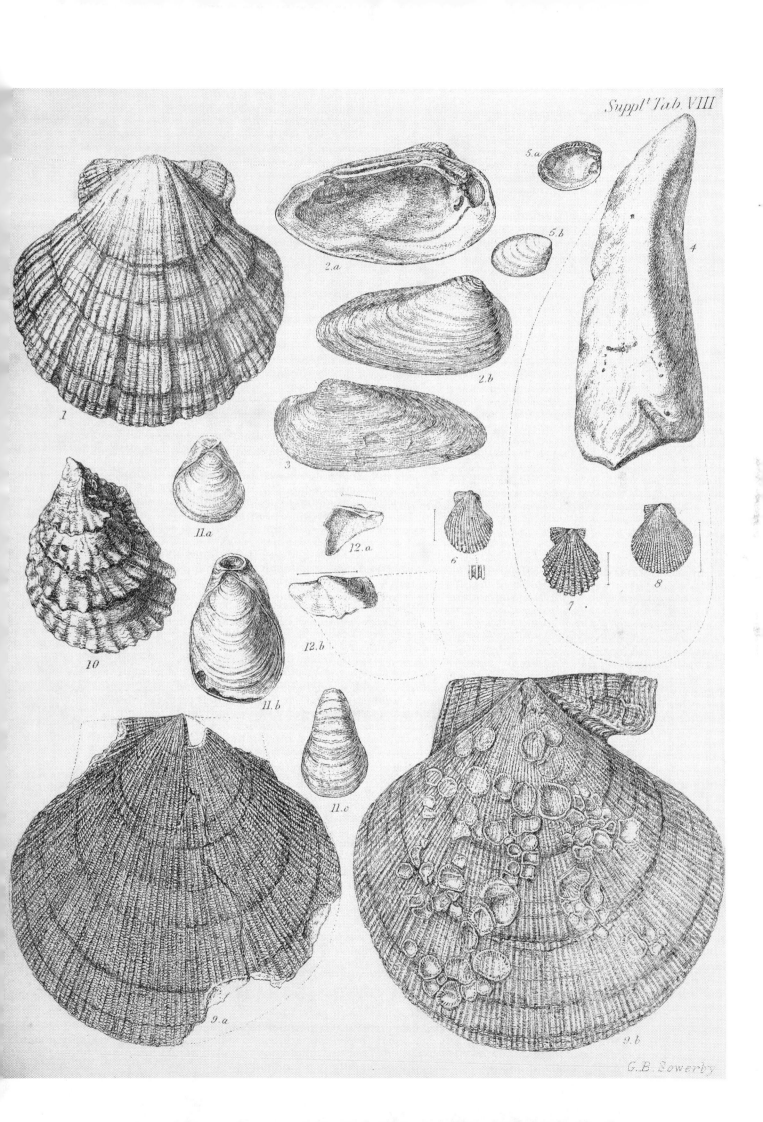

5.a

5.b

2.a

4

1

2.b

3

11.a

12.a

6

5.b

7

8

10

11.b

12.b

11.c

9.a

9.b

G.B. Sowerby

PLATE IX.

(This Plate was engraved in 1870.)

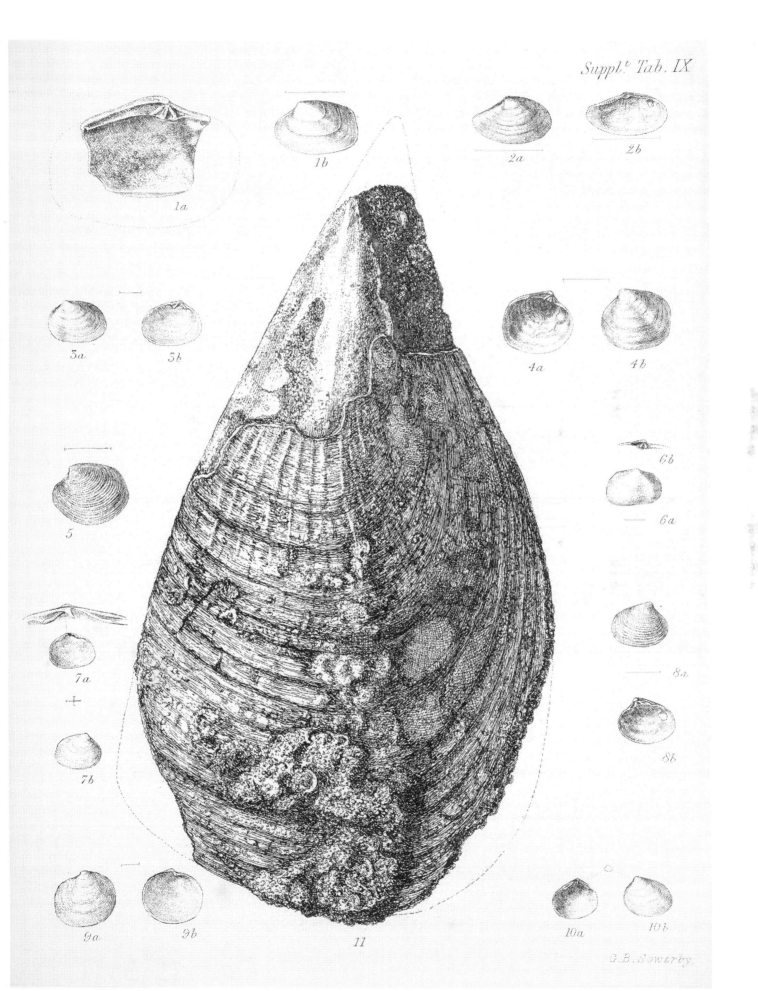

1a

1b

2a

2b

3a

3b

4a

4b

5

6b

6a

7a

8a

7b

8b

9a

9b

11

10a

10b

G.B.Sowerby

PLATE X.

(This Plate was engraved in 1872.)

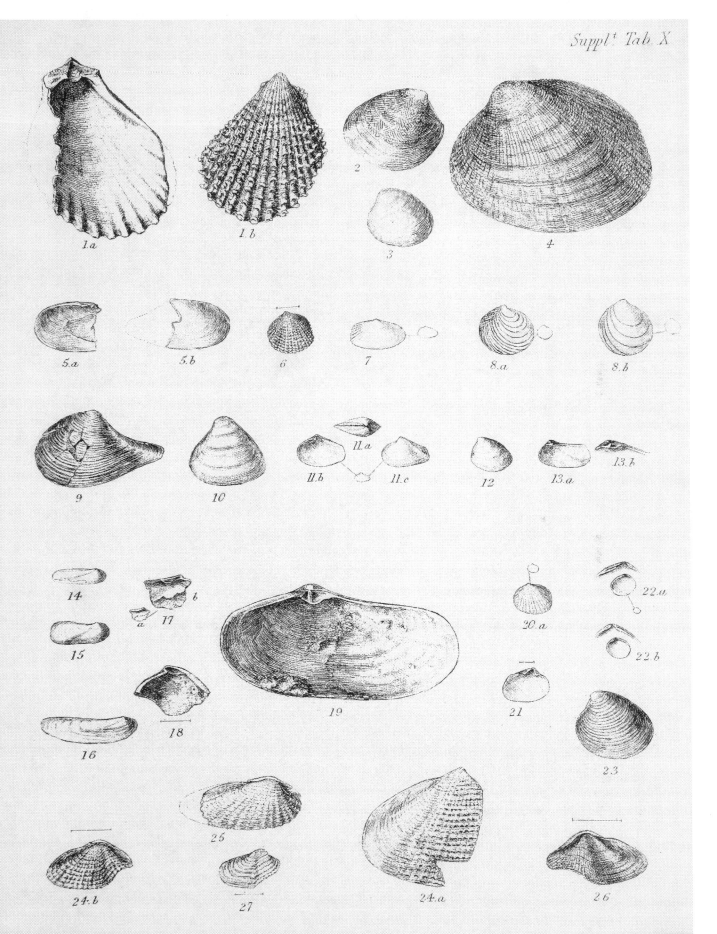

G.B. Sowerby

TABLE XI.

Fig.	Names of the shells.	Page	Localities from which the specimens figured were obtained.
1, c.	*Lingula Dumortieri* . . .	172	Coralline, Sutton.
a, b.	Magnified representations of the same.		
2, a—c.	*Rhynchonella psittacea* . .	171	Fluvio-marine Crag, Bramerton.
3, a—e.	*Terebratulina caput-serpentis*	169	Cor. Crag, Sutton.
	All magnified figures of young shells.		
4, a—d.	*Argiope cistellula*	170	Cor. Crag, Sutton.
5, a—g.	*Terebratula grandis* . . .	168	Cor. Crag, near Orford.
a.	Interior of the smaller valve, showing the apophysary system.		} Cor. Crag, Ramsholt. The small specimens from Sutton.
6.	*Discina fallens*	172	Cor. Crag, Sutton.
	The line indicates its magnitude.		

(This Plate was engraved in 1867.)

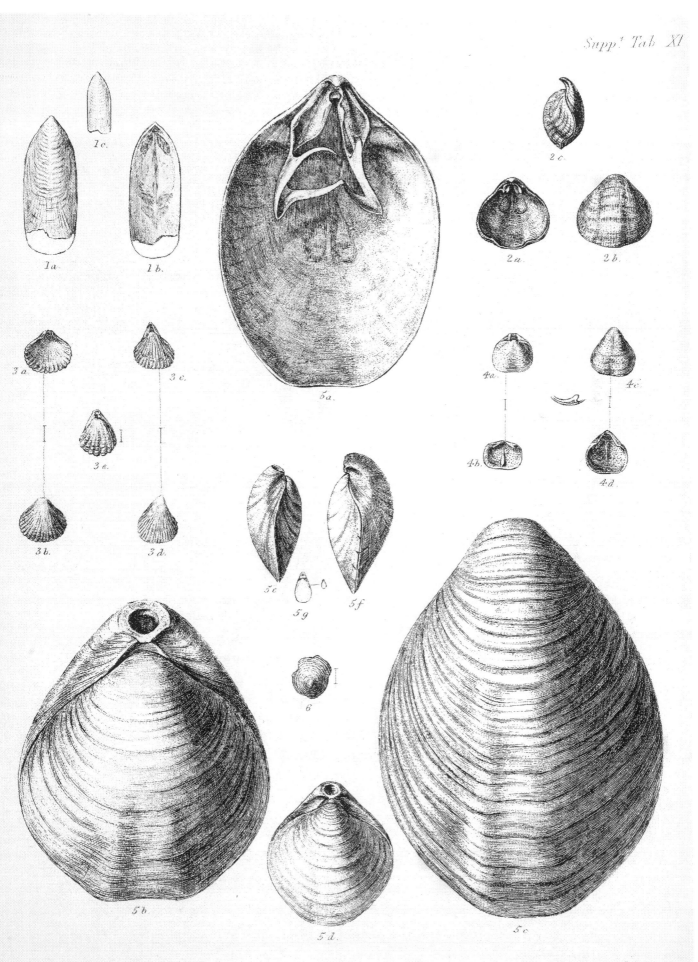

1c.

1a.

1b.

2c.

2a.

2b.

3a.

3c.

3e.

3b.

3d.

5a.

4a.

4c.

4b.

4d.

5e.

5g.

5f.

6

5b.

5d.

5c.

G. B. Sowerby

ADDENDUM PLATE.

Fig.	Names of the shells.	Page	Localities from which the specimens figured were obtained.
1.	*Scalaria semicostata*	. 183	Red Crag, probably derivative.
2.	*Pleurotoma striolata* . . .	179	Cor. Crag, near Orford.
3, *a, b*.	— *tereoides* . . .	178	Cor. Crag, Sutton.
4.	— *Bertrandi* . .	179	Red Crag, Butley.
5.	*Scalaria communis*	183	Recent.
6.	*Aporrhais Serresianus?* . .	180	Cor. Crag, near Orford.
7.	*Lachesis Anglica*	175	Cor. Crag, near Orford.
8, *a. b*.	*Pleurotoma clathrata?* . .	178	Cor. Crag, Sutton.
9.	*Murex insculptus*	176	Red Crag, Waldringfield, probably derivative.
10.	*Cancellaria umbilicaris?* . .	182	Red Crag, Waldringfield, probably derivative.
11.	*Buccinum Tomlinei* . . .	175	Red Crag, Waldringfield, probably derivative.
12.	*Dentalium entalis*	187	Cor. Crag, near Orford.
13.	*Trophon elegans·*	177	Cor. Crag, Sutton.
14.	*Chemnitzia Jeffreysii?* . .	184	Cor. Crag, Sutton.
15.	*Odostomia albella*	184	Cor. Crag, Sutton.
16.	*Trophon Norvegicus* . . .	177	Red Crag, Sutton.
17.	*Cerithium perversum?* . .	181	Cor. Crag, Sutton.
18.	*Odostomia dentiplicata* . .	185	Cor. Crag, Sutton.
19.	*Columbella Borsoni?* . . .	174	Red Crag, Walton Naze.
20.	— *minor* . . .	174	Cor. Crag, near Orford.
21.	*Menestho Britannica* . . .	185	Cor. Crag, Sutton.
22.	*Clausilia Pliocena*	188	Cor. Crag, Sutton.
23.	*Avicula phalænoides* . . .	188	Cor. Crag, near Orford.
24.	*Nassa pusillina,*var. *variabilis*	176	Red Crag, Butley.
25.	*Scacchia lata*	189	Cor. Crag, Sutton.
26.	*Bulla utriculus?*	187	Cor. Crag, near Orford.
27.	*Thracia dissimilis*	189	Cor. Crag, near Orford.

(This Plate was engraved in 1873.)

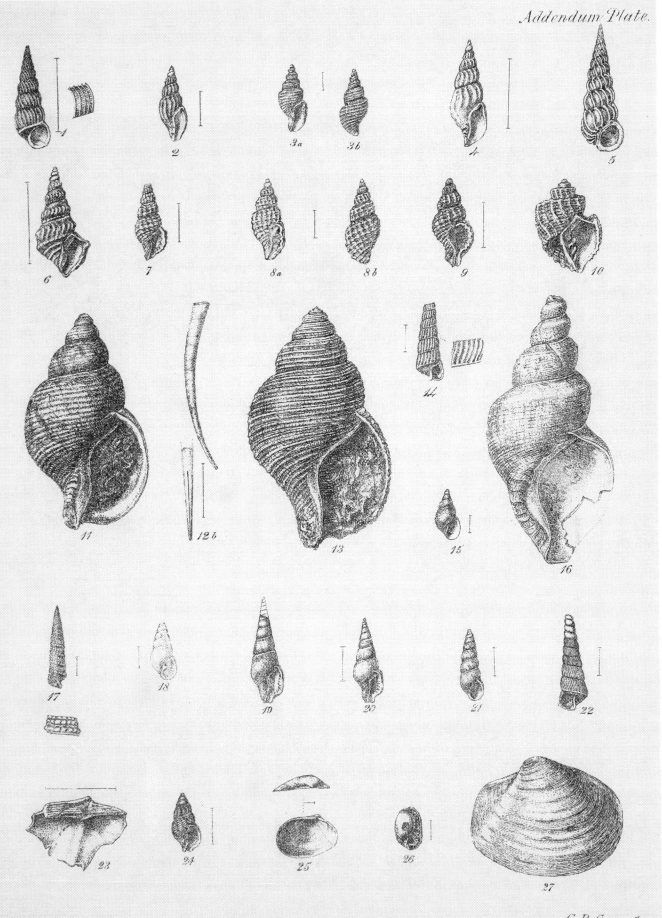

G.B.Sowerby.

Printed in the United States
By Bookmasters